Advances in Geometric Modeling

Advances in Geometric Modeling

Dr. Muhammad Sarfraz

*King Fahd University of Petroleum
and Minerals, Saudi Arabia*

John Wiley & Sons, Ltd

Other Wiley Editorial Offices

John Wiley & Sons Inc., 111 River Street, Hoboken, NJ 07030, USA

Jossey-Bass, 989 Market Street, San Francisco, CA 94103-1741, USA

Wiley-VCH Verlag GmbH, Boschstr. 12, D-69469 Weinheim, Germany

John Wiley & Sons Australia Ltd, 33 Park Road, Milton, Queensland 4064, Australia

John Wiley & Sons (Asia) Pte Ltd, 2 Clementi Loop #02-01, Jin Xing Distripark, Singapore 129809

John Wiley & Sons Canada Ltd, 22 Worcester Road, Etobicoke, Ontario, Canada M9W 1L1

Wiley also publishes its books in a variety of electronic formats. Some content that appears in print may not be
available is electronic books.

A catalogue record for this book is available from the British Library

ISBN 0-470-859377

Typeset in 10/12pt Times Roman by TechBooks, New Delhi, India
Printed and bound in Great Britain by Biddles Ltd, Kings Lynn, Norfolk
This book is printed on acid-free paper responsibly manufactured from sustainable forestry
in which at least two trees are planted for each one used for paper production.

Contents

Preface

Geometric Modeling (GM) plays a significant role in the construction, design and manufacture of various objects. In addition to its critical importance in the traditional fields of automobile and aircraft manufacturing, shipbuilding, and general product design, GM methods have also proven to be indispensable in a variety of modern industries, including computer vision, robotics, medical imaging, visualization, textile, designing, painting, and other media.

This book aims to provide a valuable resource, which focuses on interdisciplinary methods and affiliate research in the area. It aims to provide the user community with a variety of advanced geometric modeling techniques and the applications necessary for various real life problems. It also aims to collect and disseminate information in various disciplines including:

- Computer Graphics
- Computer Vision
- Computer Aided Geometric Design
- Geometric Algorithms
- Visualization
- Shape Abstraction and Modeling
- Computational Geometry
- Solid Modeling
- Shape Analysis and Description
- Reverse Engineering
- Multiresolution and Diffusion
- Texture Mapping
- CAD/CAM
- Industrial Applications

The major goal of this book is to stimulate discussion and provide a source where researchers and practitioners can find the latest developments in the field of geometric modeling. Due to the speed of scientific development, there is a great deal of thirst among the worldwide scientific community to be equipped with state of the art theory and practice to get their problems solved in diverse areas of various disciplines. Although a large amount of work has been done by researchers already, a tremendous interest still increases everyday due to complicated problems being faced in academia and industry.

This book has over twenty-two chapters focussing on new advances in the area. These contributions are meant for geometric modeling issues including:

- Reviewed literature
- New techniques
- Applications

The book is useful for researchers, computer scientists, practicing engineers, and many others who seek state of the art techniques and applications in geometric modeling. It is also useful for undergraduate students as well as graduate students in the areas of computer science, engineering, and mathematics.

The book is based on state of the art articles on geometric modeling techniques and applications. It will be a good reference book for the computer graphics and geometric modeling community worldwide. The book will be specifically of interest to people in the following industries or academic fields:

- Computer Science (especially Computer Graphics)
- Computer Aided Geometric Design
- Computer Vision
- Image Processing
- Toon Rendering
- Virtual Reality
- Body Simulation
- Engineering Disciplines
- Mathematical Sciences
- Font Industry
- Software Industry
- Information Visualization
- Manufacturing Industry

The editor is thankful to the contributors for their valuable efforts towards the completion of this book. A lot of credit is also due to the various experts who reviewed the chapters and provided helpful feed back. The editor is happy to acknowledge the support of King Fahd University of Petroleum and Minerals towards the compilation of this book. This book editing project has been funded by King Fahd University of Petroleum and Minerals under Project #ICS/ADV.MODEL/259.

M. Sarfraz

1

Polygonal Subdivision Curves for Computer Graphics and Geometric Modeling

Ahmad H. Nasri

American University of Beirut, Department of Computer Science,
PO Box 11-0236, Riad El Solh, Beirut 1107 2020, Lebanon
Email: anasri@aub.edu.lb

A polygonal subdivision curve is the limit of subdivision of a polygonal complex. Such a complex, whose shape depends on the scheme involved, is simply a control mesh that converges to a curve rather than a surface. The use of polygonal complexes has been motivated by the curve interpolation problem in subdivision surfaces. This is one of the major interpolation constraints that is partially addressed in the literature. The definition of curves by polygonal complexes carries with it cross derivative information which can be naturally embodied in the mesh of a subdivision surface. This chapter gives an overview of such complexes, their polygonal subdivision curves and their applications for computer graphics and geometric modeling.

1. Introduction

Undoubtedly, subdivision surfaces are gradually becoming state of the art in both computer graphics and geometric modeling. There are several reasons for their popularity which include simplicity, and the ability to produce globally smooth surfaces from arbitrary topological meshes.

Simply, a recursive division surface S is defined by a pair (P_0, R) where P_0 is an initial configuration, and R is a refinement procedure. The configuration consists of a set of vertices, edges and faces in the 3D space. This is often referred to as a polyhedron or polygonal mesh. The refinement procedure is a set of rules applied to a configuration to generate another with

Advances in Geometric Modeling. Edited by M. Sarfraz
© 2003 John Wiley & Sons, Ltd   ISBN: 0-470-85937-7

more vertices, edges and smaller faces than the initial one. At each level i of the refinement process, the polygonal mesh P_{i-1} is taken as input to the refinement R, which produces another polygonal mesh P_i, which may itself be taken as input to the next refinement step and so on. If R satisfies certain conditions then the sequence of meshes P_i converges to a smooth surface at the limit.

This simple idea has sparked the imagination of the graphics and geometric modeling communities. For the former, it provides a definition of models in a multiresolution fashion with as much accuracy as needed. The possibility of migrating from coarser to finer meshes and vice versa is an attractive one. For the geometric modeling community, subdivision not only provides a definition of surfaces but is also a mean of interrogation. Efficient and robust algorithms for intersecting, sectioning and interrogating such curves and surfaces can be developed. Extensive research is still ongoing and results have revealed that modifying or generating new rules is a source of flexibility to achieve various modifications of the limit surface. Interestingly enough, subdivision surfaces have found their use in multiresolution and wavelets, fluid flow, movie production, mesh processing (including fairing, compression, etc.). More details can be found in [25–26], the site http://www.subdivision.org, and the references.

Polygonal complexes have recently proven their usefulness in Computer-Aided Geometric Design (CAGD) and in computer graphics [10–11]. Put simply, a polygonal complex is a control mesh whose limit under subdivision is a curve rather than a surface. At first, one may wonder how a control mesh instead of a control polygon could be useful to define a curve. One advantage of the former is simply that polygonal complexes permit the definition of curves that carry with them cross derivatives information, this is explained later.

In the context of any given subdivision scheme, the definition of a polygonal complex requires the following information:

1. Its topological and geometrical nature.
2. The required modifications (if any) to the subdivision rules of the scheme used. This should not however involve modifying the subdivision coefficients themselves.

One particular advantage of the use of these complexes is that no modification of the subdivision coefficients is ever required in order to achieve curve interpolation constraints. This can be useful in various applications such as curve generation, curve interpolation, trimming and other computer graphics applications.

This chapter is structured as follows: Section 2 outlines two basic subdivision schemes, Section 3 gives the necessary background about Bézier curves (and surfaces) and their conversions to B-spline ones. Polygonal subdivision curves are introduced in Section 4 with examples in Doo-Sabin and Catmull-Clark schemes. The limits of these complexes are also given in Section 5 and the applications are discussed in Section 6. Finally, conclusions and future work are drawn in Section 7.

2. Two Basic Schemes of Subdivision Surfaces

Since their introduction in 1974, several subdivision schemes have emerged. From among these subdivision schemes, we outlined two pioneering ones: the quadratic Doo-Sabin [4] and the cubic Catmull-Clark [2].

2.1. Catmull-Clark Subdivision Scheme

In a Catmull-Clark setting, the following rules are used to generate a refinement of a polyhedron:

1. For each old face f with n vertices $(v_i)_{1 \le i \le n}$, a new vertex v_f is generated at the centroid by:

$$v_f = \frac{1}{n} \sum_{i=1}^{n} v_i$$

2. For each old edge e having two vertices v_1 and v_2 which is shared by two faces f and g, a new vertex v_e is generated by:

$$v_e = \frac{v_1 + v_2 + v_f + v_g}{4}$$

3. For each old vertex v incident to n edges (e_i) and shared by n faces (f_i), a new vertex v_v is generated by:

$$v_v = \alpha_n \sum_{i=1}^{n} v_{ei} + \beta_n \sum_{i=1}^{n} v_{fi} + \gamma_n v$$

where v_{ei} (respectively v_{fi}) is the vertex generated from the edge e_i (respectively the face f_i), and the weights $\alpha_n, \beta_n \gamma_n$ are given by:

$$\alpha_n = \beta_n = \frac{1}{n^2}$$

$$\gamma_n = \frac{n-2}{n}$$

The refined polyhedron is obtained by joining each F-vertex of a face F_i to the E-vertices of the edges of this face. The V-vertex of a vertex V_i is also connected to the E-vertices of the edges incident to V_i.

Figure 1.1 shows an example of such a surface.

2.2. Doo-Sabin Scheme

In the Doo-Sabin scheme [4], a refined polyhedron is generated by constructing from each face with n vertices $(v_i)_{1 \le i \le n}$, a new set of vertices $(w_i)_{1 \le i \le n}$ using the following equation:

$$w_i = \sum_{j=1}^{n} v_j \alpha_{ij}$$

where the α_{ij} are given by:

$$\alpha_{ij} = \begin{cases} \dfrac{n+5}{4n} & i = j \\[2mm] \dfrac{3 + 2\cos(2\pi(i-j)/n)}{4n} & i \neq j \end{cases}$$

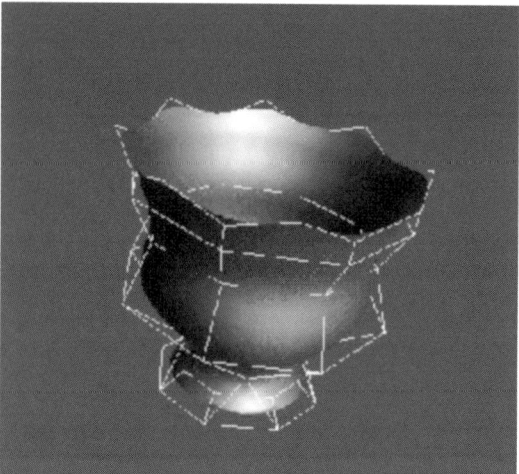

Figure 1.1   A Catmull-Clark surface of a vase and its polyhedron.

The new control mesh or polyhedron is obtained by connecting these vertices making a new F-face, E-face and V-face from each old face, edge and vertex, respectively, as shown in Figure 1.2.

3. Background

In this section, we outline the process of conversion between the Bézier and the B-spline schemes that will be employed in this paper.

3.1. The Curve Case

A cubic B-spline curve segment $s(u)$, where $0 \leq u \leq 1$, can be expressed by the following equation:

$$s(u) = U \times M_s \times P_s$$

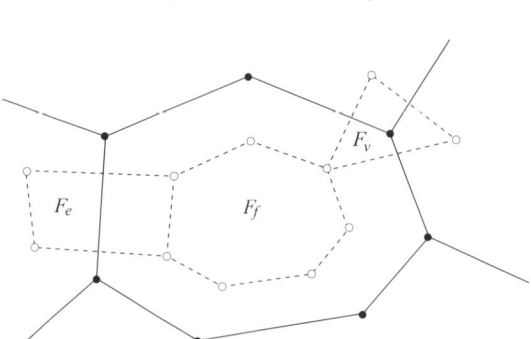

Figure 1.2   The three types of faces (dashed) in Doo-Sabin approach. F-face (F_f), E-face (F_e), V-face (F_v).

where $U = (u^3 \ u^2 \ u \ 1)$ and P_s is the matrix defining the control points p_0, p_1, p_2 and p_3:

$$P_s = \begin{pmatrix} p_0 \\ p_1 \\ p_2 \\ p_3 \end{pmatrix}$$

The matrix M_s is the B-spline basis transformation matrix and given by:

$$M_s = \frac{1}{6} \begin{pmatrix} -1 & 3 & -3 & 1 \\ 3 & 6 & 3 & 0 \\ -3 & 0 & 3 & 0 \\ 1 & 4 & 1 & 0 \end{pmatrix}$$

Similarly, a Bézier segment $b(u)$, where $0 \le u \le 1$, can be expressed by the following equation:

$$b(u) = U \times M_b \times P_b$$

where P_b is the matrix defining the control points q_0, q_1, q_2 and q_3:

$$P_b = \begin{pmatrix} q_0 \\ q_1 \\ q_2 \\ q_3 \end{pmatrix}$$

and where M_b is the Bézier basis transformation matrix:

$$M_b = \begin{pmatrix} -1 & 3 & -3 & 1 \\ 3 & -6 & 3 & 0 \\ -3 & 3 & 0 & 0 \\ 1 & 0 & 0 & 0 \end{pmatrix}$$

The conversion between Bézier and B-spline basis functions proceeds by equating the appropriate matrix forms. So, to convert from B-spline to Bézier control points, we use:

$$P_b = M_b^{-1} \times M_s \times P_s$$

Conversely, to convert from Bézier to B-spline control points, we use:

$$P_s = M_s^{-1} \times M_b \times P_b$$

3.2. The Surface Case

Following the same notation, similar matrix operations can define the conversion from a B-spline patch to a Bézier patch and vice versa. In fact, a B-spline patch $S_s(u, v)$ can be expressed by:

$$S_s(u, v) = U \times M_s \times P_s \times M_s^T \times V^T$$

where $V = (v^3 \ v^2 \ v \ 1)$ and P_s is a 4×4-mesh of data points:

$$P_s = \begin{pmatrix} p_{00} & p_{01} & p_{02} & p_{03} \\ p_{10} & p_{11} & p_{12} & p_{13} \\ p_{20} & p_{21} & p_{22} & p_{23} \\ p_{30} & p_{31} & p_{32} & p_{33} \end{pmatrix}$$

Similarly, a Bézier patch $S_b(u, v)$ can be expressed by:

$$S_b(u, v) = U \times M_b \times P_b \times M_b^T \times V^T$$

where P_b is a 4×4-mesh of data points:

$$P_b = \begin{pmatrix} q_{00} & q_{01} & q_{02} & q_{03} \\ q_{10} & q_{11} & q_{12} & q_{13} \\ q_{20} & q_{21} & q_{22} & q_{23} \\ q_{30} & q_{31} & q_{32} & q_{33} \end{pmatrix}$$

The Bézier points can be expressed in terms of the B-spline points by:

$$P_b = M_b^{-1} \times M_s \times P_s \times M_s^T \times \left(M_b^T\right)^{-1}$$

From the above expressions it can be noted that, in the conversion from a B-spline patch to a Bézier patch, the boundary curve $B = (b_{00} \ b_{01} \ b_{02} \ b_{03})$ can be expressed in terms of the following three B-spline rows $(q_{i0} \ q_{i1} \ q_{i2} \ q_{i3})_{0 \le i \le 2}$, see [14] for more details.

$$B = \frac{1}{36}(1 \quad 4 \quad 1) \times \begin{pmatrix} q_{00} & q_{01} & q_{02} & q_{03} \\ q_{10} & q_{11} & q_{12} & q_{13} \\ q_{20} & q_{21} & q_{22} & q_{23} \end{pmatrix} \times M$$

where:

$$M = \begin{pmatrix} 1 & 0 & 0 & 0 \\ 4 & 4 & 2 & 1 \\ 1 & 2 & 4 & 4 \\ 0 & 0 & 0 & 1 \end{pmatrix}$$

Obtaining the B-spline control points $S = (s_1 \ s_2 \ s_3 \ s_4)$ equivalent to B can be done through a suitable transformation performed on B (details are omitter here). Alternatively, S can be obtained directly through the following equation:

$$S = \frac{1}{6}(1 \quad 4 \quad 1) \times \begin{pmatrix} q_{00} & q_{01} & q_{02} & q_{03} \\ q_{10} & q_{11} & q_{12} & q_{13} \\ q_{20} & q_{21} & q_{22} & q_{23} \end{pmatrix}$$

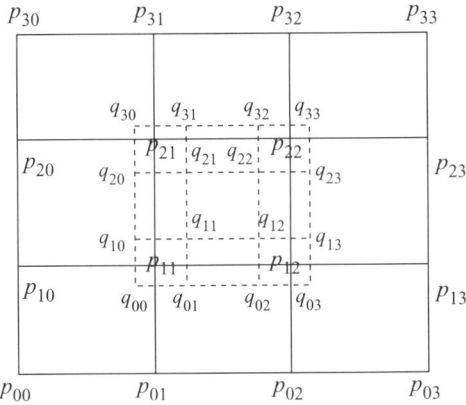

Figure 1.3 B-spline and Bézier control mesh for the same patch.

This again shows that the boundary curve of a B-spline patch depends only on the first three rows of the mesh defining the patch; the usefulness of that will be demonstrated in the following section.

4. Recursive Subdivision of Polygonal Complexes

In the context of any given subdivision scheme, the definition of a polygonal complex requires the following:

1. Its topological and geometrical information.
2. The required modifications (if any) of the rules of the subdivision scheme being used, preferably without modifying the subdivision coefficients themselves.

If the scheme has a set of rules S, a polygonal complex is then defined by the pair $(P_0, \hat{S})$ where P_0 is a control mesh and $\hat{S}$ is a slight variation of S.

In this section we give two examples of such complexes, one for Doo-Sabin and one for Catmull-Clark.

4.1. Doo-Sabin Polygonal Complexes

Topologically, a Doo-Sabin polygonal complex consists of a sequence of panels $(q_i)_{1 \leq i \leq n}$ with the property that every two panels q_j and q_{j+1} have exactly one edge in common. If the two panels q_1 and q_n do not share an edge, they are called *end*-panels [4, 11]. Each panel of a Doo-Sabin complex must be n-reflected about n mid-segments, as defined below.

ChebyChev Points: On a segment $[AB]$, m ChebyChev points $(p_i)_{i=1...m}$ can be defined as follows:

$$p_i = [(1 + \beta_i)A + (1 - \beta_i)B]/2$$

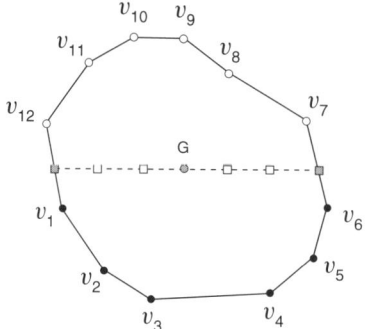

Figure 1.4   A 1-reflected panel and its mid-segment.

where:

$$\beta_i = \cos((2i-1)\pi/2m)/\cos(\pi/2m)$$

Clearly, $p_1 = A$ and $p_m = B$.

A panel f with $n(n = 2m)$ vertices $(v_i)_{1 \le i \le n}$ is called singly reflected or (1-reflected) about the segment e joining the midpoints $v_1 v_n$ and $v_m v_{m+1}$, if its vertices are symmetrically distributed about the ChebyChev points (c_i) defined on e. This means that two vertices v_i and v_{n+1-i} are symmetric about (c_i); these are called *opposite vertices*. The edges $v_1 v_n$ and $v_m v_{m+1}$ are called *contact* edges. For an example, see Figure 1.4.

N-reflected Panels: Consider a regular n-gon $(n = 2m)$ inscribed in a circle of radius r (the radius r can be chosen arbitrarily) and centered at the origin, as depicted in Figure 1.5. Its

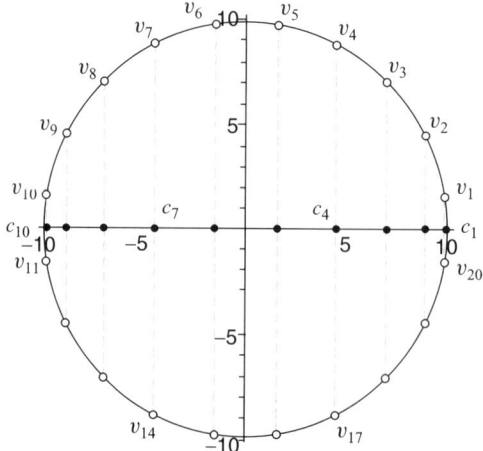

Figure 1.5   The vertices of the original n-reflected panel with one mid-segment.

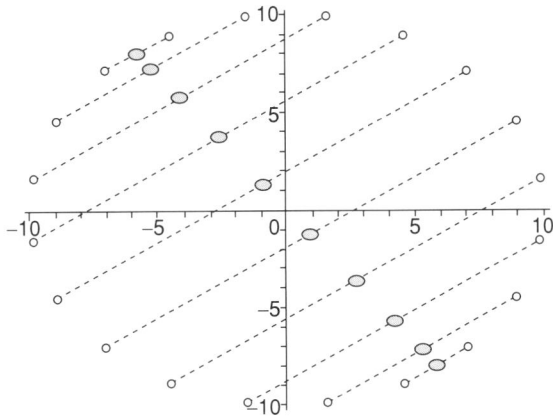

Figure 1.6 An n-reflected panel and its corresponding mid-segment.

vertices v_k are defined by:

$$v_k = \left(r \cos \left(\frac{(2k-1)\pi}{n} \right), r \sin \left(\frac{(2k-1)\pi}{n} \right) \right)$$

and $k = 1 \cdots n$ [1]. The radius r can be chosen arbitrarily.

Let A_1 and B_1 be the midpoints of the $v_1 v_n$ and $v_m v_{m+1}$, respectively. Their x coordinates x_A and x_B are given by:

$$x_A = -x_B = r \cos \frac{\pi}{n}.$$

It is then easy to show that every two vertices v_i and v_{n+1-i} are symmetric about the ChebyChev point c_i defined on $A_1 B_1$. Note that $A_1 = c_1$ and $B_1 = c_m$. The panel F formed by the set of vertices (v_k) is then called singly reflected or 1-reflected about $A_1 B_1$.

It is easy to show that F is also 1-reflected about n mid-segments $(A_j B_j)_{j=1 \cdots n}$, where A_j and B_j are the midpoints of $v_{j-1} v_j$ and $v_{m+j-1} v_{m+j}$. We will refer to this face as the **original** n-reflected panel. Figure 1.6 shows an example of such a panel.

Definition: An n-reflected panel can then be defined as a panel with $n = 2m$ vertices v_k that are symmetric about the ChebyChev points defined on its n mid-segments $(e_j)_{1 \leq j \leq n}$. Where each e_j joins the midpoints (indices are considered mod n) of $v_j v_{j+1}$ and $v_{m+j-1} v_{m+j}$ and whose vertices are obtained by an affine map (such as rotation, translation, scaling) of the original n-reflected one.

We further define the *mid-polygon* of a polygonal complex as the piecewise control polygon whose vertices are the midpoints of the contact edges. In the case of an open curve, the centroids of the end-panels are the end-points of this mid-polygon.

Figure 1.7 shows an example of a Doo-Sabin polygonal complex, its corresponding mid-polygon and its limit curve.

The modification to the Doo-Sabin subdivision rules (when applied to polygonal complexes) requires just the omission of the V-face rule [11].

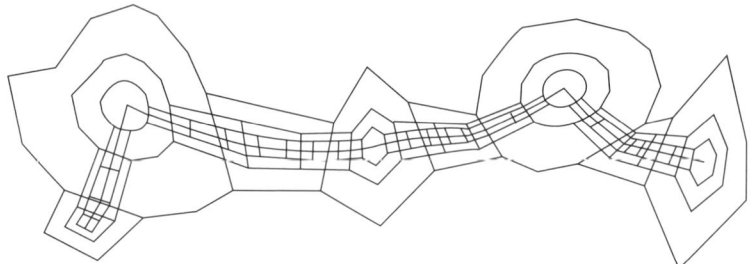

Figure 1.7   A Doo-Sabin polygonal complex with two successive subdivisions and its limit curve.

4.2. Catmull-Clark Polygonal Complexes

In the Catmull-Clark setting, the simple form of a polygonal complex P_0 consists of a double sequence of panels so that each panel has three shared edges. In the case of a polygonal complex strip, each of the four *corner* panels has exactly two shared edges. Accordingly, a Catmull-Clark polygonal complex CC can be defined by three rows (t_i), (m_i) and (b_i), as indicated in Figure 1.8. The modified rules $\hat{S}$ is the same as S, keeping in mind that m_0, (t_i), (b_i) and m_n are all boundary vertices.

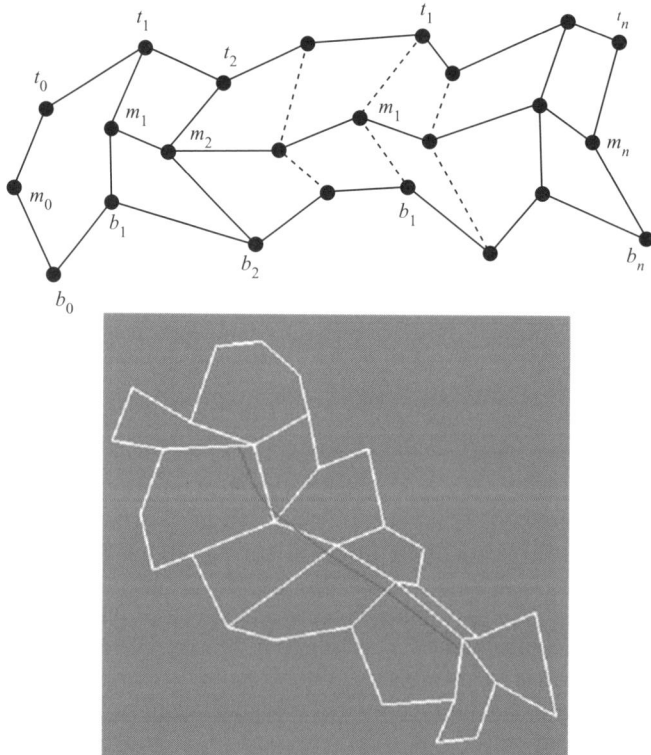

Figure 1.8   A simple (top) and a general (bottom) Catmull-Clark polygonal complex and its limit curve.

A general form of this Catmull-Clark complex permits the inclusion of non 4-sided panels in its defining mesh. In such a case, a single level of Catmull-Clark subdivision will be sufficient to generate a rectangular structure similar to the simple form suggested above. Figure 1.8 shows examples of such a complex in both forms.

The same subdivision scheme is used here without modification.

5. Polygonal Subdivision Curves

The limit of a polygonal complex is a curve, called *polygonal subdivision curve*. It is not difficult to define such a polygonal complex in a given scheme, but the challenge is to predict the limit curve and its relation to the complex. The polygonal complexes (as described above) converge to quadratic B-spline in the case of the Doo-Sabin scheme, and to the cubic B-spline in the Catmull-Clark setting. The following sections discuss these issues.

5.1. Polygonal Doo-Sabin Curves

In [11], it was shown that a 1-reflected panel remains invariant under subdivision. This means that a 1-reflected panel with a mid-segment e generates another 1-reflected panel whose mid-segment is a Chaikin subdivision of e. This property carries to an n-reflected panel with n mid-segment (e_i), where its subdivided one is n-reflected, with n mid-segments that are Chaikin's subdivision of (e_i), respectively. Consequently, a Doo-Sabin polygonal complex Q_0 with a mid-polygon M_0 converges to the piecewise quadratic B-spline curve c of the control polygon M_0, such that c interpolates the centroid of each panel of Q_0. Each piece of the limit curve starts and ends at the centroid of a panel and the continuity at the joint depends on the contact edges of the corresponding panel. If two contact edges are opposite then the two pieces join with C^1 continuity, otherwise with C^0 only. Figure 1.7 shows a Doo-Sabin complex and its corresponding limit curve.

5.2. Polygonal Catmull-Clark Curves

In the simple form, the limit curve of a cubic polygonal complex CC (defined by three rows (t_i), (m_i) and (b_i)) can be determined piecewise.

In fact, let us consider the first piece of this complex defined by $(t_i)_{0 \leq i \leq 3}$, $(m_i)_{0 \leq i \leq 3}$ and $(b_i)_{0 \leq i \leq 3}$. Assuming that these rows are the bottom three rows of a B-spline patch, we know from the previous section that this converges to a Bézier patch with one of its boundary curves defined by these rows. It can be deduced that the limit of this part of the complex converges to a Bézier curve defined by $(p_0 \quad p_1 \quad p_2 \quad p_3)$ and given by:

$$\frac{1}{36}(1 \quad 4 \quad 1) \times \begin{pmatrix} t_0 & t_1 & t_2 & t_3 \\ m_0 & m_1 & m_2 & m_3 \\ b_0 & b_1 & b_2 & b_3 \end{pmatrix} \times M$$

Similarly, the second piece of the limit curve defined by $(p_3 \quad p_4 \quad p_5 \quad p_6)$ is given by:

$$\frac{1}{36}(1 \quad 4 \quad 1) \times \begin{pmatrix} t_1 & t_2 & t_3 & t_4 \\ m_1 & m_2 & m_3 & m_4 \\ b_1 & b_2 & b_3 & b_4 \end{pmatrix} \times M$$

and so on. However, remembering what we said above, we can obtain the B-spline curve corresponding to the whole patch directly and in a single operation:

$$S = \frac{1}{6}(1 \quad 4 \quad 1) \times \begin{pmatrix} t_0 & t_1 & \cdots & t_{n-1} \\ m_0 & m_1 & \cdots & m_{n-1} \\ b_0 & b_1 & \cdots & b_{n-1} \end{pmatrix}$$

To conclude, a simple CC polygonal complex converges to a piecewise Bézier Curve which can be converted to a B-spline curve. In other words, the limit of a CC complex is a B-spline curve determined by the above operation. Figure 1.8 shows a polygonal complex with two subsequent divisions and its limit curve.

In the case of a general complex, the one-step refined mesh will form a simple complex whose limit can be similarly determined as above.

6. Applications of Polygonal Subdivision Curves

Polygonal complexes have proven to be useful in geometric modeling and computer graphics. This section gives few applications.

6.1. Free-Form Curve Generation

It may sound more complicated to define curves by a control mesh rather than a control polygon, but the use of polygonal complexes allows the curve to carry with it cross derivative information. This topic is investigated in more detail by the author in a forthcoming paper.

6.2. Curve Interpolation

This is a interesting application which has actually motivated the introduction of curve interpolation in CAGD. The idea simply consists of incorporating polygonal complexes in the polyhedron defining a subdivision surface. Subdivision of the polyhedron will generate a limit surface with the limit curves of the polygonal complexes interpolated by the surface.

The main practical issue in using the above result is how the user specifies the curve to be interpolated. Two approaches were suggested:

1. The *complex* approach: This consists of designing polygonal complexes of the curves to be interpolated [11], see Figure 1.9. These complexes are then embedded into a polyhedron to define a subdivision surface for which no further postprocessing is needed. Subdivision will be carried out in the normal way, which will automatically take the polyhedron to a limit surface interpolating the limit curves of the given complexes.
2. The *polygonal* approach: This provides a good interface to automate the process of designing such complexes through tagging control polygons on a given polyhedron. On this mesh, the system will then construct corresponding complexes that converge to the curves of these tagged polygons [12], see Figure 1.10.

Certainly, the second approach is more convenient since the construction of the *n*-reflected panels is done automatically. Its major drawback, however, is in the quality of the limit surface

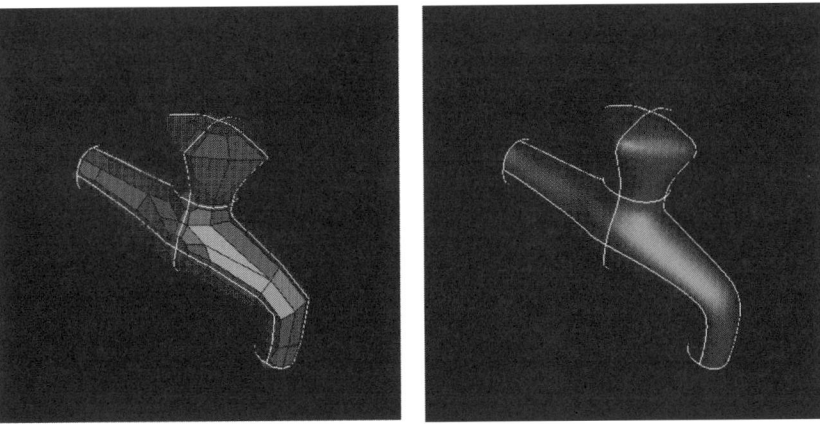

Figure 1.9   The Complex Approach: A polygonal mesh for a tap with some polygonal complexes, and its corresponding limit surface interpolating the curves defined by the complexes.

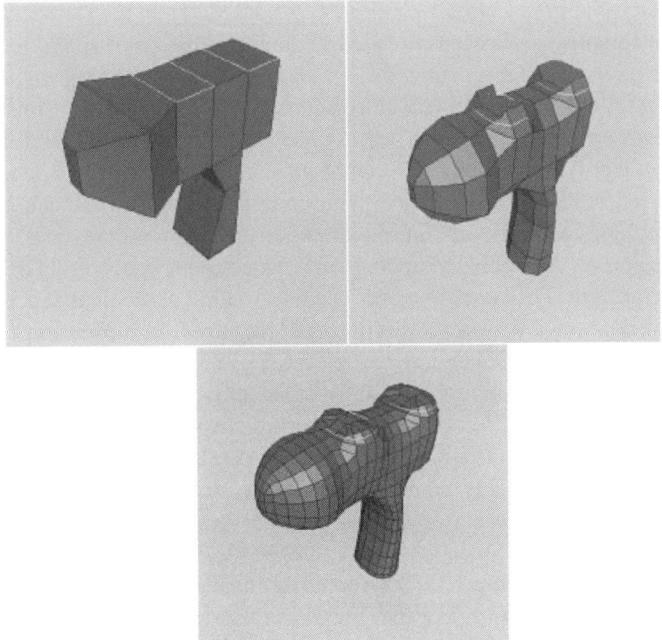

Figure 1.10   The Polygonal Approach: A hair dryer surface with two tagged control polygons, the mesh after constructing polygonal complexes, and the limit surface (bottom) before fairing.

across the interpolated curves, a problem that can be cured by smoothing polyhedral meshes with various constraints as suggested in [18].

In either of the suggested approaches, the challenging problem is the interpolation of intersecting curves. This problem has been partially addressed in the literature [8, 9, 11, 23]. In these references, no more than two intersecting curves can be interpolated by a subdivision

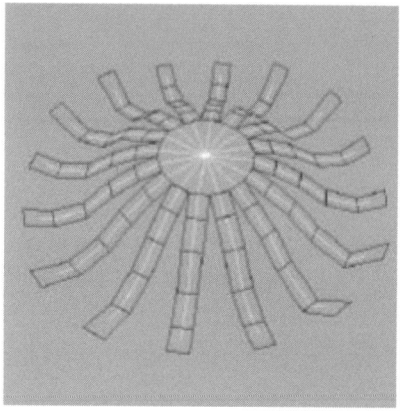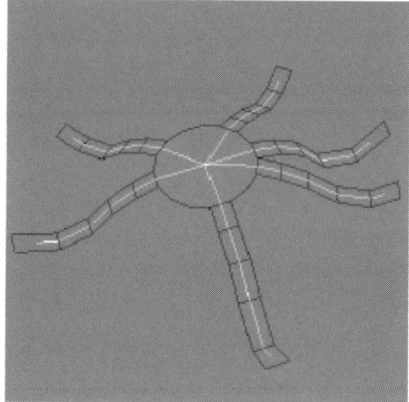

Figure 1.11   A net of 16 polygonal complexes sharing one 16-reflected panel, and a net of 6 polygonal complexes sharing the same panel.

surface such as Doo-Sabin or Catmull-Clark surfaces. Recently, in the Doo-Sabin Setting, the use of polygonal complexes has been extended to the interpolation of an *unlimited* number of curves through a point on the limit surface [19]. The idea is described below.

Each edge of an n-reflected panel can be considered as a contact edge. Consequently, with the restriction imposed on the panels of a polygonal complex being only 1-reflected (except the end-panels), up to n polygonal complexes can be attached to an n-reflected panel. Figure 1.11 (top) shows a 16-reflected panel with 16 polygonal complexes attached to it. It is also possible to connect less complexes, as shown in Figure 1.11, where only six complexes are attached to the same 16-reflected panel. Accordingly, up to n curves meeting at the centroid of an n-reflected panel can be interpolated. The continuity between these curves at the joint depends on whether the complexes are connected with opposite contact edges or not. Two complexes whose contact edges are opposite can be considered as one complex giving one C^1 curve passing through the common point. Therefore, up to $n/2$ C^1 curves can intersect at an extraordinary point. For example, in Figure 1.11 eight C^1 curves can cross the centroid of the panel.

Alternatively, a configuration of up to $n/2$ curves that can all meet with C^0 at an extraordinary point can also be achieved. This is simply because if we attach a complex to an edge, we do not attach another complex to its corresponding opposite contact edge in order to avoid the C^1 continuity with another curve. Since we have $n/2$ opposite contact edges then the number of curves is limited to $n/2$. In Figure 1.11, six curves can meet at C^0.

Theoretically, it is important to note that n can be as large as needed, hence an unlimited number of curves can be interpolated through the center of an n-reflected panel. Figure 1.12 shows a stamp interpolating several curves at one extraordinary point.

For the Catmull-Clark setting, a similar technique is under preparation by the author.

6.3. Trimming Subdivision Surfaces

Another application of the use of polygonal complexes is to trim a surface along its polygonal subdivision curve. This curve can be thought of as an edge inserted on the surface as discussed in [7]. The edge is a feature line or a shape handle along which a surface can be trimmed or

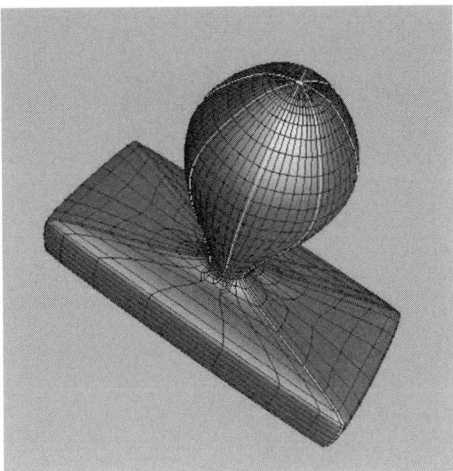

Figure 1.12   A stamp surface interpolating n curves meeting at the top.

split as indicated in Figure 1.13. This can be done by cutting the surface panel-wise rather than segment-wise, *i.e. leaving the panels of the complex to be shared by the two pieces apart.*

Continuity across the inserted edge can be controlled. A C^1 continuity can be achieved if both surfaces share the same polygonal complex. Twisting the panels of the complexes can reduce the continuity at the joint to C^0.

6.4. Lofted Recursive Subdivision Surfaces

The curve interpolation concept can be easily extended to generate lofted subdivision surfaces known as skinning. The idea consists of constructing a polygonal mesh M_0 from a given sequence of cross sections (c_i), such that the limit surface of M_0 interpolates these cross sections [16].

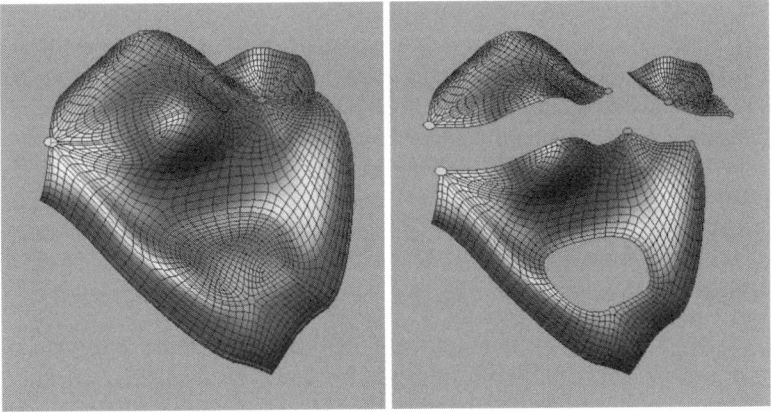

Figure 1.13   Trimming of a subdivision surface along interpolated curves. A sample surface and the same surface trimmed and then split into 3 pieces.

Figure 1.14   A lofted Catmull-Clark pawn with fourteen cross sections.

To elaborate, given a sequence of control polygons, each defining a cubic B-spline curve, we need to build a control mesh or a polyhedron whose Catmull-Clark surface will interpolate the given curves. This can be done by building a polygonal complex for each of the given curves and then connecting these complexes to get the required polyhedron. Initially, the case where all control polygons are defined by the same number of vertices has been considered. This was later extended to polygons with different number of vertices [20]. One advantage of this technique is that the interpolated curve need not be iso-parametric curves.

Figure 1.14 shows 14 cross sections of a pawn and its corresponding lofted shapes.

7. Summary

Polygonal subdivision complexes are very useful in computer graphics and geometric modeling applications. This chapter sketches on few of these applications. It is anticipated that they will play an important role not only in designing subdivision surfaces with features but also in providing shape control along such features. In computer graphics, a shape can be manipulated using the complexes while keeping the limit curves on the surface.

Future work will expand the definition of these complexes and the presented applications in various subdivision schemes.

Acknowledgements

This work was supported in part by a grant #51110-111130 from the American University of Beirut and a grant #111130022106 from the Lebanese National Council for Scientific Research. Thanks are due to A. Charrara for his valuable help in producing some of the included pictures using Geomview [5]. The author is also grateful to Dr. Malcolm Sabin for his valuable comments.

References

[1] Biermann, H., Levin, A. and Zorin, D. "Piecewise Smooth Subdivision Surfaces With Normal Control", *Computer Graphics Proceedings, ACM SIGGRAPH 2000*, pp. 113–120, 2000.

[2] Catmull, E. and Clark, J. "Recursively Generated B-Spline Surfaces On Arbitrary Tmeshes", *Seminal Graphics*, Ed. Rosalee Wolfe, ACM Press, pp. 183–188, 1998.

[3] Chaikin, G. M. "An Algorithm For High Speed Curve Generation", *Computer Graphics And Image Processing*, (3):12, pp. 346–349, 1974.

[4] Doo, D. and Sabin, M. "Behavior Of Recursive Division Surfaces Near Extraordinary Points", *Seminal Graphics*, Ed. Rosalee Wolfe, ACM Press, pp. 177–181, 1988.

[5] Levy, A., Munzner, T. and Philips, M. "Geomview 1.6.6 GL version", http://www.Geomview.org.

[6] Halstead, M., Kass, M. and De Rose, T. "Efficient, Fair Interpolation Using Catmull-Clark Surfaces", *Computer Graphics Proceedings, ACM SIGGRAPH 1993*, pp. 35–44, 1993.

[7] Habib, A. Three Approaches to Building Curves and Surfaces in Computer Aided Geometric Design, *Ph.D. Thesis*, Rice University, 1996.

[8] Levin, A. "Interpolating Nets Of Curves By Smooth Subdivision Surfaces", *Computer Graphics Proceedings*, *ACM SIGGRAPH 99*, pp. 57–64, 1999.

[9] Nasri, A. "Interpolation of Meshes of Curves by Recursive Subdivision Surfaces", presented at the *Fourth SIAM conference on Geometric Design*, Nashville, Nov. 6–9, 1995.

[10] Nasri, A. "Constructing Polygonal Complexes With Shape Handles for Curve Interpolation By Subdivision Surfaces", *Computer Aided Design*, (33), pp. 753–765, 2001.

[11] Nasri, A. "Recursive Subdivision Of Polygonal Complexes And Its Applications In CAGD", *Computer Aided Geometric Design*, (17), pp. 595–619, 2000. Presented at the fifth *SIAM Conference On Geometric Design*, Nashville, 1997.

[12] Nasri, A. "A Polygonal Approach For Interpolating Meshes of Curves by Subdivision Surfaces", *Proceedings of Geometric Modeling and Processing 2000*, Hong Kong, 10–12 April, IEEE, pp. 262–273, 2000.

[13] Nasri, A. "Designing Subdivision Surfaces Interpolating Triangular Meshes of Curves", American University of Beirut, Computer Science, Technical Report, TR-2002/4, 2002.

[14] Nasri, A. and Sabin, M. "Taxonomy of Interpolation Constraints in Recursive Subdivision Curves", *The Visual Computer*, (18):4, pp. 259–272, 2002.

[15] Nasri, A. and Sabin, M. "Taxonomy of Interpolation Constraints in Recursive Subdivision Surfaces", *The Visual Computer*, (18):5–6, pp. 382–403, 2002.

[16] Nasri, A. and Abbas, A. "Lofted Catmull-Clark Subdivision Surfaces", *Proceedings of Geometric Modeling and Processing 2002*, Japan, pp. 2002.

[17] Nasri, A. and Abbas, A. "Designing Catmull-Clark Subdivision Surfaces With Curve Interpolation Constraints", *Computers & Graphics*, (26):3, pp. 393–400, 2002.

[18] Nasri, A., Kim, T. and Lee, K. Polygonal Mesh Regularization for Subdivision Surfaces Interpolating Meshes of Curves, To appear in *The Visual Computer*, 2003.

[19] Nasri, A. "Interpolating an Unlimited Number of Curves Through an Extraordinary Point on Subdivision Surfaces". To appear in *Computer Graphics Forum*, special issue on subdivision, 2003.

[20] Nasri, A., Abbas, A. and Hasbini, I. "Skinning Subdivision Surfaces", in preparation, 2003.

[21] Reif, U. "A Unified Approach To Subdivision Algorithms Near Extraordinary Vertices", *Computer Aided Geometric Design*, (12), pp. 153–174, 1995.

[22] Peters, J. and Reif, U. "Analysis of Generalized B-spline Subdivision Algorithms", *SIAM Journal On Numerical Analysis*, (35):2, pp. 728–748, 1998.

[23] Schweitzer, J. "Analysis And Applications Of Subdivision Surfaces", *Ph.D. Thesis*, University Of Washington Seattle, 1996.

[24] Sederberg, T., Zheng, J., Sewell, D. and Sabin, M. "Non-Uniform Recursive Subdivision Surfaces", *Computer Graphics Proceedings, ACM SIGGRAPH 98*, pp. 387–394, 1998.

[25] Zorin, D. and Schröder, P. Subdivision For Modeling And Animation, *ACM SIGGRAPH Course Notes*, 2000.

[26] Warren, J. and Weimer, H. *Subdivision Methods For Geometric Design: A Constructive Approach*, Morgan Kaufmann Publishers, 2002.

2

Planar Development of Digital Free-Form Surfaces

Phillip N. Azariadis
Nickolas S. Sapidis
Department of Product and Systems Design Engineering, University of the Aegean, Ermoupolis, Syros, 84100 Greece

There are three main goals of this chapter. First, a detailed review of differential-geometry criteria for the developability of free-form surfaces is presented. Then, tools measuring the accuracy of planar developments are introduced and analyzed. These tools are a prerequisite for evaluating the numerous methods/approaches proposed for generating flat developments of digital free-form surfaces whose level of involvement in many areas of CAD/CAM and computer graphics is constantly increasing. Finally, some of the most efficient surface flattening methods are analyzed and categorized followed by a discussion of representative examples.

1. Introduction

Planar development of curved surfaces is a well known problem in the manufacturing field. Generating isometric mappings between two different surfaces has also attracted the attention of many researchers in the computer graphics community due to its application in non-distorted texture mapping. Historically, this problem first appeared in the context of the 'mapmaker' problem: mapping the surface of the Earth onto a plane. Gauss in 1828 proved that such a mapping is not possible due to the different intrinsic curvature of the two surfaces. Thus one can only aim at derivation of an approximately isometric mapping with minimal geometric distortions.

Numerous approaches have been proposed for dealing with the planar development problem of curved surfaces. The problem is trivial when the surface has zero curvature, but becomes significantly complicated when the surface is doubly-curved. In fact, in the latter case, it is proven that there is actually an infinite number of different planar developments for the same

Advances in Geometric Modeling. Edited by M. Sarfraz
© 2003 John Wiley & Sons, Ltd   ISBN: 0-470-85937-7

surface. Unfortunately, the vast majority of the products produced nowadays include free-form surfaces, which are often created from clouds of points (produced using a modern digitizing device). The selection of the right method for producing a planar development is a subtle task, since there are many issues that have to be taken into consideration, such as the production method, material properties, surface geometry and so on.

The primary purpose of this chapter is to provide an analytical description of developability criteria for surfaces and present tools for measuring the accuracy of planar developments. A classification of the many available methods for generating flat developments of free-form surfaces is also presented. The final goal of this chapter is to help the reader understand, following an intuitive approach, the various aspects of the surface flattening problem and of the related solutions.

This chapter is structured as follows: in Section 2, we review results from differential geometry related to the developability of surfaces. In Section 3, we analyze local properties of a set of affine transformations used to approximate an isometric mapping of a doubly-curved surface onto the plane. In Section 4, a classification of the available flattening methods is presented. Section 5, concludes this chapter with examples and an evaluation of current methods for obtaining adequate planar developments.

2. Criteria of Developability for Surfaces

This section reviews results from differential geometry which are applied to the development of a set of criteria for the *developability* of surfaces. The interested reader is referred to [10,19,21,32] for an in-depth analysis of the presented results. In the following, the cross product of vectors $\mathbf{x}$ and $\mathbf{y}$ is denoted by $\mathbf{x} \times \mathbf{y}$, the gross product of vectors $\mathbf{x}$, $\mathbf{y}$ and $\mathbf{z}$ by $[\mathbf{x}, \mathbf{y}, \mathbf{z}]$, and the Euclidean norm of $\mathbf{x}$ by $\|\mathbf{x}\|$.

2.1. Intrinsic Geometry of Surfaces

The surface $\mathbf{x}$ of class, C^m, $m \geq 2$, is given by a parametric equation $\mathbf{x} = \mathbf{x}(u, v)$. The differential of $\mathbf{x} = \mathbf{x}(u, v)$ at (u, v) is a 1-1 and onto linear mapping $d\mathbf{x} = \mathbf{x_u}du + \mathbf{x_v}dv$, which maps the random vector (du, dv) of the uv-plane to the surface tangent vector $\mathbf{x_u}du + \mathbf{x_v}dv$ at $\mathbf{x}(u, v)$. The First Fundamental Form of $\mathbf{x} = \mathbf{x}(u, v)$ is a second degree function of du and dv given as [21]:

$$\begin{aligned}
I(du, dv) &= d\mathbf{x} \cdot d\mathbf{x} \\
&= (\mathbf{x_u}du + \mathbf{x_v}dv) \cdot (\mathbf{x_u}du + \mathbf{x_v}dv) \\
&= (\mathbf{x_u} \cdot \mathbf{x_u})du^2 + 2(\mathbf{x_u} \cdot \mathbf{x_v})dudv + (\mathbf{x_v} \cdot \mathbf{x_v})dv^2 \\
&= E du^2 + 2F dudv + G dv^2
\end{aligned} \tag{2.1}$$

where

$$E = \mathbf{x_u} \cdot \mathbf{x_u}, \quad F = \mathbf{x_u} \cdot \mathbf{x_v}, \quad G = \mathbf{x_v} \cdot \mathbf{x_v}. \tag{2.2}$$

The coefficients E, F and G are the *First Order Fundamental Coefficients* and are functions of u and v.

At every point on $\mathbf{x} = \mathbf{x}(u, v)$ there is unit normal vector $\mathbf{N} = \dfrac{\mathbf{x_u} \times \mathbf{x_v}}{\|\mathbf{x_u} \times \mathbf{x_v}\|}$ which defines C^1 mapping $\mathbf{N}$ from the surface to the unit sphere (Gauss mapping) with respect to u and v. The differential of $\mathbf{N}$ is a vector $d\mathbf{N} = \mathbf{N_u}du + \mathbf{N_v}dv$ normal to the surface tangent plane at $\mathbf{x}(u, v)$ and it is called the *Second Order Fundamental Form* of surface $\mathbf{x}$. It is given by,

$$
\begin{aligned}
\mathrm{II}(du, dv) &= -d\mathbf{x} \cdot d\mathbf{N} \\
&= -(\mathbf{x_u}du + \mathbf{x_v}dv) \cdot (\mathbf{N_u}du + \mathbf{N_v}dv) \\
&= Ldu^2 + 2Mdudv + Ndv^2
\end{aligned}
\tag{2.3}
$$

where

$$
L = -\mathbf{x_u} \cdot \mathbf{N_u}, \quad M = -\tfrac{1}{2}(\mathbf{x_u} \cdot \mathbf{N_v} + \mathbf{x}_v \cdot \mathbf{N_u}), \quad N = -\mathbf{x_v} \cdot \mathbf{N_v}.
\tag{2.4}
$$

The coefficients L, M and N are the *Second Order Fundamental Coefficients* and can be alternatively written as:

$$
L = \mathbf{x_{uu}} \cdot \mathbf{N}, \quad M = \mathbf{x_{uv}} \cdot \mathbf{N}, \quad N = \mathbf{x_{vv}} \cdot \mathbf{N}.
\tag{2.5}
$$

Each surface point $\mathbf{P} = \mathbf{x}(u, v)$ can be characterized through the result of the scalar $\Delta = LN - M^2$ in four distinct cases:

- Elliptic point: $\Delta > 0$
- Hyperbolic point: $\Delta < 0$
- Parabolic point: $\Delta = 0$ and $L^2 + M^2 + N^2 \neq 0$
- Flat point: $\Delta = 0$ and $L = M = N = 0$.

Moreover, there are two scalars, namely κ_1 and κ_2, defined at $\mathbf{P}$ which correspond to the two main curvatures of $\mathbf{x}$ at $\mathbf{P}$ and are given as the roots of the second degree equation:

$$
\kappa^2 - 2H\kappa + K = 0
\tag{2.6}
$$

where:

$$
H = \tfrac{1}{2}(\kappa_1 + \kappa_2) = \frac{EN + GL - 2FM}{2(EG - F^2)}
\tag{2.7}
$$

is the *mean curvature* at $\mathbf{P}$ and is the average of κ_1 and κ_2. The scalar

$$
K = \kappa_1\kappa_2 = \frac{LN - M^2}{EG - F^2}
\tag{2.8}
$$

is the well-known *Gaussian curvature* at $\mathbf{P}$. Comparing Equation (2.8) with Δ we conclude that a surface point is elliptic, hyperbolic, parabolic or flat iff $K > 0$, $K < 0$ or $K = 0$, respectively.

2.2. Isometric Mappings

Let **f** be a 1-1 mapping of a surface S onto a surface S^*. Mapping **f** defines an *isometric mapping* or an *isometry* if the arc length of a normal arc $\mathbf{x} = \mathbf{x}(t)$ of S equals the length of its image $\mathbf{x}^* = \mathbf{x}^*(t) = \mathbf{f}(\mathbf{x}(t))$ on S^*. If **f** is an isometry then $\mathbf{f}^{-1}$ is an isometry from S^* onto S, too. In such a case we call S and S^* isometric. In general, it can be proved [21] that a 1-1 mapping **f** from S onto S^* is an isometry iff for each part $\mathbf{x} = \mathbf{x}(u, v)$ of S and $\mathbf{x}^* = \mathbf{f}(\mathbf{x}(u, v))$ of S^* the first order fundamental coefficients are equal:

$$E = E^*, \quad F = F^* \text{ and } G = G^*. \tag{2.9}$$

2.3. Developability Criteria for Surfaces

Two surfaces S and S^* are called *applicable* if there is a continuous family of mappings $\mathbf{f}_\lambda$, $0 \le \lambda \le 1$, from S onto S^* such that:

a. $\mathbf{f}_0(S) = S$
b. $\mathbf{f}_1(S) = S^*$
c. mappings $\mathbf{f}_\lambda$ are isometric mappings from S onto $\mathbf{f}_\lambda(S)$, for every $\lambda \in [0, 1]$.

Intuitively, one understand that S and S^* are applicable if S can be bended continuously and isometrically in such a way that the final bended surface coincides with S^*. Obviously, if S and S^* are applicable then they are isometric too.

Taking this into account, we can easily ascertain that a surface is *developable* iff it is applicable to the plane. Thus, every developable surface can be unrolled onto a plane without distortion, implying that a developable surface can be constructed through the smooth bend of a plane sheet. As a result, every developable surface is isometric to the plane and therefore it holds that $LN - M^2 = 0$ (meaning that all points of such a surface are parabolic or flat) and thus its Gaussian curvature equals zero at every point. This discussion is summarized by the following theorem:

Theorem 1. *Let S be a C^m, $m \ge 2$, class surface. The following properties are equivalent:*

  i. *The surface is developable.*
  ii. *All the surface points are parabolic or flat.*
  iii. *The Gaussian curvature at every surface point equals zero.*
  iv. *The surface is applicable to the plane.* ■

Essentially, this theorem stands as a general criterion for the developability of any surface. However, more specific criteria can be developed for certain families of surfaces like ruled or revolved surfaces.

2.3.1 Ruled Surfaces

A ruled surface is constructed from a one-parameter family of lines and its normal parametric equation is given by:

$$\mathbf{x} = \mathbf{x}(u, v) = \mathbf{n}(u) + v\mathbf{q}(u), v \in (-\infty, +\infty), \tag{2.10}$$

where, $\mathbf{n} = \mathbf{n}(u)$ is a C^m, $m \geq 2$, curve called a *directrix* and $\mathbf{q} = \mathbf{q}(u)$ is a C^m, $m \geq 2$, vector function which defines the lines' direction at each point $\mathbf{n} = \mathbf{n}(u)$, and it is called a *generatrix*. The surface normal vector is given by:

$$N = \frac{(\dot{\mathbf{n}} + v\dot{\mathbf{q}}) \times \mathbf{q}}{\sqrt{(\dot{\mathbf{n}} + v\dot{\mathbf{q}})^2 - (\mathbf{n}, \mathbf{q})^2}} \tag{2.11}$$

where, $\dot{\mathbf{n}} = \dfrac{d\mathbf{n}}{du}$ and $\dot{\mathbf{q}} = \dfrac{d\mathbf{q}}{du}$.

Theorem 2. *A ruled surface is developable iff* $[\dot{\mathbf{n}}, \mathbf{q}, \dot{\mathbf{q}}] = 0$.

Proof. A ruled surface is developable if the cross product of two normal vectors defined at arbitrary positions $v_1 \neq v_2$ of the same generatrix equals zero. It holds,

$$
\begin{aligned}
&[(\dot{\mathbf{n}} + v_1\dot{\mathbf{q}}) \times \mathbf{q}] \times [(\dot{\mathbf{n}} + v_2\dot{\mathbf{q}}) \times \mathbf{q}] \\
&= [(\dot{\mathbf{n}} + v_1\dot{\mathbf{q}}), \mathbf{q}, \mathbf{q}](\dot{\mathbf{n}} + v_2\dot{\mathbf{q}}) - [(\dot{\mathbf{n}} + v_1\dot{\mathbf{q}}), \mathbf{q}, (\dot{\mathbf{n}} + v_2\dot{\mathbf{q}})]\mathbf{q} \\
&= (v_1 - v_2)[\dot{\mathbf{n}}, \mathbf{q}, \dot{\mathbf{q}}]\mathbf{q}
\end{aligned}
\tag{2.12}
$$

which completes the proof. ∎

2.3.2 Revolved Surfaces

Let the normal parametric equation of a revolved surface be:

$$\mathbf{x} = \mathbf{x}(u, v) = (r(u)\cos v, \quad r(u)\sin v, \quad z(u)) \tag{2.13}$$

where $r = r(u)$, $z = z(u)$ are the parametric equations of the surface meridians, and u is the arc length of each meridian.

Theorem 3. *A revolved surface is developable iff* $r'' = 0$.

Proof. The first order fundamental coefficients are

$$E = 1, F = 0, G = r^2 \tag{2.14}$$
$$L = r'z'' - r''z', M = 0, N = rz' \tag{2.15}$$

Substituting into Equation (2.8) leads to the following expression for the Gaussian curvature of the surface:

$$K = K(u) = \frac{z'(r'z'' - r''z')}{r}, \tag{2.16}$$

which obviously depends only on the arc length u. Then,

$$(r')^2 + (z')^2 = 1 \Rightarrow r'r'' + z'z'' = 0 \Rightarrow z'' = -\frac{r'r''}{z'}. \tag{2.17}$$

Substituting Equation (2.17) into Equation (2.16) produces:

$$K(u) = -\frac{r''}{r}, \tag{2.18}$$

which completes the proof. ∎

Theorem 4. *A revolved surface is developable iff* $\begin{vmatrix} r'(u_1) & z'(u_1) \\ r'(u_2) & z'(u_2) \end{vmatrix} = 0$ *for every* $u_1 \neq u_2$.

Proof. The surface normal vector is,

$$\mathbf{N} = (-z'(u)\cos v, -z'(u)\sin v, r'(u)). \tag{2.19}$$

Let $\mathbf{N}_1$ and $\mathbf{N}_2$ be two surface normal vectors defined at the same meridian for $u_1 \neq u_2$. The surface will be developable iff the cross product of $\mathbf{N}_1$ and $\mathbf{N}_2$ equals to zero. This implies that:

$$\mathbf{N}_1 \times \mathbf{N}_2 = \begin{vmatrix} r'(u_1) & z'(u_1) \\ r'(u_2) & z'(u_2) \end{vmatrix} (\sin v, -\cos v, 0). \tag{2.20}$$

Taking into account that $\sin^2 v + \cos^2 v = 1$, we derive that $\mathbf{N}_1 \times \mathbf{N}_2 = \mathbf{0}$ iff $\begin{vmatrix} r'(u_1) & z'(u_1) \\ r'(u_2) & z'(u_2) \end{vmatrix} = 0$, which completes the proof. ∎

3. Evaluating Planar Developments

The above criteria for developability limit the variety of surfaces that can be isometrically unfolded onto the plane. On the other hand, the vast variety of surfaces used in today's products are doubly curved with arbitrarily complex shapes. For these surfaces, approximate, local isometric-mappings should be considered, which is the subject of this section.

The surfaces considered in the present context are approximated with an adequate mesh Φ of triangles. This implies that the mesh is allowed to have variable density depending on the local accuracy of approximation. Note that Φ contains only positive non-degenerated triangles, i.e., the vertices have a counter-clockwise order and the triangle area is always positive.

3.1. Affine Triangle Transformations

Let S be a surface given by the parametric equation $\mathbf{x} = \mathbf{x}(u, v)$ and the uv-plane P. Then, $\mathbf{x}^{-1}$ maps three-dimensional points of S onto the plane P. If $\mathbf{x}$ is an isometry then the surface S is developable. We focus on the case where $\mathbf{x}$ is not an isometry.

Since S is approximated with a finite number of triangular elements, we can also approximate $\mathbf{x}$ using the same number of local mappings between triangular elements. We assume that there is a mesh φ on plane P having equivalent topological characteristics with Φ. There is a 1-1 correspondence between the elements of Φ and φ.

Let us consider an arbitrary pair of corresponding triangles $\nabla(\mathbf{A}, \mathbf{B}, \mathbf{C})$ of Φ and $\nabla(\mathbf{a}, \mathbf{b}, \mathbf{c})$ of φ, where $\mathbf{A}, \mathbf{B}, \mathbf{C} \in \Re^3$ and $\mathbf{a}, \mathbf{b}, \mathbf{c} \in \Re^2$ are respectively the vertices of the triangles. Using the Gram-Schmidt orthogonalization, we define a local orthonormal coordinate system in each triangle as follows.

Let L be the orthonormal coordinate system with unit vectors $\mathbf{Q}_1$ and $\mathbf{Q}_2$ defined as:

$$
\begin{aligned}
\mathbf{Q}_1 &= \frac{\mathbf{B} - \mathbf{A}}{\|\mathbf{B} - \mathbf{A}\|} \\
\mathbf{Q}_2 &= \frac{(\mathbf{C} - \mathbf{A}) - ((\mathbf{C} - \mathbf{A}) \cdot \mathbf{Q}_1)\mathbf{Q}_1}{\|(\mathbf{C} - \mathbf{A}) - ((\mathbf{C} - \mathbf{A}) \cdot \mathbf{Q}_1)\mathbf{Q}_1\|}
\end{aligned}
\tag{2.21}
$$

The origin of L is set to be the point $\mathbf{A}$.

Similarly, let ℓ denote the local orthonormal coordinate system with unit vectors $\mathbf{q}_1$ and $\mathbf{q}_2$:

$$
\begin{aligned}
\mathbf{q}_1 &= \frac{\mathbf{b} - \mathbf{a}}{\|\mathbf{b} - \mathbf{a}\|} \\
\mathbf{q}_2 &= \frac{(\mathbf{c} - \mathbf{a}) - ((\mathbf{c} - \mathbf{a}) \cdot \mathbf{q}_1)\mathbf{q}_1}{\|(\mathbf{c} - \mathbf{a}) - ((\mathbf{c} - \mathbf{a}) \cdot \mathbf{q}_1)\mathbf{q}_1\|}
\end{aligned}
\tag{2.22}
$$

The origin of ℓ is the point $\mathbf{a}$. Then, the local coordinates of the two triangles are given, with respect to the two local coordinate systems, by:

$$
\begin{aligned}
\mathbf{A}^L &= (0, 0) \\
\mathbf{B}^L &= \|\mathbf{B} - \mathbf{A}\|\mathbf{Q}_1 = \left(B_x^L, 0\right) \\
\mathbf{C}^L &= ((\mathbf{C} - \mathbf{A}) \cdot \mathbf{Q}_1, (\mathbf{C} - \mathbf{A}) \cdot \mathbf{Q}_2) = \left(C_x^L, C_y^L\right)
\end{aligned}
\tag{2.23}
$$

and

$$
\begin{aligned}
\mathbf{a}^\ell &= (0, 0) \\
\mathbf{b}^\ell &= \|\mathbf{b} - \mathbf{a}\|\mathbf{q}_1 = \left(b_x^\ell, 0\right) \\
\mathbf{c}^\ell &= ((\mathbf{c} - \mathbf{a}) \cdot \mathbf{q}_1, (\mathbf{c} - \mathbf{a}) \cdot \mathbf{q}_2) = \left(c_x^\ell, c_y^\ell\right)
\end{aligned}
\tag{2.24}
$$

We define an affine transformation of triangle $\nabla(\mathbf{A}, \mathbf{B}, \mathbf{C})$ to the triangle $\nabla(\mathbf{a}, \mathbf{b}, \mathbf{c})$ by a local linear mapping $\mathbf{f}$ written in matrix form as:

$$
\begin{bmatrix} b_x^\ell & c_x^\ell \\ 0 & c_y^\ell \end{bmatrix} = \mathbf{f} \begin{bmatrix} B_x^\ell & C_x^\ell \\ 0 & C_y^\ell \end{bmatrix} \Leftrightarrow
$$

$$
\mathbf{f} = \begin{bmatrix} b_x^\ell & c_x^\ell \\ 0 & c_y^\ell \end{bmatrix} \begin{bmatrix} B_x^\ell & C_x^\ell \\ 0 & C_y^\ell \end{bmatrix}^{-1} \Leftrightarrow
$$

$$
\mathbf{f} = \begin{bmatrix} \dfrac{b_x^\ell}{B_x^L} & \dfrac{B_x^L c_x^\ell - b_x^\ell C_x^L}{B_x^L C_y^L} \\[2ex] 0 & \dfrac{c_y^\ell}{C_y^L} \end{bmatrix}
\tag{2.25}
$$

Note that the matrix $\begin{bmatrix} B_x^\ell & C_x^\ell \\ 0 & C_y^\ell \end{bmatrix}$ always has an inverse since the determinant $D = B_x^\ell C_y^\ell$ is nonzero. The set of all local mappings $\mathbf{f}$ defines an approximation of $\mathbf{x}^{-1}$. Furthermore, if the two corresponding triangles are equal then $\mathbf{f}$ is an isometry and surface S is considered, in this triangular area, locally isometric to the plane. Thus, we should focus on the study of the properties of $\mathbf{f}$, which is the subject of the next section.

3.2. Properties of Local Mappings

In this section we shall investigate both quantitative and qualitative characteristics of local mappings $\mathbf{f}$ in order to ascertain whether they define an isometry or not.

3.2.1 Mapping Points

We can simplify the used notation taking into account that, since L and ℓ are orthonormal coordinate systems, we can express the coordinates of the vertices of $\nabla(\mathbf{A}^L, \mathbf{B}^L, \mathbf{C}^L)$ and $\nabla(\mathbf{a}^\ell, \mathbf{b}^\ell, \mathbf{c}^\ell)$ in a global coordinate system W as:

$$
\begin{aligned}
\mathbf{A}^W &= (0,0) & \mathbf{a}^W &= (0,0) \\
\mathbf{B}^W &= \left(B_x^L, 0\right) & \text{and} \quad \mathbf{b}^W &= \left(b_x^\ell, 0\right) \\
\mathbf{C}^W &= \left(C_x^L, C_y^L\right) & \mathbf{c}^W &= \left(c_x^\ell, c_y^\ell\right)
\end{aligned}
\tag{2.26}
$$

In addition, we drop the use of superscripts and, in the rest of this section, we denote the triangle vertices of Equation (2.26) as $(\mathbf{A}, \mathbf{B}, \mathbf{C})$ and $(\mathbf{a}, \mathbf{b}, \mathbf{c})$ respectively. This is illustrated in Figure 2.1.

Let $\mathbf{P} = (X, Y)$ a point within triangle $\nabla(\mathbf{A}, \mathbf{B}, \mathbf{C})$. This point is mapped through $\mathbf{f}$ onto a point $\mathbf{p} = (x, y)$ within triangle $\nabla(\mathbf{a}, \mathbf{b}, \mathbf{c})$ according to $\mathbf{p} = \mathbf{fP}$. If $\mathbf{f}$ is an isometry then the squared Euclidean-distance d^2 between $\mathbf{p}$ and $\mathbf{P}$ should be zero. This distance is computed as

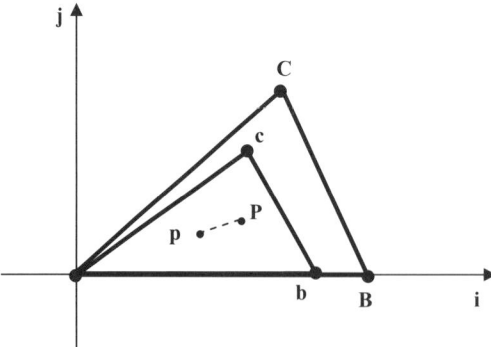

Figure 2.1   The two corresponding triangles drawn in the global coordinate system W with unit vectors $i = (1, 0)$ and $j = (0, 1)$.

follows:

$$\begin{bmatrix} x \\ y \end{bmatrix} = \mathbf{f} \begin{bmatrix} X \\ Y \end{bmatrix} \Leftrightarrow$$

$$\begin{bmatrix} X \\ Y \end{bmatrix} = \mathbf{f}^{-1} \begin{bmatrix} x \\ y \end{bmatrix} \Leftrightarrow \begin{bmatrix} X \\ Y \end{bmatrix} - \begin{bmatrix} x \\ y \end{bmatrix} = \mathbf{f}^{-1} \begin{bmatrix} x \\ y \end{bmatrix} - \begin{bmatrix} x \\ y \end{bmatrix} \Leftrightarrow \tag{2.27}$$

$$\begin{bmatrix} X - x \\ Y - y \end{bmatrix} = (\mathbf{f}^{-1} - \mathbf{I}) \begin{bmatrix} x \\ y \end{bmatrix}$$

Then,

$$d^2 = [X - x \quad Y - y] \begin{bmatrix} X - x \\ Y - y \end{bmatrix}$$

$$= [x \quad y](\mathbf{f}^{-1} - \mathbf{I})^{\mathrm{T}}(\mathbf{f}^{-1} - \mathbf{I}) \begin{bmatrix} x \\ y \end{bmatrix} \Rightarrow$$

$$d^2 = \mathbf{p}^{\mathrm{T}}\mathbf{M}\mathbf{p}, \tag{2.28}$$

where $\mathbf{M} = (\mathbf{f}^{-1} - \mathbf{I})^{\mathrm{T}}(\mathbf{f}^{-1} - \mathbf{I})$. The scalar d^2 expresses quantitatively the distortion caused by $\mathbf{f}$ regarding the mapping of $\mathbf{P}$ onto $\mathbf{p}$. Furthermore, expanding Equation (2.28) we find that the mapped point $\mathbf{p}$ belongs to an ellipse which can be computed according to the following theorem.

Theorem 5. *Any point* $\mathbf{P} = (X, Y)$ *is mapped through* f *into a point* $\mathbf{p} = (x, y)$ *which belongs to the ellipse,*

$$(m_{11} - 1)x^2 + (m_{12} + m_{21})xy + (m_{22} - 1)y^2 + 2Xx + 2Yy - X^2 - Y^2 = 0$$

where $\mathbf{M} = [m_{ij}] = (\mathbf{f}^{-1} - \mathbf{I})^{\mathrm{T}}(\mathbf{f}^{-1} - \mathbf{I}), i, j = 1, 2$ ∎

Now, we focus on estimating the error introduced by the mapping $\mathbf{f}$. For this purpose we apply $\mathbf{f}$ on a circle of unit radius centered at $(0, 0)$. This transformation will reveal more detailed characteristics of the distortion that $\mathbf{f}$ may cause if it is not an isometry.

Let us consider a circle $\mathbf{X_c}(\omega) = (X_c(\omega), Y_c(\omega)) = (\cos\omega, \sin\omega)$ of unit radius with its center lying at the origin of the global system of reference. An arbitrary circle point is mapped, through $\mathbf{f} = [f_{ij}]$, to a point $\mathbf{x_c}(\omega)$ which belongs to the ellipse,

$$\begin{bmatrix} x_c(\omega) \\ y_c(\omega) \end{bmatrix} = \begin{bmatrix} f_{11} & f_{12} \\ 0 & f_{22} \end{bmatrix} \begin{bmatrix} X_c(\omega) \\ Y_c(\omega) \end{bmatrix} \Rightarrow$$

$$x_c(\omega) = f_{11} \cos\omega + f_{12} \sin\omega$$

$$y_c(\omega) = f_{22} \sin\omega \tag{2.29}$$

Then, the squared distance between the new point $(x_c(\omega), y_c(\omega))$ and the origin is:

$$g(\omega) = x_c^2(\omega) + y_c^2(\omega)$$

$$= f_{11}^2 \cos^2\omega + \left(f_{12}^2 + f_{22}^2\right) \sin^2\omega + 2f_{11}f_{12} \cos\omega \sin\omega \tag{2.30}$$

The angle at which the unit circle suffers the maximum or minimum deformation is computed at $g'(\omega) = 0$ as:

$$\omega_1 = \frac{1}{2} \tan^{-1} \frac{2 f_{11} f_{12}}{f_{11}^2 - f_{12}^2 - f_{22}^2} \tag{2.31}$$

Substituting Equation (2.31) into Equation (2.30) the principal direction of the ellipse relative to the x-axis of the global system is given by:

$$\varphi = \tan^{-1} \left(y_c(\omega_1) \big/ x_c(\omega_1) \right) \tag{2.32}$$

The extension or shrinkage d_p of the unit circle along this direction is:

$$d_p = \sqrt{x_c^2(\omega_1) + y_c^2(\omega_1)}. \tag{2.33}$$

In a similar fashion the second component of the extension is:

$$d_q = \sqrt{x_c^2(\omega_1 + \pi/2) + y_c^2(\omega_1 + \pi/2)} \tag{2.34}$$

Having computed the lengths and the directions of the ellipse's axis, we can write the parametric equation of the ellipse as,

$$\left. \begin{array}{l} x_e(u) = d_p \cos\varphi \cos u - d_q \sin\varphi \sin u \\ y_e(u) = d_p \sin\varphi \cos u + d_q \cos\varphi \sin u \end{array} \right\}, u \in [0,\ 2\pi] \tag{2.35}$$

Based on the above analysis we can state the following theorem [2].

Theorem 6. *A circle of unit radius is transformed through f into an ellipse with half axis lengths equal to d_p and d_q, respectively, and with its major axis inclined with an angle φ relative to the x-axis of the global system of reference.* ■

Remark 1. *The ellipse components d_p and d_q can be also computed through the Singular Value Decomposition of f.*

In other words, $\mathbf{f}$ can be expressed as a matrix product $\mathbf{f} = \mathbf{R}(\theta)\mathbf{\Lambda}\mathbf{R}(\varphi)$, where $\mathbf{R}(\varphi)$ and $\mathbf{R}(\theta)$ express rotations while $\mathbf{\Lambda}$ is a diagonal matrix expressing the deformation along the two principal axis of the ellipse, i.e., $\mathbf{\Lambda} = \begin{bmatrix} d_p & 0 \\ 0 & d_q \end{bmatrix}$. Thus, d_p and d_q are the square roots of the eigenvalues of the positive matrix $\mathbf{f}\mathbf{f}^{\mathrm{T}}$.

Remark 2. *The mapping f is an isometry iff $d_p = 1$ and $d_q = 1$.*

Remark 3. *The determinant of the Jacobian of mapping f equals $d_p d_q$.*

Obviously, a trivial case is when $\mathbf{f} = \mathbf{I}_2$.

3.2.2 Measuring the Accuracy of Planar Developments

Taking into account the set of mappings $\mathbf{f}$ which approximate $\mathbf{x}^{-1}$, we wish to derive meaningful indices measuring the metric distortion during the planar development of a doubly-curved surface.

Homogeneity of distortion: The distortion should be homogeneous throughout the surface in order to avoid rapid changes in local areas of the planar development. The ratio of the minimum over the maximum value of d_p and the corresponding ratio for d_q are good measures of the distortion variation in the first and second principal directions. The ratio of $\min\{d_p d_q\}$ over $\max\{d_p d_q\}$, for all triangles, characterizes the homogeneity of the distortion along both principal directions, i.e.,

$$h = \frac{\min\{d_p d_q\}}{\max\{d_p d_q\}}. \tag{2.36}$$

Ideally, the value of h should be constant over the surface and close to the unit.

Aspect ratio: Aspect ratio should be preserved to avoid non-uniform stretching of the planar development. This distortion can be expressed as:

$$r = \frac{\min\{d_p\}}{\max\{d_q\}} \tag{2.37}$$

for all the elements of $\boldsymbol{\Phi}$.

These indices may be used to measure the accuracy of a planar development of a doubly-curved surface both locally and globally. Such applications will be given later in this chapter.

4. Methods for Approximate Planar Development of Curved Surfaces

The problem of flattening a curved surface onto a plane is important not only for manufacturing but also for the computer graphics community. In fact, the well known two-dimensional texture mapping technique is fundamentally equivalent to the planar development of three-dimensional surfaces. Thus, many attempts have been made to solve the problem of producing nearly isometric flat developments of curved surfaces both for manufacturing and texture mapping purposes. Most of these methods may be classified into three categories:

A. Methods based on the minimization of an objective function (usually called as *energy function*), assuming a degree of elasticity in the surface material.
B. Methods based on intrinsic differential-geometric properties of surfaces. These are usually employed when local accuracy in the planar development is of great importance, or when it is necessary to insert line cuttings.
C. Methods based on the approximation of the initial surface with developable surfaces which are isometrically unfolded onto the plane.

Current methods representing all categories are presented in the following subsections.

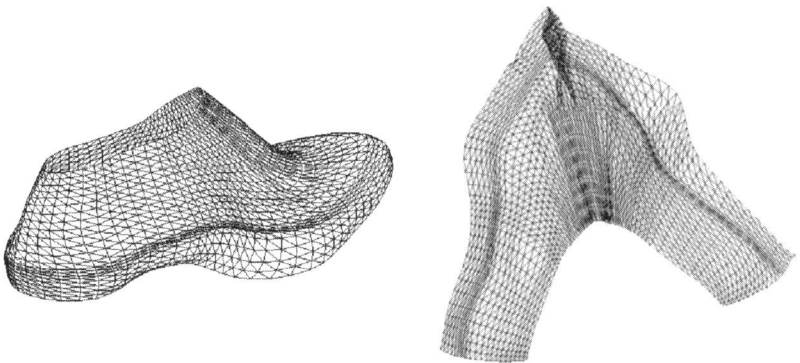

Figure 2.2   Example of applying the method of Ma and Lin to derive a planar development of a shoe
last.

4.1. Category A: Minimization of An Energy Function

These methods employ objective functions measuring the 'difference' between the planar
development and the corresponding three-dimensional surface. In other words, these functions
measure the energy needed for the planar development to be fitted onto the free-form surface or
vice-versa. The closer to zero the objective-function's value is, the better planar development
is obtained. Clearly, for a developable surface only the corresponding objective function may
obtain a zero value. Minimization of the objective function is usually achieved using a standard
optimization method, e.g., an iterative technique.

Ma and Lin [30] were the first to present a flattening technique based on optimizing an
objective function comparing the length of triangle edges of the surface mesh Φ with that of
corresponding triangle edges in the planar mesh φ. Unfortunately, this method may produce
triangles on the plane P with the wrong orientation, leading to a planar development with
overlaps (see Figure 2.2).

Maillot *et al.* [17] improve the previous method by using a new objective function, linearly
combining an energy function comparing lengths with another one based on the difference of
signed areas, which avoids definition of triangles with the wrong orientation. This is evident
in the example of Figure 2.3. An important disadvantage of both methods is that, in order for
them to converge, one must produce a good initial estimate of the planar development. Ma and
Lin propose no solution for this problem, while Maillot *et al.* offer a technique not applicable
to all surfaces.

Azariadis and Aspragathos [2] further improve the aforementioned method by modifying the
area energy function and by giving a solution to the initial guess problem. They also introduce
an algorithm for preserving, during the flattening process, either isoparametric curves [2] or
arbitrary curves [3]. The usefulness of this property has been verified by using industrial
examples. In [3], it is experimentally shown that if the mesh Φ approximates the curved
surface with a sufficient accuracy then further refinement of Φ has almost no effect on the
final result. However, estimation of the minimum size of the mesh Φ, sufficient for accurate
planardevelopment, remains an open problem.

Employing the material properties of the initial curved-surface, Shimada and Tada [28, 29]
proceeded to develop a method based on the theory of finite-elements to construct planar

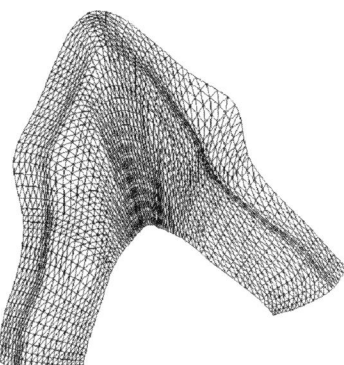

Figure 2.3   Using the method of Maillot *et al.*, it is possible to derive a planar development of the shoe last of Figure 2.2 without overlapping areas.

developments of arbitrary three-dimensional surfaces. More specifically, these researchers propose an approximation based on solving a planar stress-problem using triangular elements. The related objective function is minimized using a particular iterative process instead of a classical optimization-algorithm. An important advantage of this method is that it does not usually require an accurate initial estimate of the solution to converge. However, the method may fail if the geometry of the surface is sufficiently complex or if overlaps appear in the planar development.

Another method is proposed by Bennis *et al.* [7], where a relaxation procedure is used for the homogeneous distribution of deformation of the geodesic-curvature error in the planar development. A limitation of this technique is its strong dependence on the surface parameterization and its initialization by specification of a surface parametric-curve which is mapped onto the plane. Regarding trimmed surfaces, this method often requires decomposing the initial surface into subparts of very simple geometry.

Yu *et al.* [14] present an algorithm for optimal development of a smooth continuous curved surface onto the plane. The development process is modeled using in-plane strain from the curved surface to its planar development. Minimization of strain in the planar development is achieved by solving a constrained nonlinear programming problem. Another approach [8, 9] formulates the planar development problem using a spring-mass system and calculating the strain energy released during flattening. These authors also use a color graph to indicate areas where cutting lines should be introduced to release more strain energy.

Sheffer and de Sturler [27] introduce a method based on the observation that a triangulated planar mesh is fully defined by the mesh angles up to global scaling, rotation and translation. The authors formulate the parameterization problem in terms of the flat-mesh angles and solve it in the angle space. The method also involves constraints on angles defining a valid (continuous) planar mesh. The main part of the method is minimization of the angular distortion of the parameterization, subject to the above constraints. Recently, the authors have enhanced their flattening technique by minimizing both angular distortion and linear distortion [27]. The authors claim that this revised method avoids foldovers in the derived planar development.

4.2. Category B: Employing Intrinsic Differential-Geometric Properties of Surfaces

This category includes flattening methods that use intrinsic differential-geometric properties of surfaces like the Gaussian curvature or the geodesic curvature. Taking the characteristics of geodesic curves of a surface, Manning [22] develops a flattening method also based on an 'isometric tree', i.e., a network of surface points connected to each other with edges. Eventually, this tree is projected onto a plane using an isometric mapping.

A method based on properties of the Gaussian curvature of a surface is proposed by Hinds *et al.* [15] aiming at planar developments for apparel design. More specifically, since clothing manufacturing requires that planar developments are free of foldovers, the authors focus on developing a flattening technique that fulfils this requirement and thus, almost always, produces developments with openings, called 'radial developments'. McCartney *et al.* [23] offer another method, aiming again at the clothing industry, which handles the insertion of darts and gussets by creating appropriate openings.

Parida and Mudur [24] deal with the special case of composite materials and propose a robust flattening technique based on constraints. Azariadis and Aspragathos [1] extend this method to a general purpose surface flattening technique divided into three-stages. In the first stage, an adequate guide-strip is located on the triangulated surface. Using this guide-strip an initial planar development is derived by isometrically unfolding triangle strips onto the plane. At the final stage, foldovers and cuts are eliminated according to certain criteria. A more elaborated approach to the planar development refinement is introduced in [4] where a special genetic algorithm has been developed for global optimization under constrains.

Wolfson and Schwartz [12], and Schwartz, Shaw and Wolfson [11] used a special MDS (Multi-Dimensional Scaling) approach to flatten the curved surface using geodesic distances, and by minimizing the functional presented by Sammon in [16], which resembles the Stress-1 functional. Their method involves high computational complexity and therefore is not practical. Zigelman *el al.* [33] improved this method by introducing a new mapping method that preserves both the local and the global structure of the planar development, with minimal shearing effects.

4.3. Category C: Approximation with Developable Surfaces

These methods subdivide the initial three-dimensional surface into pieces which are approximated with developable patches. These patches are defined by a one-parameter envelope of tangent planes, which intersect pairwise and define a line in the three-dimensional space. Thus, each plane is tangent to the constructed developable patch along this straight line, which is called 'generatrix'. Subdivision of the initial surface into such patches is performed so that the surface produced is at least C^0 continuous.

A method for the construction of developable surfaces, along the lines of the above methodology, is proposed by Redont [25]. The user derives developable surfaces by specifying the orientation of the tangent plane along a geodesic. Clearly, this is not a practical method as it is very hard for any user to define appropriate orientation of the tangent planes so that the desirable developable surface is constructed.

Bodduluri and Ravani [5, 6] develop a method for the design of developable surfaces based on the concept of duality between points and planes in the three-dimensional projective space, giving a new representation for developable surfaces in the context of 'plane geometry'. The developable surface is designed using control planes (which are dual to points). Fitting is performed employing existing techniques for curve design, like Bezier or B-spline fitting.

A simple method for the approximation of an arbitrary curved surface with developable patches is proposed in [13]. More specifically, the author adopts the approach that a developable surface may be represented as an appropriate ruled-surface. On the basis of a developability condition for ruled surfaces, a simple algorithm is proposed for subdividing a surface into a set of developable ruled-surfaces within a given approximation error. The result of this algorithm is a C^0 composite-surface consisting of developable surfaces.

Many techniques have appeared which face the problem of designing developable surfaces under specific conditions related to the nature of a particular problem. For example, Sundar and Varada [31] propose a method for calculating developments of ducts, whose surface is approximated with developable surfaces. Also, Hoschek [18] proposes a method for deriving approximately developable surfaces from surfaces of revolution. Finally, Leopoldseder and Pottmann [20] present a method for designing/representing approximately developable surfaces using conic segments. Although they deal with the problem using a new approach, based on the duality between points and planes in the Euclidean space, their method may be considered as an extension of that proposed by Redont [25].

5. Application and Evaluation of Current Methods

We conclude this chapter by presenting a series of examples of planar developments of doubly-curved surfaces. First, let us briefly describe one of the most commonly used approaches for surface flattening based on an energy model.

5.1. The Modified Length-Area Energy Model

One of the most commonly used energy models for deriving planar developments of doubly-curved surfaces is the one proposed by Ma [30] and latter extended by Mailot [17]. This energy function is actually a convex combination of two energy functionals expressing the metric distortion in terms of length and signed area:

$$E(\mathbf{x}) = a E_{length}(\mathbf{x}) + (1-a)E_{area}(\mathbf{x}) \tag{2.38}$$

where $0 \le a \le 1$. The length functional is:

$$E_{length} = 2 \sum_{\mathbf{M_i} \in \Phi} \sum_{\mathbf{M_j} \in \Omega_i} \frac{(\|\mathbf{m_i} - \mathbf{m_j}\|^2 - \|\mathbf{M_i} - \mathbf{M_j}\|)^2}{\|\mathbf{M_i} - \mathbf{M_j}\|^2} \tag{2.39}$$

where $\mathbf{M_i}$ and $\mathbf{m_i}$ are vertices of triangles of Φ and $\boldsymbol{\varphi}$, respectively. Ω_i is the set of vertices of all triangle edges of Φ meeting at $\mathbf{M_i}$ ($i \ne j$). The signed-area functional is:

$$E_{area} = \sum_{M_i \in \Phi} \sum_{(j,k) \in V_i} \frac{(\det(\mathbf{m_i m_j}, \mathbf{m_i m_k}) - \|\mathbf{M_i M_j} \times \mathbf{M_i M_k}\|)^2}{\|\mathbf{M_i M_j} \times \mathbf{M_i M_k}\|} \tag{2.40}$$

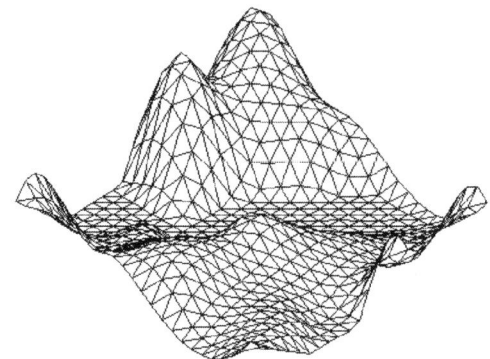

Figure 2.4   A landscape surface with high-curvature areas.

where $V_i = \{$all pairs $(j, k) : (\mathbf{M_i}, \mathbf{M_j}, \mathbf{M_k})$ define a triangle in $\mathbf{\Phi}\}$ and $i \neq j \neq k$. Equation (2.40) has been proposed in [3] in order to express the functional area with respect to mesh vertices instead of mesh triangles as it was in [17]. This modification simplifies gradient computations. The value of α is determined with respect to the complexity of the surface geometry. Usually a value around 0.5 produces acceptable results with no foldovers. However, for surfaces with high curvature areas smaller values are required.

5.2. Examples

In this section we compare the modified Length-Area Energy (LAE) algorithm [3] with the FEA algorithm [28, 29] using as a benchmark the development of the landscape shown in Figure 2.4, a significantly complex surface with high-curvature areas. This example is quite representative for texture mapping applications.

Due to the complex geometry we have to tune the parameter α of Equation (2.38) in order to avoid foldovers in the final planar development. A proper value for α is found to be 0.2 (see Figure 2.5). On the other hand, the FEA method requires no tuning; it directly produces a

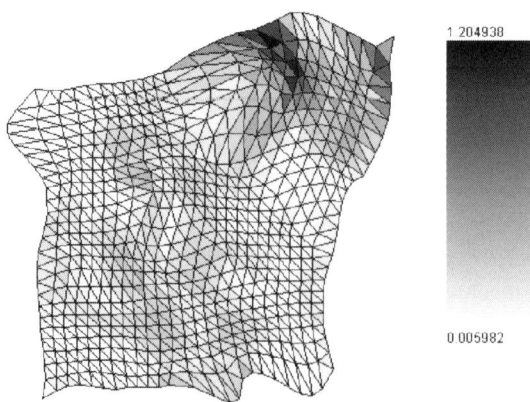

Figure 2.5   The planar development derived using the modified LAE model.

Table 2.1 Landscape and last examples: Analysis of planar developments.

Flattening Method / Surface	Homogeneity index along 1st direction $\dfrac{\min d_p}{\max d_p}$	Homogeneity index along 2nd direction $\dfrac{\min d_q}{\max d_q}$	Homogeneity index $\dfrac{\min d_p d_q}{\max d_p d_q}$	Aspect ratio $\dfrac{\min d_p}{\max d_q}$	Elapsed time (sec)
LAE / Landscape	0.216880	0.266572	0.344655	0.578145	3.365
FEA / Landscape	0.253818	0.209432	0.183896	0.485494	106.743
LAE / Last	0.687047	0.628918	0.708005	0.871228	64.894
FEA / Last	0.667407	0.698568	0.741347	0.670895	6.209

planar development without overlapping regions (Figure 2.6). Both planar developments are filled with colors indicating the areas with low/high distortion. The darker the color, the higher the distortion is in the planar development. The distortion is measured by comparing the edge lengths of triangles in the surface to those in the planar development.

Comparing the two planar developments reveals some important features. It is obvious that areas with higher distortion correspond to areas with higher curvature, as it was expected. Also, the planar development derived with the LAE model is significantly less distorted (around 1.47 times) than that obtained using FEA, which is apparent in the two color maps. An in depth analysis of the accuracy/quality of both planar developments, with respect to the indices introduced in this chapter, is presented in Table 2.1. The data in this table establish that the development based on the LAE model is better than that produced using the FEA method. The homogeneity index (Equation (2.36)) for LAE is almost twice that of FEA. This is very important for applications like texture mapping where sudden changes in the image quality are easily noticed by the human eye.

Figure 2.7 illustrates the distortion of the planar development of the shoe last derived using the LAE model. In this case, the metric distortion is much lower than the previous example

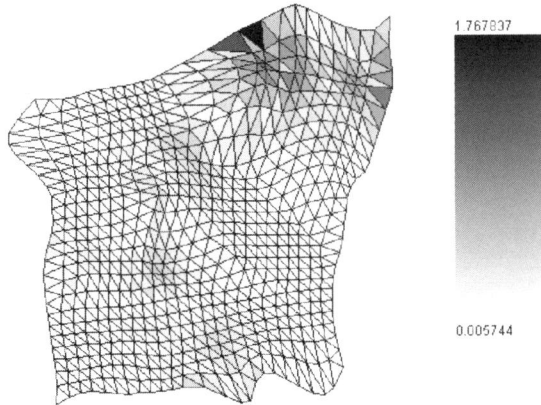

Figure 2.6 The planar development derived using the FEA flattening method.

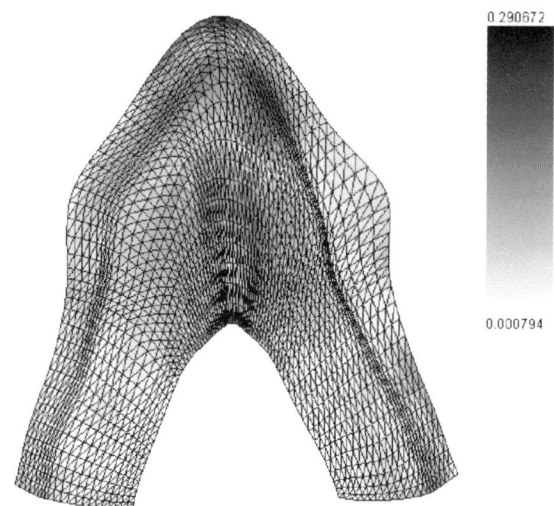

Figure 2.7 Planar development of a shoe last using the modified LAE model.

since the surface of the last does not present areas with high-curvature. The parameter α was set to the default value 0.5. The high homogeneity index h guarantees homogeneous distribution of the metric distortion within the area of the planar development. Figure 2.8, displays the corresponding planar development obtained using FEA. In this case, the distortion of the triangles is larger – more than twice that of the previous development. Also, the homogeneity index is not as high as that produced by LAE. However, in this case the execution time for FEA was significantly lower than the first example. This is due to two reasons: the algorithm

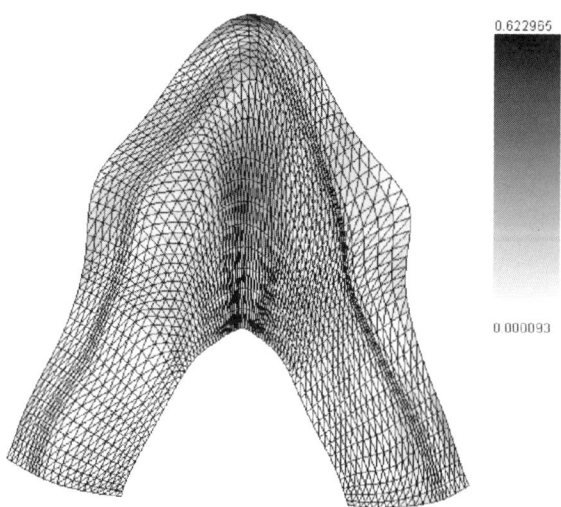

Figure 2.8 Planar development of a shoe last using the FEA method.

required fewer iterations to achieve convergence, and the triangles order was carefully designed to minimize the bandwidth of the global stiffness matrix.

6. Summary

Classical differential geometry offers a host of results to analyze developability of free-form surfaces. However, none of these can be extended to the case of *digital surfaces*, i.e., surfaces defined approximately by a triangular mesh, replacing an exact analytic representation. Digital surface descriptions are gaining in popularity as CAD/CAM and computer graphics applications are continuously shifting towards 'digital modeling and processing'. This chapter reviews related results from Differential Geometry and proceeds on to developing quality control criteria for digital flattening (Section 3). Current flattening methods are analyzed in Section 4 and categorized on the basis of fundamental characteristics. Finally, in Section 5, two state-of-the-art methods for digital flattening are evaluated using realistic examples and the criteria/metrics of Section 3.

Although many flattening methods are constantly appearing, little effort is directed towards 'numerical quality control' of planar developments. The latter has been the focal point of this chapter, aiming to assist practitioners in identifying the most appropriate method for each application.

References

[1] Azariadis P. and Aspragathos N. (1997). Design of Plane Patterns of Doubly Curved Surfaces, *Computer-Aided Design*, 29(10), 675–685.

[2] Azariadis P. and Aspragathos N. (2000). On Using Planar Developments to Perform Texture Mapping on Arbitrarily Curved Surfaces, *Computers & Graphics*, 24, 539–554.

[3] Azariadis P. and Aspragathos N. (2001). Geodesic Curvature Preservation In Surface Flattening Through Constrained Global Optimization, *Computer-Aided Design*, 33(8), 581–591.

[4] Azariadis P., Nearchou A. and Aspragathos N. (2002). An Evolutionary Algorithm for Generating Planar Developments of Arbitrarily Curved Surfaces, *Computers in Industry*, 47(3), 357–368.

[5] Bodduluri R. M. C. and Ravani B. (1993). Design of Developable Surfaces Using Duality Between Plane and Point Geometries, *Computer-Aided Design*, 25(10), 621–632.

[6] Bodduluri R. M. C. and Ravani B. (1994), Geometric Design and Fabrication of Developable Bezier and B-spline Surfaces, Transactions of the ASME, *Journal of Mechanical Design*, Vol. 116/1043.

[7] Chakib Bennis, Jean-Marc Vezien, Gerald Iglesias (1991). Piecewise Surface Flattening for Non-Distorted Texture Mapping, *Computer Graphics*, 25(4), 237–246.

[8] Wang C. C. L., Chen S.-F. and Yuen M. M. F. (2001). Surface Flattening for the Fashion Industry: a Generic Approach Using Spring-Mass System, *Computers in Industry* 1548, 1–10.

[9] Wang C. C. L., Chen S. -F. and Yuen M. M. F. (2002). Surface Flattening Based on Energy Model, *Computer-Aided Design*, 34, 823–833.

[10] Manfredo L. D. C. (1976). *Differential Geometry of Curves and Surfaces*. Prentice-Hall, Englewood Cliffs, New Jersey.

[11] Schwartz E. L., Shaw A., and Wolfson E. (1989). A Numerical Solution to the Generalized Mapmaker's Problem: Flattening Nonconvex Polyhedral Surfaces. *IEEE Trans. on Pattern Analysis and Machine Intelligence*, 11(9), 1005–1008.

[12] Wolfson E. and Schwartz E. L. (1989). Computing Minimal Distances on Polyhedral Surfaces. *IEEE Trans. on Pattern Analysis and Machine Intelligence*, 11(9), 1001–1005.

[13] Elber G. (1995). Model Fabrication Using Surface Layout Projection, *Computer-Aided Design*, 27(4), 283–291.

[14] Yu G., Patrikalakis N.M. and Maekawa T. (2000). Optimal Development of Doubly Curved Surfaces, *Computer Aided Geometric Design*, 17, 545–577.

[15] Hinds B.K., McCartney L. and Woods G. (1991). Pattern Development for 3D Surfaces, *Computer-Aided Design*, 23(8), 583–592.

[16] Sammon J.W. (1969). A Nonlinear Mapping for Data Structure Analysis. *IEEE Transactions on Computers*, 18(5), 401–409.

[17] Maillot J., Yahia H. and Verroust A. (1993). Interactive Texture Mapping, *Proc. SIGGRAPH 93*, Anaheim, California, 27–34 (1–6 August 1993).

[18] Hoschek J. (1998). Approximation of Surfaces of Revolution by Developable Surfaces, *Computer-Aided Design*, 30(10), 757–763.

[19] Karger A. and Novak J. (1985). *Space Kinematics and Lie Groups*, Gordon & Breach Science Publishers.

[20] Leopoldseder S. and Pottmann H. Approximation of Developable Surfaces With Cone Spline Surfaces, *Computer-Aided Design*, 30(7), 571–582.

[21] Lipschutz M.M. (1969). *Differential Geometry*, McGraw-Hill, USA.

[22] Manning J.R. (1980). Computerized Pattern Cutting, *Computer-Aided Design*, 12(1), 43–47.

[23] McCartney J., Hinds B.K. and Seow B.L. (1999), The Flattening of triangulated Surfaces Incorporating Darts and Gussets, *Computer-Aided Design*, 31, 249–260.

[24] Parida L. and Mudur S.P. (1993), Constraint-Satisfying Planar Development of Complex Surfaces, *Computer-Aided Design*, 25(4), 225–232.

[25] Redont P. (1989). Representation and Deformation of Developable Surfaces, *Computer-Aided Design*, 21(1), 13–20.

[26] Sheffer A. and de Sturler E. (2000). Parameterization of Faceted Surfaces for Meshing using Angle Based Flattening. *Engineering with Computers* 17(3) 326–337.

[27] Sheffer A. and de Sturler E. (2002). Smoothing an Overlay Grid to Minimize Linear Distortion in Texture Mapping, *ACM Transactions on Graphics*, 21(4), 874–890.

[28] Shimada T. and Tada Y. (1989). Development of Curved Surface Using Finite Element Method, *Proc. 1st Int. Conf. Computer-Aided Optimum Design of Structures*, Recent Advances, Springer-Verlag, 23–30.

[29] Shimada T. and Tada, Y. (1991). Approximate Transformation of an Arbitrary Curved Surface Into a Plane Using Dynamic Programming, *Computer-Aided Design*, 23(2), 153–159.

[30] Ma S.O. and Lin H. (1998). Optimal Texture Mapping, *Proc. EUROGRAPHICS 88*, 421–428.

[31] Sundar P. and Varada R. (1995). Evolution of Generic Mathematical Models and Algorithms for the Surface Development and Manufacture of Complex Ducts, *Journal of Engineering for Industry*, Vol.117, 177–185.

[32] Willmore T.J. (1972). *An Introduction to Differential Geometry*, Oxford University Press.

[33] Gil Z., Ron K., and Nahum K. (2002). Texture Mapping using Surface Flattening via Multi-Dimensional Scaling. *IEEE Transactions on Visualization and Computer Graphics*, 8(2), 198–207.

3

A Shape Preserving Representation for Rational Curves with Efficient Evaluation Algorithm

Jorge Delgado
Juan Manuel Peña

Department of Matemática Aplicada, University of Zaragoza, Pedro Cerbuna, 12
50009 Zaragoza, SPAIN.

In this chapter we consider two families of rational bases derived from two families of bases of polynomials as an analternative to the Bernstein basis which have recently been introduced. We also provide two corner cutting evaluation algorithms for rational polynomial curves with the associated linear time complexity. The two families of rational bases are formed by normalized totally positive bases. Therefore, they simultaneously satisfy efficiency and shape preservation. We also provide the corner cutting algorithm for obtaining the corresponding rational Bézier polygon (with adequate weights) from the control polygon with respect to these bases.

1. Introduction

We introduced [6] new families of bases $(c_{0,h}^n(t), \ldots, c_{n,h}^n(t))$ $(1 \leq h \leq n)$ of the space of polynomials of degree at most n on [0, 1] which provide shape preserving representations of curves. The case $h = n$ corresponds to the Bernstein basis and, when $h < n$, we obtained evaluation algorithms more efficient than the de Casteljau algorithm. The computational cost of the evaluation algorithm decreases with h and, for the cases $h = 1, 2$, we have linear complexity

Advances in Geometric Modeling. Edited by M. Sarfraz
© 2003 John Wiley & Sons, Ltd   ISBN: 0-470-85937-7

(the basis corresponding to $h = 1$ was already considered in [5]). In this chapter we extend the mentioned representations and algorithms to the rational case. Let us start by introducing some basic notations.

Let U be a vector space of real functions defined on a real interval I and $(u_0, \ldots, u_n)$ $(t \in I)$ a basis of U. If a sequence $P_0, \ldots, P_n$ of point in $\mathfrak{R}^k$ is given then we define a curve $\gamma(t) = \sum_{i=0}^{n} P_i u_i(t), t \in I$. The points $P_0, \ldots, P_n$ are called *control points* and the corresponding polygon $P_0 \cdots P_n$ is called the *control polygon* of γ. In computer aided geometric design the functions $u_0, \ldots, u_n$ are usually nonnegative and $\sum_{i=0}^{n} u_i(t) = 1 \, \forall t \in [a, b]$ (i.e. the system $(u_0, \ldots, u_n)$ is *normalized*) and in this case we say that $(u_0, \ldots, u_n)$ is a *blending system*. Let us recall that the convex hull property holds if and only if $(u_0, \ldots, u_n)$ is a blending system.

On the other hand, it is desirable for the designer to have a precise control over what happens at the ends of the curve. This leads to the endpoint interpolation and boundary tangent properties. We say that the *endpoint interpolation property* holds if the first control point always coincides with the start point of the curve and the last control point always coincides with the final point of the curve. We say that a basis $(u_0, \ldots, u_n)$ satisfying the endpoint interpolation property also satisfies the *boundary tangent property* if the segments $P_0 P_1$, $P_{n-1} P_n$ are, respectively, tangent to the curve $\gamma(t) = \sum_{i=0}^{n} P_i u_i(t)$ at the endpoints P_0, P_n.

In computer aided geometric design, the usual representation of a polynomial curve is the so-called Bernstein-Bézier form. The Bernstein-Bézier form of a polynomial $p(t)$ of degree less than or equal to n on $[0, 1]$ is given by:

$$p(t) = \sum_{j=0}^{n} \alpha_j B_j^n(t), t \in [0, 1], \qquad (3.1)$$

where $\alpha_j \in \mathfrak{R}$, and:

$$B_j^n(t) = \binom{n}{j} t^j (1 - t)^{n-j}, j = 0, \ldots, n,$$

is the corresponding Bernstein polynomial of degree n. The usual algorithm used to evaluate a polynomial in the form Equation (3.1) is the de Casteljau algorithm. But, although the Bernstein-Bézier form is the usual representation of a polynomial curve (see [7] and [8]) and presents optimal shape preserving properties (see [2]), other polynomial representations can also be useful. The computational cost of the de Casteljau algorithm for a polynomial curve of degree n is quadratic (that is, of $O(n^2)$ elementary operations), but there are other evaluation algorithms useful in design with linear computational cost (that is, of $O(n)$ elementary operations). This can lead to an important reduction of the computational cost when they are generalized for the evaluation of parametric surfaces or rational curves derived from these representations. The evaluation algorithm for the Wang-Ball basis (see [11], [10] and [3]) is a corner cutting algorithm and it also has linear complexity. In [4], a family of rational bases was derived from the Wang-Ball basis. However, in [5] we proved that these rational bases are not monotonicity preserving. In contrast, the rational bases studied in this chapter are shape preserving and, in particular, monotonicity preserving, as we shall see in Section 3.

Definition 1.1. Let $(c_{0,1}^n(t), \ldots, c_{n,1}^n(t))$, $n \geq 2$, be the system of polynomials on $[0, 1]$ defined by:

$$c_{0,1}^n(t) = (1 - t)^n,$$

$$c_{i,1}^n(t) = t(1 - t)^{n-i}, \, 1 \leq i \leq \left\lfloor \frac{n}{2} \right\rfloor - 1,$$

$$c_{i,1}^n(t) = t^i(1 - t), \, \left\lfloor \frac{n+1}{2} \right\rfloor + 1 \leq i \leq n - 1,$$

$$c_{n,1}^n(t) = t^n.$$

In addition, if n is even,

$$c_{\frac{n}{2},1}^n(t) = 1 - t^{\frac{n}{2}+1} - (1 - t)^{\frac{n}{2}+1},$$

and, if n is odd,

$$c_{\frac{n-1}{2},1}^n(t) = t(1 - t)^{\frac{n+1}{2}} + \frac{1}{2}\left[1 - t^{\frac{n+1}{2}} - (1 - t)^{\frac{n+1}{2}}\right],$$

$$c_{\frac{n+2}{2},1}^n(t) = \frac{1}{2}\left[1 - t^{\frac{n+1}{2}} - (1 - t)^{\frac{n+1}{2}}\right] + t^{\frac{n+1}{2}}(1 - t).$$

We proved in Section 3 of [6] that the family of bases from Definition 1.1 satisfies the endpoint interpolation property but doesn't satisfy the boundary tangent property. In the following definition we introduce a family of bases, which satisfies both endpoint interpolation and boundary tangent properties and also has a linear complexity evaluation algorithm (although of greater complexity).

Definition 1.2. Let $(c_{0,2}^n(t), \ldots, c_{n,2}^n(t))$, $n \geq 2$, be the system of polynomials on $[0, 1]$ that coincides with the Bernstein basis for $2 \leq n \leq 4$ and that for $n \geq 5$ is defined by:

$$c_{i,2}^n(t) = \binom{n}{i} t^i(1 - t)^{n-i}, \, 0 \leq i \leq 1,$$

$$c_{i,2}^n(t) = (m + 1 - i)t^2(1 - t)^{n-i}, \, 2 \leq i \leq \left\lfloor \frac{n}{2} \right\rfloor - 1,$$

$$c_{i,2}^n(t) = (i + 1)t^i(1 - t)^2, \, \left\lfloor \frac{n+1}{2} \right\rfloor + 1 \leq i \leq n - 2,$$

$$c_{i,2}^n(t) = \binom{n}{i} t^i(1 - t)^{n-i}, \, n - 1 \leq i \leq n.$$

In addition, if n is even,

$$c_{\frac{n}{2},2}^n(t) = \sum_{j=4}^{n+1} jt^2(1 - t)^{j-1} + 6t^2(1 - t)^2 + \sum_{j=4}^{n+1} jt^{j-1}(1 - t)^2,$$

and, if n is odd,

$$c_{\frac{n-1}{2},2}^n(t) = \frac{n+3}{2}t^2(1 - t)^{\frac{n+1}{2}} + \frac{1}{2}c_{\frac{n-1}{2},2}^{n-1}(t),$$

$$c_{\frac{n+1}{2},2}^n(t) = \frac{1}{2}c_{\frac{n-1}{2},2}^{n-1}(t) + \frac{n+3}{2}t^{\frac{n+1}{2}}(1 - t)^2.$$

Given a sequence of positive weights $\omega_0, \ldots, \omega_n$ and a system $(u_0, \ldots, u_n)$ such that $\sum_{j=0}^n \omega_j u_j(t) > 0$, we can form the system $(r_0(t), \ldots, r_n(t))$ with:

$$r_i(t) := \frac{\omega_i u_i(t)}{\sum_{j=0}^n \omega_j u_j(t)}, \, i = 0, \ldots, n. \tag{3.2}$$

Then $(r_0(t), \ldots, r_n(t))$ is a system of the corresponding space of rational functions. We say that $(r_0(t), \ldots, r_n(t))$ is a rational system associated to the system $(u_0, \ldots, u_n)$. Given control points $P_0, \ldots, P_n \in \Re^d$, the generated curve:

$$\gamma(t) = \sum_{j=0}^n P_j r_j(t)$$

is called a rational curve.

In Section 2 we provide two algorithms with linear complexity to evaluate polynomial rational curves associated to the families of rational systems derived from the two families of systems of Definition 1.1 and 1.2. In Section 3 we provide corner cutting algorithms for the conversion between the rational systems derived from the bases $(c_{0,h}^n(t), \ldots, c_{n,h}^n(t))$ and the corresonding rational Bézier systems.

2. Evaluation Algorithms for Rational Curves

This section presents two efficient evaluation algorithms for curves generated by the rational systems derived from the two systems presented in the previous section. An nth degree rational Bézier curve is given by:

$$\gamma(t) = \sum_{i=0}^n b_i \frac{\omega_i B_i^n(t)}{\sum_{i=0}^n \omega_i B_i^n(t)},$$

where $\omega_0, \ldots, \omega_n$ are positive weights. A rational Bézier curve may be evaluated by applying the de Casteljau algorithm to both numerator and denominator and finally dividing through. But there is a warning: this method, although simple and usually effective, is not numerically stable. If some of the ω_i are large, the intermediate control points are no longer in the convex hull of the original control polygon; this may result in a loss of accuracy. The usual algorithm for evaluating a rational curve is called the rational de Casteljau algorithm, and is given by:

Algorithm 2.1.
```
for i=1 to n do
```
$$\omega_i^0(t) = \omega_i$$
$$b_i^0(t) = b_i$$
```
end i;
for i=1 to n do
  for j=0 to n-1 do
```
$$\omega_j^i(t) = (1-t)\omega_j^{i-1}(t) + t\omega_{j+1}^{i-1}(t)$$

$$b_j^i(t) = (1-t)\frac{\omega_j^{i-1}(t)}{\omega_j^i(t)}b_j^{i-1}(t) + t\frac{\omega_{j+1}^{i-1}(t)}{\omega_j^i(t)}b_{j+1}^{i-1}(t)$$

```
  end j;
end i;
echo
```
$\omega_0^n(t), b_0^n(t)$;

where $\omega_0^n(t) = \sum_{i=0}^n \omega_i B_i^n(t)$ and $b_0^n(t) = \gamma(t)$. But this algorithm is computationally more expensive than the previous one. A stable and cheaper evaluation algorithm for rational curves can be deduced for the rational systems derived from the family of systems of Definition 1.1, and is given, if $\gamma(t) = \sum_{i=0}^n v_i \dfrac{\omega_i c_{i,1}^n(t)}{\sum_{i=0}^n \omega_i c_{i,1}^n(t)}$, by:

Algorithm 2.2.
```
for i=1 to n do
```
$$\omega_i^0(t) = \omega_i$$
$$V_i^0(t) = V_i$$
```
end i;
for i=1 to n-2 do
```
$$\omega_0^i(t) = (1-t)\omega_0^{i-1}(t) + t\omega_1^{i-1}(t)$$

$$v_0^i(t) = (1-t)\frac{\omega_0^{i-1}(t)}{\omega_0^i(t)}v_0^{i-1}(t) + t\frac{\omega_1^{i-1}(t)}{\omega_0^i(t)}v_1^{i-1}(t)$$

```
  if ((n+1-i)%2=1) do
    for j=1 to (n-2-i)/2 do
```
$$\omega_j^i(t) = \omega_{j+1}^{i-1}(t)$$
$$v_j^i(t) = v_{j+1}^{i-1}(t)$$
```
    end j
```
$$\omega_{\frac{n-1}{2}}^i(t) = \frac{1}{2}\omega_{\frac{n-1}{2}}^{i-1}(t) + \frac{1}{2}\omega_{\frac{n-1}{2}+1}^{i-1}(t)$$

$$v_{\frac{n-i}{2}}^i(t) = \frac{1}{2}\frac{\omega_{\frac{n-i}{2}}^{i-1}(t)}{\omega_{\frac{n-1}{2}}^i(t)}v_{\frac{n-i}{2}}^{i-1}(t) + \frac{1}{2}\frac{\omega_{\frac{n-i}{2}+1}^{i-1}(t)}{\omega_{\frac{n-i}{2}}^i(t)}v_{\frac{n-i}{2}+1}^{i-1}(t)$$

```
    for j=(n-i)/2+1 to n-1-i do
```
$$\omega_j^i(t) = \omega_j^{i-1}(t)$$
$$v_i^j(t) = v_j^{i-1}(t)$$
```
    end j
  else do
    for j=1 to (n-1-i)/2 do
```
$$\omega_j^i(t) = \omega_{j+1}^{i-1}(t)$$
$$v_j^i(t) = v_{j+1}^{i-1}(t)$$
```
    end j
```

```
for j=(n+1-i)/2 to n-1-i do
```
$$\omega_j^i(t) = \omega_j^{i-1}(t)$$
$$v_j^i(t) = v_j^{i-1}(t)$$
```
  end j
  end if;
```
$$\omega_{n-i}^i(t) = (1-t)\omega_{n-i}^{i-1}(t) + t\omega_{n+1-i}^{i-1}(t)$$
$$v_{n-i}^i(t) = (1-t)\frac{\omega_{n-i}^{i-1}(t)}{\omega_{n-i}^i(t)}v_{n-i}^{i-1}(t) + t\frac{\omega_{n+1-i}^{i-1}(t)}{\omega_{n-i}^i(t)}v_{n+1-i}^{i-1}(t)$$
```
end i;
for i=n-1 to n do
  for j=0 to n-i do
```
$$\omega_j^i(t) = (1-t)\omega_j^{i-1}(t) + t\omega_{j+1}^{i-1}(t)$$
$$v_j^i(t) = (1-t)\frac{\omega_j^{i-1}(t)}{\omega_j^i(t)}v_j^{i-1}(t) + t\frac{\omega_{j+1}^{i-1}(t)}{\omega_j^i(t)}v_{j+1}^{i-1}(t)$$
```
  end j
end i;
echo ω₀ⁿ(t), v₀ⁿ(t);
```
echo $\omega_0^n(t)$, $v_0^n(t)$;

then, as the following result states, $\omega_0^n(t) = \sum_{i=0}^n \omega_i c_{i,1}^n(t)$ and $v_0^n(t) = \gamma(t)$.

Theorem 2.3. if $\gamma(t) = \sum_{i=0}^n v_i \dfrac{\omega_i c_{i,1}^n(t)}{\sum_{i=0}^n \omega_i c_{i,1}^n(t)}$, performing Algorithm 2.2 we have that:

$$\omega_0^n(t) = \sum_{i=0}^n \omega_i c_{i,1}^n(t) \quad \text{and} \quad v_0^n(t) = \gamma(t).$$

Proof. Let us denote by $\Lambda_k(t) = (\alpha_{ij}^k)_{i=0,\dots,k-1;j=0,\dots,k}$ the matrix given, if k is odd, by:

$a_{ij}^k = 1 - t$ for $i = j \in \{0, k-1\}$,

$a_{ij}^k = t$ for $i + 1 = j \in \{1, k\}$,

$a_{ij}^k = 1$ for $i = j - 1 \in \{1, \dots, (k-3)/2\}$ or $i = j \in \{(k+1)/2, \dots, k-2\}$,

$a_{ij}^k = 1/2$ for $i = (k-1)/2$ and $j \in \{(k-1)/2, (k+1)/2\}$,

$a_{ij}^k = 0$ in otherwise,

and, if k is even, by:

$a_{ij}^k = 1 - t$ for $i = j \in \{0, k-1\}$,

$a_{ij}^k = t$ for $i + 1 = j \in \{1, k\}$,

$a_{ij}^k = 1$ for $i = j - 1 \in \{1, \dots, k/2 - 1\}$ or $i = j \in \{k/2, \dots, k-2\}$,

$a_{ij}^k = 0$ in otherwise.

Taking into account that,

$$\Lambda_1(t) \cdots \Lambda_n(t) = \left(c_{0,1}^n(t), \dots, c_{n,1}^n(t) \right), \tag{3.3}$$

we can observe that,

$$\sum_{i=0}^{n} \omega_i c_{i,1}^n(t) = \Lambda_1(t) \cdots \Lambda_n(t) \begin{pmatrix} \omega_0 \\ \vdots \\ \omega_n \end{pmatrix}. \tag{3.4}$$

Using the previous algorithm we can check that,

$$\Lambda_1(t) \cdots \Lambda_n(t) \begin{pmatrix} \omega_0 \\ \vdots \\ \omega_n \end{pmatrix} = \omega_0^n(t). \tag{3.5}$$

Then, by Equations (3.4) and (3.5), we have that,

$$\omega_0^n(t) = \sum_{i=0}^{n} \omega_i c_{i,1}^n(t). \tag{3.6}$$

Now, taking into account Equation (3.3), we can deduce that,

$$\sum_{i=0}^{n} v_i \omega_i c_{i,1}^n(t) = \Lambda_1(t) \cdots \Lambda_n(t) \begin{pmatrix} \omega_0 & & \\ & \ddots & \\ & & \omega_n \end{pmatrix} \begin{pmatrix} v_0 \\ \vdots \\ v_n \end{pmatrix}. \tag{3.7}$$

Now, let us denote by $\overline{\Lambda}_k^n(t) = (\overline{a}_{ij}^{k,n})_{i=0,\ldots,k-1;j=0,\ldots,k}$ the matrix given, if k is odd, by:

$$\overline{a}_{ij}^{k,n} = (1-t)\frac{\omega_j^{n-k}(t)}{\omega_j^{n+1-j}(t)} \quad \text{for} \quad i = j \in \{0, k-1\},$$

$$\overline{a}_{ij}^{k,n} = t\frac{\omega_j^{n-k}(t)}{\omega_{j-1}^{n+1-j}(t)} \quad \text{for} \quad i+1 = j \in \{1, k\},$$

$$\overline{a}_{ij}^{k,n} = 1 \quad \text{for} \quad i = j-1 \in \{1, \ldots, (k-3)/2\} \quad \text{or} \quad i = j \in \{(k+1)/2, \ldots, k-2\},$$

$$\overline{a}_{ij}^{k,n} = \frac{1}{2}\frac{\omega_j^{n-k}(t)}{\omega_j^{n+1-k}(t)} \quad \text{for} \quad i = (k-1)/2 \quad \text{and} \quad j \in \{(k-1)/2, (k+1)/2\},$$

$$\overline{a}_{ij}^{k,n} = 0, \quad \text{otherwise},$$

and, if k is even, by:

$$\overline{a}_{ij}^{k,n} = (1-t)\frac{\omega_j^{n-k}(t)}{\omega_j^{n+1-k}(t)} \quad \text{for} \quad i = j \in \{0, k-1\},$$

$$\overline{a}_{ij}^{k,n} = t\frac{\omega_j^{n-k}(t)}{\omega_{j-1}^{n+1-k}(t)} \quad \text{for} \quad i+1 = j \in \{1, k\},$$

$$\overline{a}_{ij}^{k,n} = 1 \quad \text{for} \quad i = j-1 \in \{1, \ldots, k/2-1\} \quad \text{or} \quad i = j \in \{k/2, \ldots, k-2\},$$

$$\overline{a}_{ij}^{k,n} = 0 \text{ in otherwise}.$$

We can easily check that:

$$\Lambda_1(t)\cdots\Lambda_n(t)\begin{pmatrix}\omega_0 & & \\ & \ddots & \\ & & \omega_n\end{pmatrix}\begin{pmatrix}v_0 \\ \vdots \\ v_n\end{pmatrix} = \omega_0^n(t)\overline{\Lambda}_1^n(t)\cdots\overline{\Lambda}_n^n(t)\begin{pmatrix}v_0 \\ \vdots \\ v_n\end{pmatrix}. \tag{3.8}$$

Using the previous algorithm we can check that:

$$\overline{\Lambda}_1^n(t)\cdots\overline{\Lambda}_n^n(t)\begin{pmatrix}v_0 \\ \vdots \\ v_n\end{pmatrix} = v_0^n(t). \tag{3.9}$$

Now, from Equations (3.7), (3.8) and (3.9) we can deduce that:

$$\sum_{i=0}^n v_i\omega_i c_{i,1}^n(t) = \omega_0^n(t)v_0^n(t).$$

Hence, and using Equation (3.6) we get:

$$v_0^n(t) = \sum_{i=0}^n v_i\frac{\omega_i c_{i,1}^n(t)}{\omega_0^n(t)} = \sum_{i=0}^n v_i\frac{\omega_i c_{i,1}^n(t)}{\sum_{i=0}^n \omega_i c_{i,1}^n(t)} = \gamma(t),$$

and the result follow.

Another stable method with linear complexity for rational curves formed by polynomials of degree greater than or equal to 4 using the rational systems is derived from the family of systems of Definition 1.2, and is given, if $\gamma(t) = \sum_{i=0}^n v_i\dfrac{\omega_i c_{i,2}^n(t)}{\sum_{i=0}^n \omega_i c_{i,2}^n(t)}$, by:

Algorithm 2.4.
```
for i=1 to n do
```
$$\omega_i^0(t) = \omega_i$$
$$V_i^0(t) = V_i$$
```
end i;
for i=1 to n-4 do
```
$$\omega_0^i(t) = (1-t)\omega_0^{i-1}(t) + t\omega_i^{i-1}(t)$$

$$v_0^i(t) = (1-t)\frac{\omega_0^{i-1}(t)}{\omega_0^i(t)}v_0^{i-1}(t) + t\frac{\omega_1^{i-1}(t)}{\omega_0^i(t)}v_1^{i-1}(t)$$

$$\omega_1^i(t) = (1-t)\omega_1^{i-1}(t) + t\omega_2^{i-1}(t)$$

$$v_1^i(t) = (1-t)\frac{\omega_1^{i-1}(t)}{\omega_1^i(t)}v_1^{i-1}(t) + t\frac{\omega_2^{i-1}(t)}{\omega_1^i(t)}v_2^{i-1}(t)$$

```
   if ((n+1-i)%2=1) do
      for j=2 to (n-2-i)/2 do
```
$$\omega_j^i(t) = \omega_{j+1}^{i-1}(t)$$
$$v_j^i(t) = v_{j+1}^{i-1}(t)$$
```
      end j
```
$$\omega_{\frac{n-i}{2}}^i(t) = \frac{1}{2}\omega_{\frac{n-i}{2}}^{i-1}(t) + \frac{1}{2}\omega_{\frac{n-i}{2}+1}^{i-1}(t)$$

$$v^i_{\frac{n-i}{2}}(t) = \frac{1}{2}\frac{\omega^{i-1}_{\frac{n-i}{2}}(t)}{\omega^i_{\frac{n-i}{2}}(t)}v^{i-1}_{\frac{n-i}{2}}(t) + \frac{1}{2}\frac{\omega^{i-1}_{\frac{n-i}{2}+1}(t)}{\omega^i_{\frac{n-i}{2}}(t)}v^{i-1}_{\frac{n-i}{2}+1}(t)$$

```
for j=(n-i)/2+1 to n-2-i do
```
$$\omega^i_j(t) = \omega^{i-1}_j(t)$$
$$v^i_j(t) = v^{i-1}_j(t)$$
```
end j
else do
  for j=2 to (n-1-i)/2 do
```
$$\omega^i_j(t) = \omega^{i-1}_{j+1}(t)$$
$$v^i_j(t) = v^{i-1}_{j+1}(t)$$
```
  end j
  for j=(n+1-i)/2 to n-2-i do
```
$$\omega^i_j(t) = \omega^{i-1}_j(t)$$
$$v^i_j(t) = v^{i-1}_j(t)$$
```
  end j
end if
```
$$\omega^i_{n-1-i}(t) = (1-t)\omega^{i-1}_{n-1-i}(t) + t\omega^{i-1}_{n-i}(t)$$

$$v^i_{n-1-i}(t) = (1-t)\frac{\omega^{i-1}_{n-1-i}(t)}{\omega^i_{n-1-i}(t)}v^{i-1}_{n-1-i}(t) + t\frac{\omega^{i-1}_{n-i}(t)}{\omega^i_{n-1-i}(t)}v^{i-1}_{n-i}(t)$$

$$\omega^i_{n-i}(t) = (1-t)\omega^{i-1}_{n-i}(t) + t\omega^{i-1}_{n+1-i}(t)$$

$$v^i_{n-i}(t) = (1-t)\frac{\omega^{i-1}_{n-i}(t)}{\omega^i_{n-i}(t)}v^{i-1}_{n\ i}(t) + t\frac{\omega^{i-1}_{n+1-i}(t)}{\omega^i_{n-i}(t)}v^{i-1}_{n+1-i}(t)$$

```
end i;
for i=n-3 to n do
  for j=0 to n-i do
```
$$\omega^i_j(t) = (1-t)\omega^{i-1}_j(t) + t\omega^{i-1}_{j+1}(t)$$

$$v^i_{n-i}(t) = (1-t)\frac{\omega^{i-1}_j(t)}{\omega^i_j(t)}v^{i-1}_j(t) + t\frac{\omega^{i-1}_{j+1}(t)}{\omega^i_j(t)}v^{i-1}_{j+1}(t)$$

```
  end j
end i;
echo ω₀ⁿ(t), v₀ⁿ(t);
```
```
echo $\omega^n_0(t)$, $v^n_0(t)$;
```

Analogously to the proof of Theorem 2.3 we can prove that
$\omega^n_0(t) = \sum_{i=0}^{n} \omega_i c^n_{i,2}(t)$ and $v^n_0(t) = \gamma(t)$.

Remark 2.5. If we perform Algorithm 2.2 with $n \geq 3$, then we get:

$$\omega^i_{\frac{n-1-i}{2}}(t) = \omega^i_{\frac{n+1-i}{2}}(t) \quad \text{and} \quad v^i_{\frac{n-1-i}{2}}(t) = v^i_{\frac{n+1-i}{2}}(t) \quad \text{for all } i \in \{1, \dots, n-2\}$$

such that $(n-1-i)/2$ is even, so $\omega^i_{\frac{n-i}{2}}(t) = \omega^{i-1}_{\frac{n-i}{2}}(t)$ and $v^i_{\frac{n-i}{2}}(t) = v^{i-1}_{\frac{n-i}{2}}(t)$, for all $i \in \{1, \ldots, n-2\}$ such that $(n-i)/2$ is even if n is even, or for all $i \in \{2, \ldots, n-2\}$ such that $(n-i)/2$ is even if n is odd. Analogously, if we perform Algorithm 2.4 with $n \geq 5$, then we get $\omega^i_{\frac{n-1-i}{2}}(t) = \omega^i_{\frac{n+1-i}{2}}(t)$ and $v^i_{\frac{n-1-i}{2}}(t) = v^i_{\frac{n+1-i}{2}}(t)$ for all $i \in \{1, \ldots, n-4\}$ such that $(n-1-i)/2$ is even, so $\omega^i_{\frac{n-i}{2}}(t) = \omega^{i-1}_{\frac{n-i}{2}}(t)$ and $v^i_{\frac{n-i}{2}}(t) = v^{i-1}_{\frac{n-i}{2}}(t)$ for all $i \in \{1, \ldots, n-4\}$ such that $(n-i)/2$ is even if n is even, or for all $i \in \{2, \ldots, n-4\}$ such that $(n-1)/2$ is even if n is odd. Taking these facts into account, one can check that, in Algorithm 2.2, when n is odd, the number of sum is $4n$, the number of multiplications is $8n$ and the number of divisions is $4n$, and if n is even the number of sums is $4n - 2$, the number of multiplications is $8n - 4$ and the number of divisions is $4n - 2$. Then, in Algorithm 2.4, when n is odd, the number of sums is $8n - 10$, the number of multiplications is $16n - 20$ and the number of divisions is $8n - 10$, and if n is even the number of sums is $8n - 12$, the number of multiplications is $16n - 24$ and the number of divisions is $8n - 12$. In contrast with the $(n+1)n$ sums, the $2(n+1)n$ multiplications and the $(n+1)n$ divisions of Algorithm 2.1 (de Casteljau algorithm for rational curves).

3. Total Positivity and Conversion to the Rational Bernstein-Bézier Form

The *collocation matrix* of $(u_0, \ldots, u_n)$ at $t_0 < t_1 < \cdots < t_m$ in I is given by:

$$M \begin{pmatrix} u_0, \ldots, u_n \\ t_0, \ldots, t_m \end{pmatrix} = (u_j(t_i))_{i=0,\ldots,m; j=0,\ldots,n}. \tag{3.10}$$

A matrix is *totally positive* (TP) if all its minors are nonnegative and a system of functions is TP when all its collocation matrices Equation (3.10) are TP. In the case of a Normalized Totally Positive (NTP) basis one knows that the curve imitates the shape of its control polygon, due to the variation diminishing properties of TP matrices. In fact, shape preserving representations are associated with NTP bases (see [2] and [9]). In particular, by Theorem 2.6 of [1] they are monotonicity preserving.

Let us consider Equation (3.2). Since $\omega_0, \ldots, \omega_n > 0$ and $\sum_{j=0}^n \omega_j u_j(t) > 0$, if $(u_0, \ldots, u_n)$ is NTP then $(r_0, \ldots, r_n)$ also is NTP.

In this section, we shall present a generalization of the families of Definitions 1.1 and 1.2. considered in [6], where we showed that these families of systems are formed by NTP bases. Thus, the rational bases (with positive weights) derived from them are also NTP bases of the corresponding space of rational functions and so satisfy shape preserving properties. Finally, we shall provide for each rational basis derived an equivalent rational Bernstein basis and between them a conversion corner cutting algorithm.

Now, let us present the families of generalized systems.

Definition 3.1. Let $(c^n_{0,h}(t), \ldots, c^n_{n,h}(t))$, $n \geq 2$ and $h \geq 1$, be the system of polynomials on $[0, 1]$ that coincides with the Bernstein basis for $2 \leq n \leq 2h$ and that for $n \geq 2h + 1$ is defined by:

$$c_{i,h}^n(t) = \binom{n}{i} t^i (1-t)^{n-i}, 0 \leq i \leq h-1,$$

$$c_{i,h}^n(t) = \binom{n-1+h-i}{h-1} t^h (1-t)^{n-i}, h \leq i \leq \left\lfloor \frac{n}{2} \right\rfloor - 1,$$

$$c_{i,h}^n(t) = \binom{i+h-1}{i} t^i (1-t)^h, \left\lfloor \frac{n+1}{2} \right\rfloor + 1 \leq i \leq n-h,$$

$$c_{i,h}^n(t) = \binom{n}{i} t^i (1-t)^{n-i}, n+1-h \leq i \leq n.$$

In addition, if n is even,

$$c_{\frac{n}{2},h}^n(t) = \sum_{j=2h}^{n+h-1} \binom{j}{h-1} t^h (1-t)^{j-h+1} + \binom{2h}{h} t^h (1-t)^h$$

$$+ \sum_{j=2h}^{n+h-1} \binom{j}{j-h+1} t^{j-h+1}(1-t)^h,$$

and, if n is odd,

$$c_{\frac{n-1}{2},h}^n(t) = \binom{\frac{n-1}{2}+h}{h-1} t^h (1-t)^{\frac{n+1}{2}} + \frac{1}{2} c_{\frac{n-1}{2},h}^{n-1}(t),$$

$$c_{\frac{n+1}{2},h}^n(t) = \frac{1}{2} c_{\frac{n-1}{2},h}^{n-1}(t) + \binom{\frac{n-1}{2}+h}{\frac{n+1}{2}} t^{\frac{n+1}{2}}(1-t)^h.$$

Remark 3.2. Let us observe that the systems $(c_{0,h}^n(t), \ldots, c_{n,h}^n(t))$ for $h = 1$ and 2 coincide with the systems of Definition 1.1 and Definition 1.2 respectively for all $n \geq 2$.

Now, we shall show, for all positive weights $\omega_0, \ldots, \omega_n$, a corner cutting algorithm from all rational from all rational systems $\left(\dfrac{\omega_0 c_{0,h}^n(t)}{\sum \omega_i c_{i,h}^n(t)}, \ldots, \dfrac{\omega_n c_{n,h}^n(t)}{\sum \omega_i c_{i,h}^n(t)} \right)$ (for all $n \geq 2$ and $h \geq 1$) to rational Bernstein systems $\left(\dfrac{\overline{\omega}_0 B_0^n(t)}{\sum \overline{\omega}_i B_i^n(t)}, \ldots, \dfrac{\overline{\omega}_n B_n^n(t)}{\sum \overline{\omega}_i B_i^n(t)} \right)$. Let us start with the even case:

Theorem 3.3. if:

$$\gamma(t) = \sum_{i=0}^{2m} v_i \frac{\omega_i c_{i,h}^{2m}(t)}{\sum_{i=0}^{2m} \omega_i c_{i,h}^{2m}(t)}, m \geq 1, h \geq 1,$$

where $\omega_0, \ldots, \omega_{2m}$ are positive weights and $(c_{0,h}^{2m}(t), \ldots, c_{2m,h}^{2m}(t))$ is given in Definition 3.1, then, performing the following algorithm:

```
for i=0 to h-1 do
```
$$\overline{\omega}_i = \omega_i$$
$$b_i = v_i$$
```
end i;
```
$$\overline{\omega}_n = \omega_n;$$
$$b_n = v_n;$$
```
for i=2m+1-h to 2m do
```
$$\overline{\omega}_i = \omega_i$$
$$b_i = v_i$$
```
end i;
for i=h to 2m-h do
  if i≠m do
```
$$\omega_i^{h-1} = \omega_i$$
$$v_i^{h-1} = v_i$$
```
  end if
end i;
for i=h to m do
```
$$\omega_m^i = \omega_m$$
$$v_m^i = v_m$$
```
end i;
for i=to m-1 do
  for j=m-1 to i step -1 do
```
$$\omega_j^i = \frac{i}{2m - j + i}\omega_j^{i-1} + \frac{2m - j}{2m - j + 1}\omega_{j+1}^i$$

$$v_j^i = \frac{i}{2m - j + i}\frac{\omega_j^{i-1}}{\omega_j^i}v_j^{i-1} + \frac{2m - j}{2m - j + 1}\frac{\omega_{j+1}^i}{\omega_j^i}v_{j+1}^i$$

$$\omega_{2m-j}^i = \frac{2m - j}{2m - j + i}\omega_{2m-j-1}^i + \frac{i}{2m - j + i}\omega_{2m-j}^{i-1}$$

$$v_{2m-j}^i = \frac{2m - j}{2m - j + i}\frac{\omega_{2m-j-1}^i}{\omega_{2m-j}^i}v_{2m-j-1}^i + \frac{i}{2m - j + i}\frac{\omega_{2m-j}^{i-1}}{\omega_{2m-j}^i}v_{2m-j}^{i-1}$$

```
  end j
```
$$\overline{\omega}_i = \omega_i^i$$
$$b_i = v_i^i$$
$$\overline{\omega}_{2m-i} = \omega_{2m-i}^i$$
$$b_{2m-i} = v_{2m-i}^i$$
```
end i;
```
echo $(\overline{\omega}_0, \ldots, \overline{\omega}_{2m}), (b_0, \ldots, b_{2m});$

we have that $\sum_{i=0}^{2m} \omega_i c_{i,h}^{2m}(t) = \sum_{i=0}^{2m} \overline{\omega}_i B_i^{2m}(t)$ and

$$\gamma(t) = \sum_{i=0}^{2m} b_i \frac{\overline{\omega}_i B_i^{2m}(t)}{\sum_{i=0}^{2m} \overline{\omega}_i B_i^{2m}(t)}.$$

Proof. Let us denote $\lambda_j^i := \dfrac{2m - j}{2m - j + 1}$ and $\lambda_{2m-j}^i := \dfrac{2m - j}{2m - j + i}$ for all $i \in \{1, \ldots, m - 1\}$ and $j \in \{1, \ldots, m - 1\}$. Now, let us consider the matrices $A_k^{2m} := (a_{ij}^{k,2m})_{0 \leq i,j \leq 2m}$ for $k \in \{h, \ldots, m - 1\}$ defined by:

$$a_{ij}^{k,2m} = 1 \quad \text{for} \quad i = j \in \{0, \ldots, k - 1, m, 2m + 1 - k, \ldots, 2m\},$$

$$a_{ij}^{k,2m} = 1 - \lambda_i^{j+1-k} \quad \text{for} \quad i = j \in \{k, \ldots, m - 1\},$$

$$a_{ij}^{k,2m} = \lambda_i^{j-k} \quad \text{for} \quad i = j - 1 \in \{k, \ldots, m - 1\},$$

$$a_{ij}^{k,2m} = 1 - \lambda_i^{2m+1-j-k} \quad \text{for} \quad i = j \in \{m + 1, \ldots, 2m - k\},$$

$$a_{ij}^{k,2m} = \lambda_i^{2m-j-k} \quad \text{for} \quad i = j + 1 \in \{m + 1, \ldots, 2m - k\},$$

$$a_{ij}^{k,2m} = 0 \text{ in otherwise.}$$

By the results of Section 3 of [6] we have:

$$\left(c_{0,h}^{2m}(t), \ldots, c_{2m,h}^{2m}(t)\right) = \left(B_0^{2m}(t), \ldots, B_{2m}^{2m}(t)\right) A_h^{2m}(t) \ldots A_{m-1}^{2m}(t). \tag{3.11}$$

Now, taking into account the alorithm, we can easily check that:

$$A_h^{2m} A_{h+1}^{2m} \cdots A_{m-1}^{2m} \begin{pmatrix} \omega_0 \\ \vdots \\ \omega_m \end{pmatrix} = \begin{pmatrix} \overline{\omega}_0 \\ \vdots \\ \overline{\omega}_m \end{pmatrix}. \tag{3.12}$$

Then, by Equations (3.11) and (3.12) we can deduce that:

$$\sum_{i=0}^{2m} \omega_i c_{i,h}^{2m}(t) = \sum \overline{\omega}_i B_i^{2m}(t). \tag{3.13}$$

Now, from Equation (3.11) we also can deduce that:

$$\left(c_{0,h}^{2m}(t), \ldots, c_{2m,h}^{2m}(t)\right) \begin{pmatrix} \omega_0 & & \\ & \ddots & \\ & & \omega_{2m} \end{pmatrix} \begin{pmatrix} v_0 \\ \vdots \\ v_{2m} \end{pmatrix}$$

$$= \left(B_0^{2m}(t), \ldots, B_{2m}^{2m}(t)\right) A_h^{2m} \ldots A_{m-1}^{2m} \begin{pmatrix} \omega_0 & & \\ & \ddots & \\ & & \omega_{2m} \end{pmatrix} \begin{pmatrix} v_0 \\ \vdots \\ v_{2m} \end{pmatrix}. \tag{3.14}$$

Now, let us consider the matrices $\overline{A}_k^{2m} := (\overline{a}_{ij}^{k,2m})_{0 \leq,j \leq 2m}$ for $k \in \{h, \ldots, m - 1\}$ defined by:

$$\overline{a}_{ij}^{k,2m} = 1 \quad \text{for} \quad i = j \in \{0, \ldots, k - 1, m, 2m + 1 - k, \ldots, 2m\},$$

$$\overline{a}_{ij}^{k,2m} = \left(1 - \lambda_i^{j+1-k}\right) \frac{\omega_i^{j-k}}{\omega_i^{j+1-k}} \quad \text{for} \quad i = j \in \{k, \ldots, m - 1\},$$

$$\overline{a}_{ij}^{k,2m} = \lambda_i^{j-k} \frac{\omega_{i+1}^{j-k}}{\omega_i^{j-k}} \quad \text{for} \quad i = j-1 \in \{k, \ldots, m-1\},$$

$$\overline{a}_{ij}^{k,2m} = \left(1 - \lambda_i^{2m+1-j-k}\right) \frac{\omega_i^{2m-j-k}}{\omega_i^{2m+1-j-k}} \quad \text{for} \quad i = j \in \{m+1, \ldots, 2m-k\},$$

$$\overline{a}_{ij}^{k,2m} = \lambda_i^{2m-j-k} \frac{\omega_{i-1}^{2m-j-k}}{\omega_i^{2m-j-k}} \quad \text{for} \quad i = j+1 \in \{m+1, \ldots, 2m-k\},$$

$$\overline{a}_{ij}^{k,2m} = 0 \text{ in otherwise.}$$

We can easily check that:

$$A_h^{2m} \cdots A_{m-1}^{2m} \begin{pmatrix} \omega_0 & & \\ & \ddots & \\ & & \omega_{2m} \end{pmatrix} \begin{pmatrix} v_0 \\ \vdots \\ v_{2m} \end{pmatrix} = \begin{pmatrix} \overline{\omega}_0 & & \\ & \ddots & \\ & & \overline{\omega}_{2m} \end{pmatrix} \overline{A}_h^{2m} \cdots \overline{A}_{m-1}^{2m} \begin{pmatrix} v_0 \\ \vdots \\ v_{2m} \end{pmatrix} \tag{3.15}$$

Now, using the algorithm, we can derive:

$$\overline{A}_h^{2m} \cdots \overline{A}_{m-1}^{2m} \begin{pmatrix} v_0 \\ \vdots \\ v_{2m} \end{pmatrix} = \begin{pmatrix} b_0 \\ \vdots \\ b_{2m} \end{pmatrix}. \tag{3.16}$$

Then, by Equations (3.14), (3.15) and (3.16) we have:

$$\sum_{i=0}^{2m} v_i \omega_i c_{i,h}^{2m}(t) = \sum_{i=0}^{2m} b_i \overline{\omega}_i B_i^{2m}(t).$$

Finally, from Equation (3.13) and the previous formula we can deduce that:

$$\gamma(t) = \sum_{i=0}^{2m} v_i \frac{\omega_i c_{i,h}^{2m}(t)}{\sum_{i=0}^{2m} \omega_i c_{i,h}^{2m}(t)} = \sum_{i=0}^{2m} b_i \frac{\overline{\omega}_i B_i^{2m}(t)}{\sum_{i=0}^{2m} \overline{\omega}_i B_i^{2m}(t)}$$

and the result follows.

In the following result we consider the odd case and it can be proved analogously to the previous result.

Theorem 3.4. If:

$$\gamma(t) = \sum_{i=0}^{2m+1} v_i \frac{\omega_i c_{i,h}^{2m+1}(t)}{\sum_{i=0}^{2m+1} \omega_i c_{i,h}^{2m+1}(t)}, \quad m \geq 1, h \geq 1,$$

where $\omega_0, \ldots, \omega_n$ are positive weights and $(c_{0,h}^{2m+1}(t), \ldots, c_{2m+1,h}^{2m+1}(t))$ is given in Definition

3.1, then, performing the following algorithm:

```
for i=0 to h-1 do
```
$$\overline{\omega}_i = \omega_i$$
$$b_i = v_i$$
```
end i;
for i=2m+2-h to 2m+1 do
```
$$\overline{\omega}_i = \omega_i$$
$$b_i = v_i$$
```
end i;
for i=h to 2m+1-h do
```
$$\omega_i^{h-1} = \omega_i$$
$$v_i^{h-1} = v_i$$
```
end i;
for i=h to m do
```

$$\omega_m^i = \frac{i + \frac{1}{2}(m+1)}{m+1+i}\omega_m^{i-1} + \frac{\frac{1}{2}(m+1)}{m+1+i}\omega_{m+1}^{i-1}$$

$$v_m^i = \frac{i + \frac{1}{2}(m+1)}{m+1+i}\frac{\omega_m^{i-1}}{\omega_m^i}v_m^{i-1} + \frac{\frac{1}{2}(m+1)}{m+1+i}\frac{\omega_{m+1}^{i-1}}{\omega_m^i}v_{m+1}^{i-1}$$

$$\omega_{m+1}^i = \frac{\frac{1}{2}(m+1)}{m+1+i}\omega_m^{i-1} + \frac{i + \frac{1}{2}(m+1)}{m+1+i}\omega_{m+1}^{i-1}$$

$$v_{m+1}^i = \frac{\frac{1}{2}(m+1)}{m+1+i}\frac{\omega_m^{i-1}}{\omega_{m+1}^i}v_m^{i-1} + \frac{i + \frac{1}{2}(m+1)}{m+1+i}\frac{\omega_{m+1}^{i-1}}{\omega_{m+1}^i}v_{m+1}^{i-1}$$

```
for j=m-1 to i step-1 do
```

$$\omega_j^i = \frac{i}{2m+1-j+i}\omega_j^{i-1} + \frac{2m+1-j}{2m+1-j+i}\omega_{j+1}^i$$

$$v_j^i = \frac{i}{2m+1-j+i}\frac{\omega_j^{i-1}}{\omega_j^i}v_j^{i-1} + \frac{2m+1-j}{2m+1-j+i}\frac{\omega_{j+1}^i}{\omega_j^i}v_{j+1}^i$$

$$\omega_{2m+1-j}^i = \frac{2m+1-j}{2m+1-j+i}\omega_{2m-j}^i + \frac{i}{2m+1-j+i}\omega_{2m+1-j}^{i-1}$$

$$v_{2m+1-j}^i = \frac{2m+1-j}{2m+1-j+i}\frac{\omega_{2m-j}^i}{\omega_{2m+1-j}^i}v_{2m-j}^i + \frac{i}{2m+1-j+i}\frac{\omega_{2m+1-j}^{i-1}}{\omega_{2m+1-j}^i}v_{2m+1-j}^{i-1}$$

```
end j
```
$$\overline{\omega}_i = \omega_i^i$$
$$b_i = v_i^i$$
$$\overline{\omega}_{2m+1-i} = \overline{\omega}_{2m+1-i}^i$$
$$b_{2m+1-i} = v_{2m+1-i}^i$$
```
end i;
```
$$\text{echo } (\overline{\omega}, \ldots, \overline{\omega}_{2m+1}), (b_0, \ldots, b_{2m+1});$$

we have that $\sum_{i=0}^{2m+1} \omega_i c_{i,h}^{2m+1}(t) = \sum_{i=0}^{2m+1} \overline{\omega}_i B_i^{2m+1}(t)$ and

$$\gamma(t) = \sum_{i=0}^{2m+1} b_i \frac{\overline{\omega}_i B_i^{2m+1}(t)}{\sum_{i=0}^{2m+1} \overline{\omega}_i B_i^{2m+1}(t)}.$$

4. Summary

In this chapter we have shown an alternative to the usual rational Bernstein-Bézier representation. Our representations also present shape preserving properties and are more efficient than the rational de Casteljau algorithm. We also provide a stable conversion from our representations to the rational Bernstein-Bézier representation.

References

[1] Carnicer J.M., García-Esnaola M. and Peña J.M. (1996) Convexity of rational curves and total positivity. *J. Comput. and Appl. Math.* 71, 365–382

[2] Carnicer J.M. and Peña J.M. (1993) Shape preserving representations and optimality of the Bernstein basis. *Advances in Computational Mathematics* 1, 173–196

[3] Dejdumrong N. and Phien H.N. (2000) Efficient algorithms for Bezier curves. *Comput. Aided Geom. Design* 17, 247–2502.

[4] Dejdumrong N., Phien H.N., Le Tien H. and Lay K.M. (2001) Rational Wang-Ball curves. *Internat. J. Math. Ed. Sci. Tech.* 32, no. 4, 247–250.

[5] Delgado J. and Peña J.M. (2002) Monotonicity preservation of some polynomial and rational representations. In: *Information Visualization*. IEEE Computer Society, Los Alamitos (CA), 57–62

[6] Delgado J. and Peña J.M. A shape preserving representation with an evaluation algorithm of linear complexity. To appear in *Comput. Aided Geom. Design*.

[7] Farin G. (1996) *Curves and Surfaces for Computer Aided Geometric Design* (4th edition). San Diego, Academic Press.

[8] Hoscheck J. and Lasser D. (1993) *Fundamentals of Computer Aided Geometric Design*. Wellesley, AK Peters.

[9] Peña J.M. (1999) *Shape preserving representations in Computer Aided Geometric Design*. Nova Science Publishers, Commack (New York)

[10] Shi-Min H., Guo-Zhao W. and Tong-Guang J. (1996) Properties of two types of generalized Ball curves. *Computer-Aided Design* 28, 125–133

[11] Wang G.J. (1987) Ball curve of high degree and its geometric properties. *Appl. Math.: A journal of Chinese Universities* 2, 126–140

4

Piecewise Power Basis Conversion of Dynamic B-spline Curves and Surfaces

Deok-Soo Kim

Department of Industrial Engineering, Hanyang University
17 Haengdang-Dong, Seongdong-Ku, Seoul, 133-791, Korea.

Joonghyun Ryu

Samsung SDS Co., LTD.
159-9 HIGH-TECH Center, Gumi-Dong, Bundang-Gu, Seongnam, Korea.

B-spline is one of the most popular representations for curves and surfaces in CAGD and computer graphics. Although the recursive form of B-spline curves and surfaces is frequently used, the piecewise power basis form is often preferred when the speed of evaluation is important. Existing approaches handle only the conversion problem for B-spline curves and surfaces where the positions of control points are fixed. In this chapter we present an efficient algorithm for transforming a B-spline curve into a piecewise power basis form. Extension to the same problem for surfaces is also discussed. Experiments show that the presented algorithm significantly outperforms conventional approaches when curves or surfaces change their shapes dynamically by moving their control points.

1. Introduction

The evaluation of points and tangent vectors on parametric curves or surfaces [10–21] is one of the frequently required geometric calculations in CAGD and computer graphics. For shading curved surfaces, a typical graphics API often employs the evaluation of points and normal vectors [24]. The deformation of an object represented by parametric surfaces can be realized via moving some control points and re-evaluating the surface [23]. Once a parametric curve

---

Advances in Geometric Modeling. Edited by M. Sarfraz
© 2003 John Wiley & Sons, Ltd   ISBN: 0-470-85937-7

or surface is represented in power basis form, a point evaluation can be made much faster than represented on a Bernstein basis, since Horner's rule can be applied. However, it should be noted that there is the issue of numerical stability for a power basis form [5].

The piecewise power basis form of a B-spline curve can also be utilized for quickly computing characteristic points such as cusps and inflection points. Note that subdivision at these characteristic points can facilitate fast computation of intersection points between parametric curves [7]. Moreover, IGES supports a power basis form curve as an entity type 112.

Due to the relative advantages of an implicit representation of curves or surfaces over parametric ones in some geometric calculations such as a point inclusion test, a curve or a surface in parametric form is often transformed into one in implicit form [1]. The implicitization process, which uses a resultant, usually demands curves or surfaces to be represented in power basis form [22]. Since this operation is computationally expensive, the reduction of computation time should not be disregarded.

There are a few conventional approaches for the piecewise power basis conversion of B-spline curves or surfaces. Since a polynomial represented in an arbitrary basis can be transformed into power basis form via Taylor expansion, a conversion can be done by evaluating an appropriate number of derivatives and factorial functions for each knot span or knot cell [8]. Applying knot refinement and basis conversion to each piece of B-spline curves or surfaces can also yield piecewise power basis representations of given B-spline curves or surfaces [2, 3, 4, 5, 6, 9].

In this chapter, we discuss an algorithm for piecewise power basis conversion of B-spline curves and surfaces. Based on the presented algorithm, the fast evaluation of B-spline curves and surfaces is possible especially when they change their shapes dynamically.

The core part of the algorithm is to split and compute B-spline basis functions into a set of polynomials in power basis form in each knot span. The main idea of this algorithm is to locate appropriate linear polynomials in the segments of B-sline basis functions and to compute all segments of a B-spline basis function in power basis form, defined as *truncated basis functions*, via unfolding the recurrence formula of a B-spline basis function.

Once truncated basis functions in a knot span are obtained in power basis form, the polynomial curves in power basis form of a knot span can be easily obtained by a linear combination between control points and corresponding truncated basis functions. Repeating this operation for each knot span, all of the polynomials of a B-spline curve are transformed into a set of polynomials in power basis representation. In addition, the conversion of a tensor product B-spline surface into a set of polynomials in power basis representation could be performed through a similar procedure.

The structure of this chapter is organized as follows: Section 2 presents a procedure for splitting B-spline basis functions into a set of polynomial segments in power basis form. In Sections 3 and 4, the conversion of a B-spline curve and surface with moving control points is discussed. Experiments for comparing the presented algorithm with the conventional approaches are provided in Section 5. The conclusion of this chapter is given in Section 6. Note that a supplementary pseudo-code is provided in the appendix.

2. Splitting B-spline Basis Functions

For splitting a B-spline basis function into a set of independent polynomials in power basis form, let us look into its recursive representation. A B-spline curve of degree p with knot

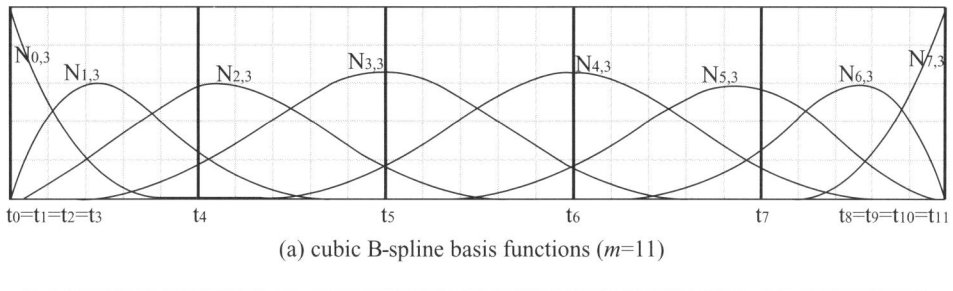

(a) cubic B-spline basis functions ($m=11$)

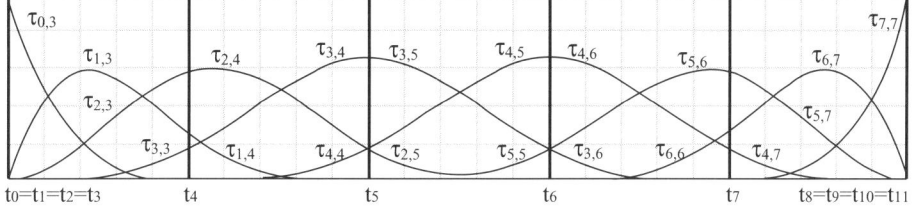

(b) Truncated basis functions corresponding to the basis functions in (a)

Figure 4.1 B-spline basis functions and the corresponding truncated basis functions.

vector of size $(m + 1)$ is given as follows [9]:

$$\mathbf{C}(t) = \sum_{i=0}^{m-p-1} \mathbf{P}_i N_{i,p}(t), \tag{4.1}$$

where $\mathbf{P}_i$ and $N_{i,p}(t)$ are control points and a B-spline basis function of degree p defined on the knot vector $\mathbf{U} = \{0, \ldots, 0, t_{p+1}, \ldots, t_{m-p-1}, 1, \ldots 1\}$, $t_i \leq t_{i+1}$, respectively. $N_{i,p}(t)$ is also defined as:

$$N_{i,p}(t) = \frac{t - t_i}{t_{i+p} - t_i} N_{i,p-1}(t) + \frac{t_{i+p+1} - t}{t_{i+p+1} - t_{i+1}} N_{i+1,p-1}(t),$$

$$N_{i,0}(t) = \begin{cases} 1 & if \qquad t_i \leq t < t_{i+1}, \\ 0 & otherwise, \end{cases} \tag{4.2}$$

where $i = 0, 1, \ldots, m - p - 1$. Figure 4.1(a) shows cubic B-spline basis functions and Figure 4.1(b) shows all of the polynomial segments of the basis functions. In this figure, each polynomial segment is denoted by $\tau_{i,w}$ defined in Definition 1.

2.1. Representation of B-spline Basis Functions

The presented algorithm splits a B-spline basis function into a set of polynomial segments and computes them in power basis form. Such polynomials are defined as follows:

Definition 1. A truncated basis function, $\tau_{i,w}(t)$, $i = w - p, w - p + 1, \ldots, w$, $w = p, p + 1, \ldots, m - p - 1$, is a non-zero polynomial segment of $N_{i,p}(t)$ in $[t_w, t_{w+1})$.

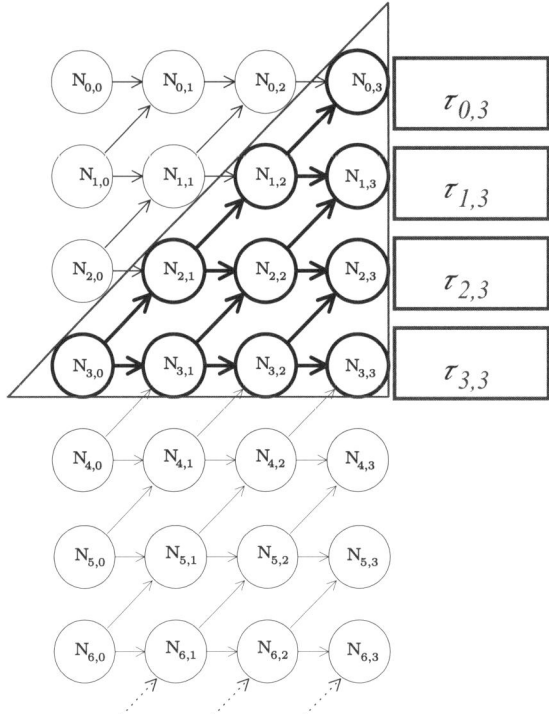

Figure 4.2   Graph representation for B-spline basis functions.

Figure 4.1(b) shows that there are four truncated basis functions in each knot span for cubic B-spline basis functions. In general, a B-spline basis function of degree p is split into $(p + 1)$ truncated basis functions in each knot span.

It is known that the recursive computation of Equation (4.2) follows a triangular scheme as illustrated in Figure 4.2. For example, the computation of $N_{1,3}(t)$ requires both $N_{1,2}(t)$ and $N_{2,2}(t)$ in addition to two corresponding linear polynomial terms. Again, the computation of both $N_{1,2}(t)$ and $N_{2,2}(t)$ also needs another two basis functions with two linear polynomial terms, respectively.

To compute and rearrange truncated basis functions in power basis form, the repeation form is used instead of the recursive form of Equation (4.2). The procedure for computing truncated basis functions can be captured from Figure 4.3, which can be regarded as a directed graph. The root node in Figure 4.3 corresponds to a knot span and four leaf nodes correspond to four truncated basis functions in the knot span. In this figure, the intermediate arrows can be interpreted as linear polynomials shown in Equation (4.2).

For the time being, let us focus on a knot span $[t_3, t_4)$. Then, $N_{3,0}(t)$ in Figure 4.2 corresponds to a knot span $[t_3, t_4)$ and $N_{i,3}(t), i = 0, \ldots, 3$, in the triangle corresponds to four truncated basis functions $\tau_{i,3}(t), i = 0, \ldots, 3$, in the knot span. (In Figure 4.2, $N_{i,3}(t)$ and $\tau_{i,3}(t)$ are denoted as $N_{i,3}$ and $\tau_{i,3}$, respectively, for a notational convenience.)

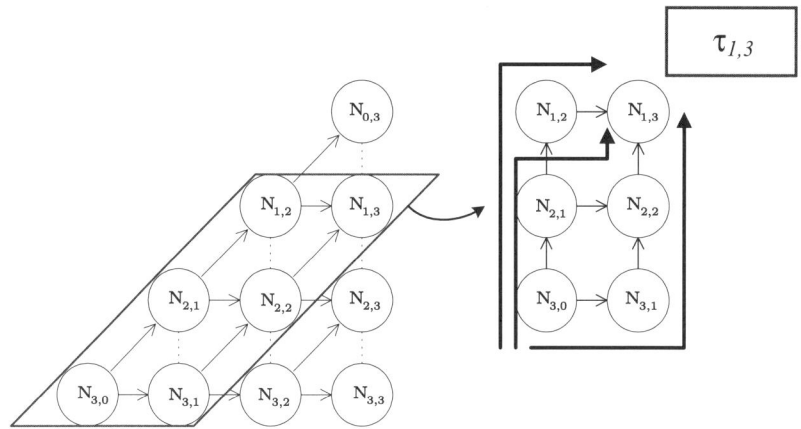

Figure 4.3   A directed graph and paths for computing $\tau_{1,3}(t)$.

For the convenience of discussion, let us rewrite Equation (4.2) as follows:

$$N_{i,p} = h_{i,p}(t)N_{i,p-1} + v_{i,p}(t)N_{i+1,p-1},\tag{4.3}$$

where $h_{i,p}(t) = \dfrac{t - t_i}{t_{i+p} - t_i}$ and $v_{i,p}(t) = \dfrac{t_{i+p+1} - t}{t_{i+p+1} - t_{i+1}}$.

Then, an intermediate arrow in Figure 4.2 corresponds to either $h_{i,p}(t)$ or $v_{i,p}(t)$. Hence, if linear polynomials, $h_{i,p}(t)$ and $v_{i,p}(t)$ necessary to compute $\tau_{i,w}(t)$ are found, it is possible to obtain the desired truncated basis function. Furthermore, given two nodes of an edge (shown as an arrow) the linear term corresponding to the edge can be computed from the subscripts of both nodes. (The edge between $N_{i,j}(t)$ and $N_{k,l}(t)$ is $h_{i,j+1}(t)$ if $k = i, l = j + 1$ or $v_{i-1,j+1}(t)$ if $k = i - 1, l = j + 1$.) We call such a linear polynomial an edge value.

It turns out that $h_{i,p}(t)$'s and $v_{i,p}(t)$'s can be collected from the edges on the paths between $\tau_{i,w}(t)$ and $N_{w,0}(t)$. The left side of Figure 4.3 is a directed graph, which is the subset of Figure 4.2. In the case of $\tau_{1,3}(t)$, for example, the corresponding paths are illustrated in the right side of Figure 4.3 as rectangular grids.

In this case, all necessary paths are three and each path includes three edges and four nodes.

2.2. Direct Expansion of B-spline Basis Functions

To compute and rearrange the truncated basis functions in power basis form it is necessary to find all paths between each root node, representing a knot span, and appropriate leaf nodes, representing corresponding truncated basis functions, from the directed graph in Figure 4.2. Hence, it is necessary to enumerate all paths for each truncated basis function.

This problem can be formulated as an enumeration of all possible string of size p where the alphabets are only 0 and 1. The number of 0's and 1's are the number of horizontal and vertical grids, respectively. For example, Figure 4.4 illustrates the enumeration of three paths for $\tau_{1,3}(t)$ shown as three 0–1 sequences. Simple pseudo-code for this task is provided in the appendix.

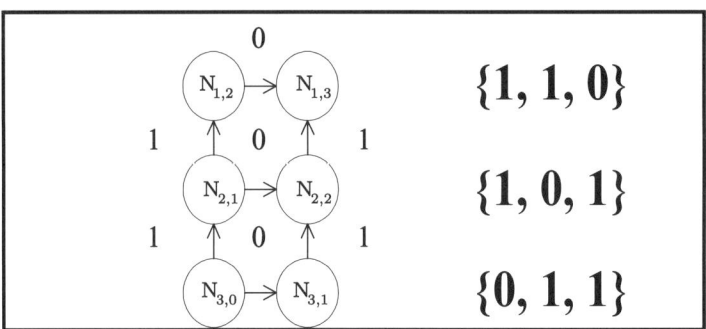

Figure 4.4   Path enumerations.

Once three paths between $\tau_{1,3}(t)$ and $N_{3,0}(t)$ are enumerated, $\tau_{1,3}(t)$ can be obtained by multiplying the appropriate edge values of each path and summing up the results of all paths. Figure 4.5 illustrates edge values showing up in the process of computing $\tau_{1,3}(t)$, and Equation (4.4) provides two truncated basis functions in power basis form, $\tau_{1,3}(t)$ and $\tau_{0,3}(t)$:

$$
\begin{aligned}
\tau_{1,3}(t) \\
&= v_{2,1}(t)v_{1,2}(t)h_{1,3}(t) + v_{2,1}(t)h_{2,2}(t)v_{1,3}(t) + h_{3,1}(t)v_{2,2}(t)v_{1,3}(t) \\
&= \frac{t_4 - t}{t_4 - t_3}\frac{t_4 - t}{t_4 - t_2}\frac{t - t_1}{t_4 - t_1} + \frac{t_4 - t}{t_4 - t_3}\frac{t - t_2}{t_4 - t_2}\frac{t_5 - t}{t_5 - t_2} + \frac{t - t_3}{t_4 - t_3}\frac{t_5 - t}{t_5 - t_3}\frac{t_5 - t}{t_5 - t_2}. \quad (4.4)
\end{aligned}
$$

$$
\tau_{0,3}(t) = v_{2,1}(t)v_{1,2}(t)h_{1,3}(t) = \frac{t_4 - t}{t_4 - t_3}\frac{t_4 - t}{t_4 - t_2}\frac{t - t_1}{t_4 - t_1}.
$$

Note that both $\tau_{1,3}(t)$ and $\tau_{0,3}(t)$ have as many polynomial terms as the number of paths possible for a truncated basis function. Each polynomial term also consists of as many linear terms as

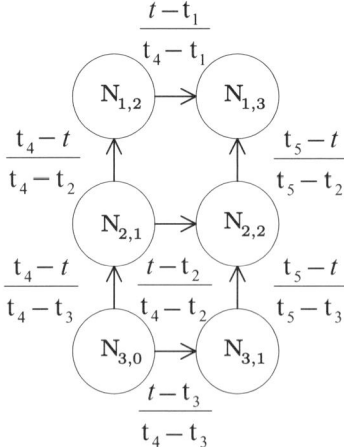

Figure 4.5   Edge values appear in the computation of $\tau_{1,3}(t)$.

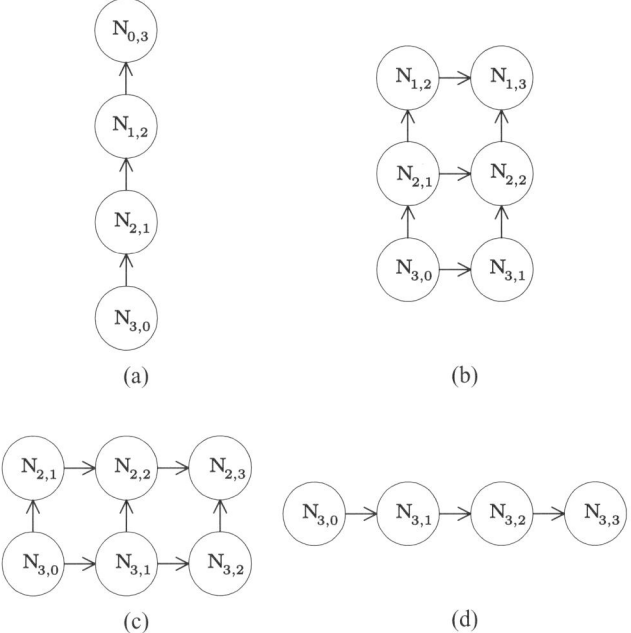

Figure 4.6 Grid paths for computing $\tau_{0,3}(t)$, $\tau_{1,3}(t)$, $\tau_{2,3}(t)$ and $\tau_{3,3}(t)$.

the degree of a corresponding B-spline curve. Note that $\tau_{0,3}(t)$ consists of only one polynomial term since there is only one path.

If this procedure is repeated for each knot span, all truncated basis functions in each knot span are obtained by expanding the multiplications and summations between linear terms directly. For example, four truncated basis functions in $[t_3, t_4)$ exist and the numbers of paths for $\tau_{0,3}(t)$, $\tau_{1,3}(t)$, $\tau_{2,3}(t)$ and $\tau_{3,3}(t)$ are one, three, three and one, respectively. This fact can be verified in Figure 4.6, which shows the paths necessary for computing four truncated basis functions.

3. Direct Expansion of a Dynamic B-spline Curve

3.1. A Static B-spline Curve

If all truncated basis functions of an arbitrary knot span, $[t_i, t_{i+1})$, are obtained, a B-spline curve in the knot span can be represented as Equation (4.5). Note that this equation is a polynomial in power basis form since $\tau_{j,i}$ are already in power basis form.

$$\mathbf{C}_i(t) = \sum_{j=i-p}^{i} \tau_{j,i}(t)\mathbf{P}_j. \tag{4.5}$$

In the case of cubic B-spline curve, for example, the truncated basis functions in $[t_3, t_4)$ are $\tau_{0,3}(t)$, $\tau_{1,3}(t)$, $\tau_{2,3}(t)$ and $\tau_{3,3}(t)$. Then, a set of polynomial curves of a given B-spline curve can be represented in power form as Equation (4.6).

$$C(t) = \sum_{i=p}^{m-p-1} C_i(t)N_{i,0} \tag{4.6}$$

Note that $N_{i,0}$ plays the role of a filter for an appropriate polynomial curve so that the portions of the polynomial curve exterior to the corresponding knot span are suppressed to null.

Until now, this problem has been solved using Taylor expansion (denoted as *TE*-approach) or knot refinement followed by basis conversion (denoted as *KR*-approach). In the case of TE-approach, after a number of hodographs are computed depending on the degree of B-spline curve, the hodographs are evaluated and a number of factorial functions are computed to solve the conversion problem. KR-approach settles the same problem via piecewise knot refinement followed by the basis conversion of Bernstein basis to power basis, which actually performs a number of matrix multiplications.

It turns out that the presented approach is not computationally superior to other conventional approaches when the control points of a B-spline curve are fixed. Equation (4.6) shows that the computation required is not less than the existing approaches. The presented algorithm shows an especially quadratic-like increase whereas KR-approach and TE-approach show only linear-like increases with respect to the degrees of curves. This result is due to the fact that there are 2^p paths for computing $(p+1)$ truncated basis functions in each knot span, when the degree of a B-spline curve is p.

However, it is quite different when the curve changes its shape continuously by moving some control points.

3.2. A Dynamic B-spline Curve

The presented algorithm outperforms the conventional approaches when curves or surfaces change their shapes dynamically. Let us define such a dynamic B-spline curve as follows:

Definition 2. A *dynamic* B-spline curve, $C_d(t)$ is a B-spline curve with more than one control point moving.

Thus, $C_d(t)$ can be represented by Equation (4.7).

$$C_d(t) = \sum_{i \in I} N_{i,p}P_i + \sum_{j \in J} N_{j,p}\tilde{P}_j. \tag{4.7}$$

where I and J are the index sets of fixed control points P_i and moving control points $\tilde{P}_j$, respectively. $N_{i,p}$ denotes a basis function.

A simple, but naive, approach to transform $C_d(t)$ to a set of polynomials in power form is to recalculate power basis forms of all curve segments in every knot span whenever the curve changes its shape. This method is obviously unsatisfactory since it wastes computing time for the knot spans where the shape has not been changed.

The efficiency of the algorithm can be improved by locating the knot spans of curve segments that changed by moved control points. KR-approach performs knot refinement and basis conversion for the target knot spans. In the case of TE-approach, reapplying Taylor expansion to the target knot spans can also only recalculate the coefficients of new polynomials where the curve has changed its shape. The derivative information and factorial evaluation are needed for each coefficient of the polynomial.

However, the presented algorithm, denoted as the *DE*-approach, shows different computational behavior. Whether control points are moving or not, all truncated basis functions remain unchanged. It turns out that the computational gain of the presented algorithm for a dynamic curve is more significant than those of others if $C(t)$ is provided (as Equation (4.6)) as a preprocessing tool for a dynamic curve.

Let $C(t)$ be a B-spline curve before any control point moves and $C_d(t)$ be a dynamic curve counterpart of $C(t)$. Then, $C_d(t)$ can be now rewritten as the following equation using difference vectors, starting at the positions of old control points and ending at those of new control points:

$$C_d(t) = \sum_{k \in K} N_{k,p}(t)P_k + \sum_{j \in J} N_{j,p}(t)D_j, \tag{4.8}$$

where $K \equiv I \bigcup J$. That is, P_k, $k \in K$, is all control points of $C(t)$, and $D_j = (\tilde{P}_j - P_j)$ corresponds to the displacement of the moving control point.

Thus, from Equation (4.8), it can be found that $C_d(t)$ can be obtained by the summation of original curve $C(t)$ and difference vectors multiplied by the corresponding basis functions. It is necessary to locate knot spans that are affected by the second term of Equation (4.8) so that the transformation can be done more efficiently.

On the other hand, $C_d(t)$ can also be divided into two groups: the first group is the set of curve segments whose shapes are fixed, and the second is the set of curve segments whose shapes are changed by moving control points. Hence the following equation holds:

$$C_d(t) = \sum_{m \in M} C_m(t)N_{m,0}(t) + \sum_{n \in N} \tilde{C}_n(t)N_{n,0}(t), \tag{4.9}$$

where M and N are index sets for knot spans of curve segments whose shapes are fixed and changed by moving control points, respectively. In addition, $\tilde{C}_n(t)$ can be again rewritten by using truncated basis functions as follows, since $\tilde{C}_n(t)$ may also have both fixed and moving control points:

$$\tilde{C}_n(t) = \sum_{q \in Q} \tau_{q,n}(t)P_q + \sum_{r \in R} \tau_{r,n}(t)\tilde{P}_r, \tag{4.10}$$

where Q and R are index sets for fixed and moving control points for $\tilde{C}_n(t)$, respectively. Therefore,

$$\tilde{C}_n(t) = \sum_{s \in S} \tau_{s,n}(t)P_s + \sum_{r \in R} \tau_{r,n}(t)(\tilde{P}_r - P_r), \tag{4.11}$$

where $S \equiv Q \bigcup R$ and $|S| = p + 1$. P_s, $s \in S$, is all control points of $C_n(t)$ before they move. Thus, Equation (4.11) can be rewritten as Equation (4.12) using difference vector and truncated basis function:

$$\tilde{C}_n(t) = C_n(t) + \sum_{r \in R} \tau_{r,n}(t)D_r, \tag{4.12}$$

where $\mathbf{D}_r = \tilde{\mathbf{P}}_r - \mathbf{P}_r$ is a difference vector whose value is the displacement of the moving control point. Thus, for a particular knot span, a changed curve segment in power form, $\tilde{\mathbf{C}}_n(t)$ can be obtained by summing the original polynomial $\mathbf{C}_n(t)$ in the form of Equation (4.5) and difference vectors multiplied by the corresponding truncated basis functions. Performing the operation in Equation (4.12) for the knot spans that are influenced by the moving control points completes the desired transformation for $\mathbf{C}_d(t)$.

4. Extension to a Dynamic Surface

The presented idea can be easily extended to the conversion problem of a B-spline surface. The computational gain for a dynamic B-spline surface is more significant compared to its counterpart of curve. A B-spline surface of degree $p \times q$ is defined as Equation (4.13) [9].

$$\mathbf{S}(s, t) = \sum_{i=0}^{n-p-1} \sum_{j=0}^{m-q-1} \mathbf{P}_{i,j} N_{i,p}(s) N_{j,q}(t) \quad \text{for } 0 \le t \le 1, 0 \le s \le 1, \tag{4.13}$$

where $\mathbf{P}_{i,j}$, $N_{i,p}(t)$ and $N_{j,q}(t)$ are control points and B-spline basis functions of degree p and q on knot vectors $\mathbf{S} = \{0, \ldots, 0, s_{p+1}, \ldots, s_{n-p-1}, 1, \ldots 1\}$, $s_i \le s_{i+1}$ and $\mathbf{T} = \{0, \ldots, 0, t_{q+1}, \ldots, t_{m-q-1}, 1, \ldots 1\}$, $t_i \le t_{i+1}$, respectively.

A similar application of the presented algorithm to a B-pline surface yields the desired power basis representation of the surface. For a particular knot cell, $[s_i, s_{i+1}) \times \lfloor t_j, t_{j+1})$, the result will have the form as Equation (4.14).

$$\mathbf{S}_{i,j}(s, t) = \sum_{k=i-p}^{i} \sum_{l=j-q}^{j} \mathbf{P}_{k,l} \tau_{k,l,i,j}(s, t) \quad \text{for } [s_i, s_{i+1}) \times \lfloor t_j, t_{j+1}), \tag{4.14}$$

where $\tau_{k,l,i,j}(s, t) = \tau_{k,i}(s) \tau_{l,j}(t)$ is the multiplication between two truncated basis functions. Thus, applying Equation (4.14) repeatedly for each knot cell completes the transformation of a B-spline surface into a set of polynomial surfaces in power basis representation as follows:

$$\mathbf{S}(s, t) = \sum_{i=p}^{n-p-1} \sum_{j=q}^{m-q-1} \mathbf{S}_{i,j}(s, t) N_{i,0}(s) N_{j,0}(t). \tag{4.15}$$

A dynamic B-spline surface can be similarly defined as follows.

Definition 3. A dynamic B-spline surface, $\mathbf{S}_d(s, t)$ can be defined as a B-spline surface with more than one control point moving.

Thus, $\mathbf{S}_d(s, t)$ can be represented by Equation (4.16).

$$\mathbf{S}_d(s, t) = \sum_{(i,j) \in I} \sum \mathbf{P}_{i,j} N_{i,p}(s) N_{j,q}(t) + \sum_{(k,l) \in J} \sum \tilde{\mathbf{P}}_{k,l} N_{k,p}(s) N_{l,q}(t), \tag{4.16}$$

where I and J are index sets of ordered indices for fixed and moving control points, respectively. Then, by following the same process as the conversion problem of a dynamic B-spline curve,

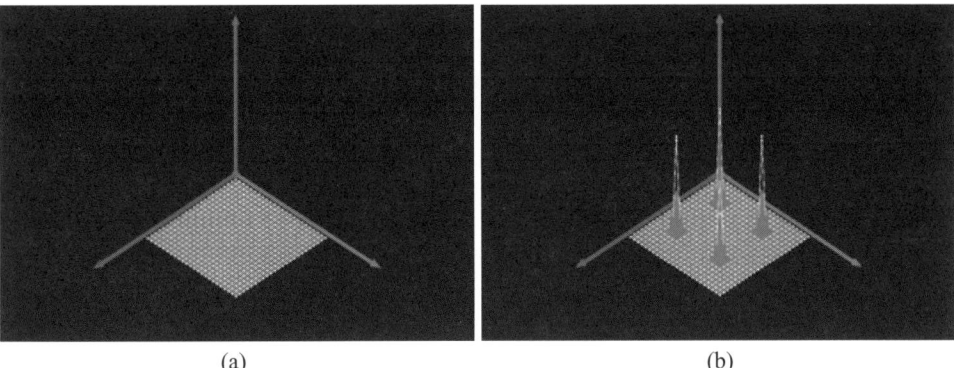

(a)   (b)

Figure 4.7 Example of the evaluations of a dynamic B-spline surface.

Equation (4.17) can be obtained.

$$
\begin{aligned}
\tilde{\mathbf{S}}_{k,l}(s,t) &= \sum_{(i,j)\in S}\sum \mathbf{P}_{i,j}\,\tau_{i,j,k,l}(s,t) + \sum_{(m,n)\in R}\sum \mathbf{D}_{m,n}\,\tau_{m,n,k,l}(s,t) \\
&= \mathbf{S}_{k,l}(s,t) \qquad\qquad\quad + \sum_{(m,n)\in R}\sum \mathbf{D}_{m,n}\,\tau_{m,n,k,l}(s,t).
\end{aligned}
\tag{4.17}
$$

where $S \equiv Q \bigcup R$ and $|S| = (p+1) \times (q+1)$. $\mathbf{P}_{i,j}$, $(i,j) \in S$, is all control points of $\mathbf{S}_{i,j}(s,t)$ before they move. $\mathbf{D}_{m,n} = (\tilde{\mathbf{P}}_{m,n} - \mathbf{P}_{m,n})$ are difference vectors whose values are the displacements of moving control points.

Thus, for the entire knot cells that the surface shape was influenced by moving control points, performing the operation of Equation (4.17) can yield the desired power basis form of a B-spline surface under an assumption that a B-spline surface is given as Equation (4.15) via pre-processing.

Figure 4.6 illustrates the evaluation process of 44,100 points on a cubic dynamic B-spline surface with four moving control points. Figure 4.7a shows a static B-spline surface evaluated through Equation (4.15) and Figure 4.7b shows that the point evaluation is performed via Equation (4.17).

5. Experiments and Discussion

Transformation of a dynamic B-spline curve into a set of piecewise polynomial curves in power form through the presented algorithm consists of two steps:

- pre-processing in Equation (4.6), and
- the operation in Equation (4.12) for the entire knot spans where a curve changed its shape.

Suppose that one of the control points of a B-spline curve of degree p is moving. While TE and KR-approaches have much more computational burdens, the presented algorithm takes only $(p+1)$ multiplications and $(p+1)$ additions for a knot span since the corresponding truncated basis functions are fixed. Thus, the presented algorithm need only $O(p^2)$ arithmetic operations.

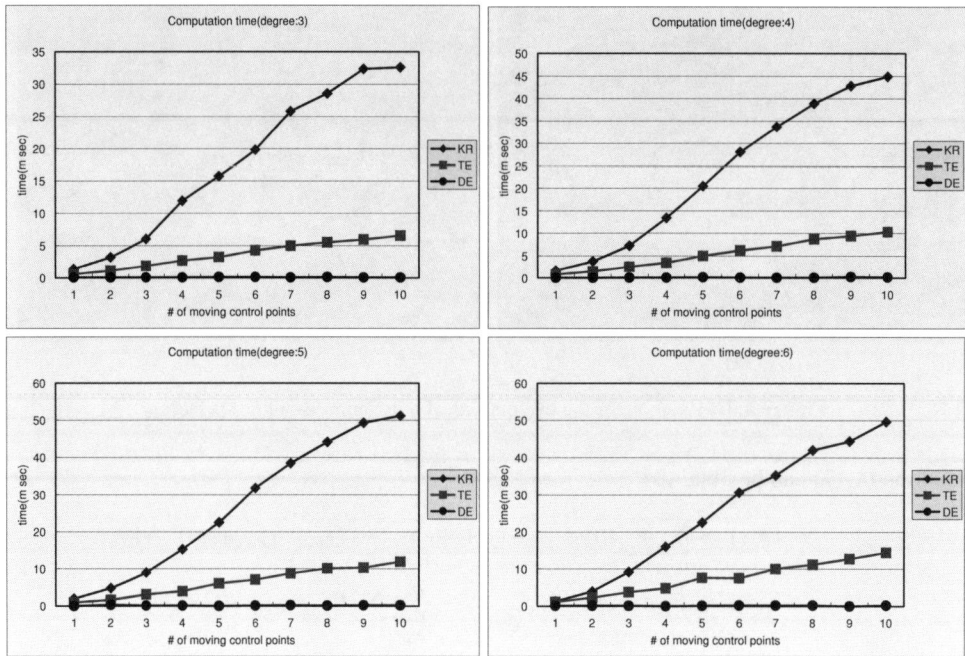

Figure 4.8   Computation time vs. number of moving control points.

For $(p + 1)$ knot spans that are influenced by a control point, TE-approach need to recalculate new polynomial curve segments. In this approach, each knot span has $(p + 1)$ coefficients to be computed, and each coefficient needs to evaluate a factorial function and a derivative. If Horner's rule is used, a derivative evaluation requires $O(p)$ operations. Otherwise, $O(p^2)$ operations are needed. Factorial evaluation also needs $O(p)$ operations. Thus, TE-approach requires $O(p^3)$ or $O(p^4)$ operations. In the case of KR-approach, knot refinement and basis conversion from Bernstein basis to power basis should also be performed for $(p + 1)$ knot spans.

In Figure 4.8 the computation time for each approach is provided. The presented algorithm, denoted by *DE*, outperforms other approaches and the computational gain of DE algorithm gets

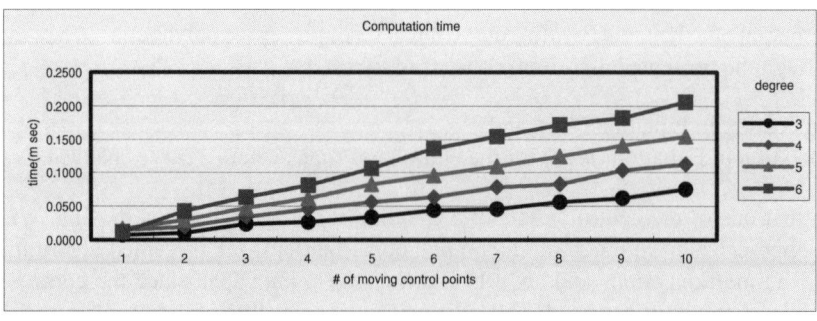

Figure 4.9   Computation time vs. number of moving control points for degree 3, 4, 5, 6.

significant as the degree of curve and the number of control points increase. The computational behavior of DE algorithm itself is provided separately in Figure 4.9. Note that the computation time is linear with respect to the number of moving control points.

6. Summary

B-spline like representations [10, 16, 21] are for surface modeling schemes in CAGD and computer graphics fields. It is often necessary to manipulate B-spline representations in power basis form. Once B-spline curves or surfaces are transformed into a power basis representation, fast evaluations are available and the computation of inflection points or cusps on a curve can be also accelerated.

In this chapter, we have provided a new conversion algorithm for transforming a B-spline curves or surfaces into a set of power basis polynomial curves or surfaces. In particular, the presented algorithm focuses on the conversion problem when curves or surfaces change their shapes continuously by moving some control points. In this case, the speed of computation is more important than other issues.

Experiments show that the presented algorithm is computationally efficient compared to conventional approaches. Thus, the presented algorithm is useful for intersection problems, the blending between B-spline curves, and the visualization of deforming B-spline curves or surfaces. In addition, the same problem for rational B-spline curves or surfaces can be handled with ease using the algorithm presented in this chapter.

References

[1] Bloomenthal, J. (1997) *Introduction to implicit surfaces*. Morgan Kaufmann Publishers Inc.

[2] Boehm, W. and Prautzsch, H. (1985) The insertion algorithm. *Computer-Aided Design* 12, 58–59.

[3] Boehm, W. (1985) On the efficiency of knot insertion algorithms. *Computer Aided Geometric Design* 2, 141–143.

[4] Cohen, E., Lyche, T. and Riesenfeld, R. (1980) Discrete B-splines and subdivision techniques in computer-aided geometric design and computer graphics. *Computer Graphics and Image Processing* 14, 87–111.

[5] Farin, G. (1997) *Curves and surfaces for computer-aided geometric design*, 3rd Ed., Academic Press.

[6] Goldman, R.N. (1990) Blossoming and knot insertion algorithm for B-spline curves. *Computer Aided Geometric Design* 7, 69–81.

[7] Kim, D.-S., Lee, S.-W., and Shin, H. (1998) A cocktail algorithm for planar Bezier curve intersections. *Computer Aided Design* 30, 1047–1051.

[8] Lasser, D. and Hoschek, J. (1993) *Fundamentals of computer aided geometric design*. A. K. Peters.

[9] Piegl, L. and Tiller, W. (1995) *The NURBS book*, Springer.

[10] Sarfraz, M. (2003). Weighted nu splines with local support basis functions, *Advances in Geometric Modeling*, Ed.: M. Sarfraz, John Wiley & Sons, Ltd. 101–118.

[11] Sarfraz, M. (2003). Optimal curve fitting to digital data. *International Journal of WSCG*, Vol. 11(1).

[12] Sarfraz, M. (2003). Curve fitting for large data using rational cubic splines. *International Journal of Computers and Their Applications*, Vol. 10(3).

[13] Sarfraz, M., and Razzak, M. F. A. (2003). A web based system to capture outlines of arabic fonts. *International Journal of Information Sciences*, Elsevier Science Inc., Vol. 150(3–4), 177–193.

[14] Sarfraz, M., and Razzak, M. F. A. (2002). An algorithm for automatic capturing of font outlines. *International Journal of Computers & Graphics*, Elsevier Science, Vol. 26(5), 795–804.

[15] Sarfraz, M. (2002) Fitting curves to planar digital data. *Proceedings of IEEE International Conference on Information Visualization IV'02-UK*: IEEE Computer Society Press, USA, 633–638.

[16] Sarfraz, M. (1992) A C^2 Rational cubic alternative to the NURBS. *Computers and Graphics* 16(1), 69–77.

[17] Sarfraz, M. (1992) Interpolatory rational cubic spline with biased, point and interval tension. *Computers and Graphics* 16(4), 427–430.

[18] Sarfraz, M. (1993) Designing of curves and surfaces using rational cubics. *Computers and Graphics* 17(5), 529–538.

[19] Sarfraz, M. (1995) Curves and surfaces for CAD using C^2 Rational Cubic Splines. *Engineering with Computers*, 11(2), 94–102.

[20] Gregory, J.A., Sarfraz, M. *et al.* (1994) Interactive curve design using C^2 rational splines. *Computers and Graphics* 18(2), 153–159.

[21] Sarfraz, M. (1994) Cubic spline curves with shape control. *Computers and Graphics* 18(4), 707–713.

[22] Sederberg, T.W. (1983) Implicit and parametric curves and surfaces for computer aided geometric design, *Ph. D. Thesis*, Purdue University.

[23] Watt, A. and Watt, M. (1992) *Advanced animation and rendering techniques*, Addison Wesley.

[24] Woo, M., Neider, J., and Davis, T. (1996) *OpenGL programming guide*, 2nd. Addison Wesley.

Appendix: Pseudo-code for Path Enumerations

The following pseudo-code performs path enumerations via lexicographic ordering of zero-one sequences. If there is a k numbers of 1's, then there are $\binom{p}{k}$ numbers of paths, which can be obtained by enumerating lexico graphically ordered positions of 1's. The size of each sequence is p (the degree of B-spline curve) and each sequence has n_h zeros and n_v ones ($n_h + n_v = p$). For example, all paths for a cubic B-spline curve in a knot span are $\{1, 1, 1\}$, $\{1, 1, 0\}$, $\{1, 0, 1\}$, $\{0, 1, 1\}$, $\{1, 0, 0\}$, $\{0, 1, 0\}$, $\{0, 0, 1\}$, and $\{0, 0, 0\}$. Hence, the number of all paths in a knot span for a curve of degree p is as follows:

$$\sum_{j=w-p}^{w} \frac{p!}{(w-j)!(p-w+j)!} = \sum_{k=0}^{p} \frac{p!}{k!(p-k)!} = \sum_{k=0}^{p} \binom{p}{k} = 2^p \qquad (4.18)$$

The number of all paths in a knot span increases rapidly as the degree of a curve gets higher. However, note that a path enumeration is needed only once for a given B-spline curve regardless of knot vector size. Once all paths for a knot span are determined, simple changes of subscripts will produce correct paths for an arbitrary knot span. Note that 2-dimensional array Index[][] keeps the position of ones.

```
Initialize Index  Index[0][l]=l+1 for l=0, ..., nv-1


for i=1 to  ( p )  do
              ( nv )
    for j=0 nv-1 do
       if(Index[i-1][nv-j-1]<p-j)
          for k=0 to nv-j-2 do
             Index[i][k]= Index[i-1][k];
          end k;
          Index[i][nv-j-1]= Index[i-1][nv-j-1]+1;
          for k=nv-j to nv-1 do
             Index[i][k]= Index[i][k-1]+1;
          end k;
          break;
       end if
    end j;
end i;
```

5

Computational Methods for Geometric Processing of Surfaces: Blending, Offsetting, Intersection, Implicitization

Andres Iglesias

*Department of Applied Mathematics and Computational Sciences, University of
Cantabria, Avda. de los Castros s/n, E-39005, Santander, Spain.*

*The aim of this chapter is to provide the interested reader with a general overview of some
relevant problems in geometric processing of surfaces. To this end, the chapter offers a classified
bibliography with more than 160 selected references on computational methods for surface
blending, offsetting, intersection and implicitization. The literature referenced in this chapter
is not intended to comprise a totally exhaustive review on these topics, but it still includes
enough comments and references to be useful for the reader.*

1. Introduction

Geometric processing is defined as the calculation of geometric properties of already constructed curves, surfaces and solids [12]. In its most comprehensive meaning, this term includes
all the algorithms that are applied to already existing geometric entities [35]. As pointed out
in [12] since geometric processing is intrinsically *hard* there is neither a unified approach nor
'key developments' such as the Bezier technique [39, 108] for design. On the contrary, the
literature on geometric processing is much more dispersed amongst different sources. The aim
of this chapter is to provide the interested reader with a general overview of some relevant
problems in geometric processing of surfaces. To this end, the chapter offers a non-exhaustive
classified bibliography with more than 160 selected references on computational methods for

Advances in Geometric Modeling. Edited by M. Sarfraz
© 2003 John Wiley & Sons, Ltd   ISBN: 0-470-85937-7

surface blending (Section 2), offsetting (Section 3), intersection (Section 4) and implicitization (Section 5). Finally, we should remark that the chapter has been influenced by both the extraordinary magnitude of this task and the limitations of space. Some references have had to be omitted and many explanations have been reduced to the minimum. Due to these reasons, the literature referenced in this chapter is not intended to comprise a totally exhaustive survey on these topics. However, we hope that we have included enough comments and references to make the chapter useful for our readers.

2. Blending Surfaces

We use the term *blending* to mean the construction of connecting curves and surfaces and the rounding off of sharp corners or edges. Although we can easily distinguish between interior and exterior blending, it is better to classify blending methods according to their mathematical definition [146]. Thus, we talk about *superficial blending* to indicate that no explicit mathematical formula is available. Superficial blending appears in the production process [27, 152, 155] in procedures such as rounding off a corner or edge with radius r. The blend described by additional surfaces connecting smoothly to some given surfaces is usually referred to as *surface blending*, while *volumetric blending* is used to mean the combination of objects in a solid modeling system (see, for instance, [46, 54, 146, 156]). Finally, some authors also consider *polyhedral blending* for those cases in which the objects to be blended are defined by polyhedra. In this case, blending is performed through surface blending of the resulting patches [76, 126] or by recursive subdivision [20, 97–99].

Especially interesting for many applied purposes is surface blending, which can be given in either implicit or parametric form. Early works on unbounded implicit blending were considered in [14] and [117] for molecule design and free-form modeling, respectively. Field functions, such as those used to define soft objects [15, 159], are inspired by the same ideas. Implicit surface blending was applied to C^1 data interpolation by Bajaj [9] while C^2 data interpolation was later analyzed in [51] by considering rational cubic splines [50, 128] as an alternative to NURBS [127] with applications to design [129, 130]. Other remarkable works in this area are [63, 85, 95, 104, 118, 119]. A nice paper on blending of algebraic surfaces can be found in [151].

Analogously, a number of methods for parametric blending have been described, ranging from *interactive methods* [11, 104] to *automatic methods* based on calculation of intersections of offset surfaces to the two given surfaces [79, 104]. The core of these methods is the calculation of the trim lines. In the interactive methods such lines are defined interactively by means of a sequence of points on the surfaces which are inverted onto the parameter planes, fitted through B-splines [11] and then mapped back onto the surfaces. These methods are highly flexible and they do not require the existence of any intersection curve for the surfaces to be blended. On the contrary, some automatic methods assume that the surfaces intersect [11]. In some special cases, it is even possible to obtain an analytic expression for the blending surface. This is the case for B-spline trim curves: the resulting blending surface can be defined as a B-spline tensor product surface [38, 109, 120]. Blending of tensor product B-spline or Bezier surfaces are analyzed, for example, in [11, 28, 45, 77]. A good survey on blending methods for parametric surfaces can be found in [156]. A more recent survey can also be found in [147].

Finally, some works have been published in the last few years where emphasis is put on blending for pairs of implicit and parametric surfaces [54, 55] allowing G^1 continuity (that is, tangent plane continuity) [56]. A solution to the problem of maintaining the G^1 continuity of the blended curves by adjusting the positions of the junction points of the curve segments was proposed in [69].

3. Offset Surfaces

Offsetting is a geometric operation which expands a given object into a similar object to a certain extent. For instance, offset of curves and surfaces are also curves and surfaces (respectively) at a constant distance d from the given initial curve or surface. Offsetting has various important applications [78, 122]. For example, if the inner surface of a piece is taken as the reference surface, the outer surface can be mathematically described by an offset surface corresponding to a distance equal to the thickness of the material. Offsets also appear in cutter-path generation for numerical control machine tools [68, 125, 139]: pieces of a surface can be cut, milled or polished using a laser-controlled device to follow the offset [27, 103, 144]. In the case of curves, they can be seen as the envelope corresponding to moving the center of a circle of radius d along the initial curve. This allows us to define both the inside and outside offset curves, with applications in milling [153]. Finally, they are fundamental tools (among others) in the constant-radius rounding and filleting of solids or in tolerance analysis [67, 70, 73], for definition of tolerance zones, etc.

In this chapter we restrict ourselves to offset surfaces. For the case of curves, several methods for the computation of their offsets are described and compared in [34]. As pointed out in [107], offsetting general surfaces is generally quite complicated, and an offset surface is often approximated [33]. Approximations of offset surfaces by bicubic spline surfaces are described in [40]. However, some of these approximations become inaccurate near its self-intersecting area [3, 107]. In [24] a marching method to compute the self-intersection curve of an offset surface is proposed. The same problem was solved in [3] for the case of offset surfaces of a uniform bicubic B-spline surface patch and in [92, 143] for Bezier surface patches, by using differential geometry and ray tracing, respectively. Subsequently, some approaches to remove loops appearing in the generation of tool paths for two-and-a-half axis pocket machining by using a pairwise intersection, the Voronoi diagram method and the Invariant-Gaus-Bonet offset were described in [53, 57, 121], respectively.

We should note, however, that offset curves and surfaces lead to several practical problems. Depending on the shape of the initial curve, its offset can come closer than d to the curve, thus causing problems with *collisions*, for instance, when steering a tool. These collision problems also arise in other applications, such as path-planning for robot motions, a key problem in the current industry. To avoid this, we need to remove certain segments of the curve which start and end at self-intersections [57, 123]. Special methods for the case of interior offsets (as used in milling holes or *pockets*) can be found in [57] and [105]. Other alternatives include a method to compute the offset of the scallop hull (a generalization of the convex hull [2, 57]) in linear time, i.e., in time that is linearly proportional to the number of segments of the input curve [29].

For surfaces the scenario is, by large, much more complicated: singularities at a point can arise when the distance d of the smallest value of the principal curvature is attained at the point. In addition, these singularities can be of many different types: cusps, sharp edges or self-intersections [40]. Finally, the set of rational curves and surfaces is not closed under offsetting [37]. Therefore, considerable attention has been paid to identify the curves and surfaces which admit rational offsets [42, 107, 111]. Special cases for surfaces whose offsets are rational include, for example, planes, spheres, circular cylinders and cones, torus and cyclides. The case of polynomial and rational curves with rational offsets is analyzed in [86]. In [87] the author showed that the offsets of paraboloids, ellipsoids and hyperboloids can be rationally parameterized. On the contrary, cylinders and cones do not have any rational offset except for the particular cases of parabolic cylinders, and cylinders and cones of revolution. In [114] it has been shown that offsets of nondevelopable rational ruled surfaces can be rationally parameterized in the whole space, although special care must be put on the case of finite patches, since the offset of a rational patch may not be expressed as a rational patch. An interesting approach for computing offsets of NURBS curves and surfaces is given in [110]. The smoothness of the offset surfaces is discussed in [59]. We also recommend [91] for a more recent overview of offset curves and surfaces.

Other recent developments are *geodesic offsets* [102] (the locus of points at a constant distance measured from an arbitrary curve C on a surface along the geodesic curve orthogonal to C) and *general offsets*, first introduced in [17] and extended in [112]. Both kinds of offsets exhibit interesting applications in manufacture. For example, geodesic offset curves are used to generate tool paths on a part for zig-zag finishing using 3-axis machining with ball-end cutter so that the scallop-height (the cusp height of the material removed by the cutter) will become constant [131, 140]. This leads to a significant reduction in size of the cutter location data and hence in the machining time. On the other hand, not only ball-end but also cylindrical and toroidal cutters are used in 3-axis NC machining. When the center of the ball-end cutter moves along the offset surface, the reference point on the cylindrical and toroidal cutters move along the general offset. Finally, a nice application of general offsets to collision problems is described in [113].

4. Intersection of Surfaces

In many applications, computation of the intersections of curves and surfaces is required. Among them, we quote smooth blending of curves and surfaces (Section 2), the construction of contour maps to visualize surfaces [48] (intersection with series of parallel planes, cylinders, cones, etc), Boolean operations on solid bodies for constructive solid geometry models of the objects, in manufacturing [12] (slicing operations for rapid prototyping, determination of collisions) and determination of self-intersections in offset curves and surfaces (Section 3).

A significant body of literature exists on the calculation of intersections of two surfaces (see, for example, the excellent reviews on this topic in [64] (Chapter 12) or [103] (Chapter 5) and the references therein. We also recommend the survey in [115]. Earlier references can be found in [36]). Basically, they can be classified into analytical and numerical methods. Analytical methods seek exact solutions by finding some function describing the intersection curves [22, 101]. The obvious advantage of these methods is that they are unaffected by robustness and efficiency limitations. The simplest example for the case of surfaces is that of an implicitparametric surface intersection. In this case, we can insert the parametric formula into the implicit form to get a nonlinear system of four algebraic equations in five variables for the intersection. If both the implicit and the parametric surfaces have a low degree (as is usual in many practical cases) we can obtain an implicit curve in the parametric domain [82, 115] but it still remains a problem for high degree surfaces. In other cases, solutions for this equation can be achieved through numerical methods [56] and differential geometry [4]. Other proposals include a combination of algebraic and analytical methods [41], hybrid algorithms combining subdivision (based on the divide-and-conquer methodologies), tracing and numerical methods (mainly Newton's method) [81], etc. Unfortunately, they exhibit a substantial loss of accuracy making them unsuitable for practical applications. Finally, there is a family of methods known as marching methods based on generating a sequence of points of an intersection curve branch by stepping from a point on such a curve in a direction determined by some local differential geometry analysis [8, 13, 80]. These methods are globally incomplete since they require starting points for every branch of the solution. Motivated by this, great effort has been devoted to the determination of such starting points by using hodographs [137], elimination methods [21] or by two local methods, namely iterative optimization and Moore-Penrose pseudo-inverse method [1].

In the case of a couple of implicit surfaces to be intersected, the analytical method yields a system of two equations in three variables. Some methods to obtain solutions to this problem in the case of quadric surfaces (clearly, a problem with relevance in CAD/CAM of mechanical parts) can be found in [43, 96, 132, 138, 154]. Other methods to solve this problem through a combination of geometric and analytic methods are given in [100]. Also, Bajaj *et al.* proposed a method for solving this problem that does not need to know the functions explicitly [8]. In [56] the concept of *numerical implicitization* was proposed. This author realized that in tracing the intersection curve of two implicit surfaces we only need to know both the values and the gradients of the implicit function at the intersection points. The practical implications of this is that we can implicitize parametric surfaces numerically so that algorithms for implicit-implicit surface intersection can be successfully applied to initial parametric surface intersection problems.

Finally, the intersection of a couple of parametric surfaces can be reduced to the implicit-parametric case by implicitizing the parametric equations of one surface (see Section 5 of this chapter for more details on implicitization). Other techniques for parametric surfaces can be found in [1, 13, 37, 44, 60, 137]. However, there has been no known algorithm that can compute the intersection curve of two arbitrary rational surfaces accurately, robustly and efficiently [64]. In addition, it is known that two surface patches intersect in a curve whose degree is much higher than the parametric degree of the two patches. Thus, two bicubic patches intersect in a curve of degree 324! Fortunately, the situation is better when we restrict the domain of input surfaces to simple surfaces (planes, quadrics and tori, i.e., the so-called *CSG primitives*) [23, 75, 84, 96, 116, 138]. These surfaces are important in conventional solid modeling systems for industry, since they can represent a large number of mechanical parts of a car, ship, plane, etc. Another algorithm to deal with general ruled surfaces (which can be classified as an algebraic method because it consists of reducing the surface intersection problem to a simpler problem of computing an implicit curve in the parametric domain) has been described in [58].

As the reader can see, the analytical methods require many different algorithms designed *ad hoc* for each kind of surface involved. While for low degree surfaces there is a number of efficient methods, more difficulties appear for high-order surfaces. Furthermore, the analytical methods cannot deal with non-algebraic surfaces and hence numerical methods are usually applied instead. At their turn, numerical methods can be classified into several categories [115]: *subdivision methods*, which divide the objects to be intersected into many pieces and check for intersections of the pieces [12, 19, 30, 47, 52, 74, 83, 97, 98, 164]. The goal of these methods is to find a polygonal approximation of the intersection curves by using face splitting of the surfaces, and it is mostly applied to create characters and other complicated shapes in computer animation [31]; *discretization methods*, which reduce the degrees of freedom by discretizing the surface representation in several ways, such as contouring [32, 106, 141] or parameter discretization [12, 66], *marching methods* which require at least a point on the intersection curve, called the starting point, to generate a sequence of points on the intersection curve using local geometry of the intersecting surfaces [8, 13, 80], *hybrid methods*, which combine subdivision and numerical methods [142, 163], *differential methods* such as the second order boundary algebraic-differential approach in [48, 49], etc.

Other recent developments include the possibility of handling intersection singularities and loops [8, 25, 62, 88–90, 136, 137]. Intersections of offsets of parametric surfaces are analyzed in [150]. This problem is often of great interest: for instance, a blend surface (see Section 2) of two surfaces can be constructed by moving the center of a sphere of given radius along the intersection curve of two surfaces that are offset from the base surfaces by the radius of the sphere.

5. Implicitization

In the last few years, implicit representations are being used more frequently in CAGD, allowing a better treatment of several problems. For instance, the point classification problem is easily solved with the implicit representation: it consists of a simple evaluation of the implicit functions. This is useful in many applications, such as solid modeling for mechanical parts, for example, where points must be defined inside or outside the boundaries of an object, or for calculating intersections of free-form curves and surfaces. The problem of computing the parametric-implicit (parametric-parametric) surface intersection is very often reduced to an implicit-implicit (implicitparametric) surface intersection problem (see Section 4). Through implicit representation, the problem is reduced to a trivial sign test. Other advantages are that the class of implicit surfaces is closed under such operations such as offsetting, blending and bisecting. In other words, the offset (see Section 3) of an algebraic curve (surface) is again an algebraic curve (surface) and so on. In addition, the intersection (see Section 4) of two algebraic surfaces is an algebraic curve. Furthermore, the implicit representation offers surfaces of desired smoothness with the lowest possible degree. Finally, the implicit representation is more general than the rational parametric one [60]. All these advantages explain why the implicit equation of a geometric object is of importance in practical problems.

Implicitization is the process of determining the implicit equation of a parametrically defined curve or surface. One remarkable fact is that this parametric-implicit conversion is *always* possible [26, 133]. Therefore, for any parametric curve or surface there exists an implicit polynomial equation defining exactly the same curve or surface. The corresponding algorithm for curves is given in [134] and [135]. In addition, a parametric curve of degree n has an implicit equation also of degree n. Further, the coefficients of this implicit equation are obtained from those of the parametric form by using only multiplication, addition and subtraction, so conversion can be performed through symbolic computation, with no numerical error introduced. Implicitization algorithms also exist for surfaces [93, 94, 134, 135]. However, a triangular parametric surface patch of degree n has an implicit equation of degree n^2. Similarly, a tensor product parametric patch of degree (m, n) has an implicit equation of degree $2mn$. For example, a bicubic patch has an implicit equation of degree *18* with *1330* terms!

In general, the implicitization algorithms are based on *resultants*, a classical technique [124], Grobner basis techniques [18] and on the Wu-Ritt method [157]. Resultants provide a set of techniques [72] for eliminating variables from systems of nonlinear equations. However, the derived implicit equation may have extraneous factors: for example, surfaces can exhibit additional sheets. On the other hand, symbolic computation required to obtain the implicit expression exceeds the resources in space and time, although parallel computation might, at least partially, solve this problem. On the other hand, given an initial set of two or three polynomials defining the parametric curve or surface as a basis for an ideal [60], the Grobner basis will be such that it contains the implicit form of the curve or surface. In the rational case, additional polynomials are needed to account for the possibility of base points [71]. Finally, the Wu-Ritt method consists of transforming the initial set into a triangular system of polynomials. This transformation involves rewriting the polynomials using pseudo-division and adding the remainders to the set. The reader is referred to [72] and [157] for more details.

With respect to implementation, hybrid symbolic/numerical methods have been proposed in [94]. Also, in [61] atractive speed-ups for Grobner based implicitization using numerical and algebraic techniques have been obtained.

On the other hand, we remark that implicitization can be seen as a particular case of conversion between different curve or surface forms (see, for example, [148, 149]). See also [65] (and references therein) for a survey on approximate conversion between Bezier and B-spline surfaces, which are also applied to offsets.

Finally, we remark that during the last few years great interest has been put on the applications of implicit surfaces for CAGD and computer graphics. We must quote (among many others) the seminal work of Bajaj and collaborators on implicit A-patches (algebraic patches) [5–7, 10, 162] and the work of B. Wyvill *et al.* on implicit surfaces for defining *soft objects* [15, 145, 158–161]. The interested reader is referred to [16] for a nice introduction to implicit surfaces with applications to computer graphics and design.

6. Summary

In this chapter a general overview of some relevant computational methods for surface blending, offsetting, intersection and implicitization is given. However, the chapter must not be understood as a survey with an exhaustive bibliography on these topics, which is beyond the scope of this work. Our only aim is to provide the interested reader with a comprehensive and soft introduction to the geometric processing of surfaces. To this end, more than 160 selected and commented references have been included.

References

[1] Abdel-Malek K., Yeh H.J. (1997) On the determination of starting points for parametric surface intersections. *CAD* 29, 21–35

[2] Anderson R.O. (1978) Detecting and eliminating collision in NC machining. *CAD* 10, 231–237

[3] Aomura S., Uehara T. (1990) Self-intersection of an offset surface. *CAD* 22, 417–422

[4] Asteasu C. (1988) Intersection of arbitrary surfaces. *CAD* 20(9), 533–538

[5] Bajaj C. (1992) Surface fitting with implicit algebraic surface patches. In: Hagen H. (ed.) Topics in Surface Modeling. *SIAM*, 23–52

[6] Bajaj C. (1997) Implicit surface patches. In: Bloomenthal J. (ed.) *Introduction to Implicit Surfaces*, Morgan Kaufman Publishers, San Francisco, CA, 98–125

[7] Bajaj C., Chen J. *et al.* (1995) Modeling with cubic A-patches. *ACM Transactions on Graphics* 14(2), 103–133

[8] Bajaj C., Hoffmann C.M. *et al.* (1988) Tracing surface intersections, *CAGD* 5, 285–307

[9] Bajaj C., Ihm, I. *et al.* (1993) Higher order interpolation and least squares approximation using implicit algebraic surfaces. *ACM Transactions on Graphics* 12(4), 327–347

[10] Bajaj C., Xu G. (1999) A-splines: local interpolation and approximation using G^k-continuous piecewise real algebraic curves. *CAGD* 16, 557–578

[11] Bardis L., Patrikalakis N.M. (1989) Blending rational B-spline surfaces. *Proceedings of EUROGRAPHICS'89*, 453–462

[12] Barnhill R.E. (1992) *Geometric Processing for Design and Manufacturing.* SIAM, Philadelphia

[13] Barnhill R.E., Kersey S.N. (1990) A marching method for parametric surface/surface intersection. *CAGD* 7, 257–280

[14] Blinn J.F. (1982) A generalization of algebraic surface drawing. *ACM Transactions on Graphics* 1, 235–256

[15] Bloomenthal J., Bajaj C. *et al.* (1997) *Introduction to Implicit Surfaces.* Morgan Kaufmann Publishers, San Francisco, CA

[16] Bloomenthal J., Wyvill B. (1990) Interactive techniques for implicit modeling. *Computer Graphics* 24(2), 109–116

[17] Brechner E.L. (1992) General tool offset curves and surfaces. In: Barnhill R.E. (1992) *Geometric Processing for Design and Manufacturing.* SIAM, 101–121

[18] Buchberger B. (1985) Grobner bases: an algorithmic method in polynomial ideal theory. In: Rose N.K. (ed.) *Multidimensional Systems Theory*, Reidel Publishing Co., 184–232

[19] Carlson W.R. (1982) An algorithm and data structure for 3D object synthesis using surface patch intersections. *Computer Graphics* 16, 255–263

[20] Catmull E.E., Clark J. (1978) Recursively generated B-spline surfaces on arbitrary topological meshes. *CAD* 10, 350–355

[21] Chandru V., Dutta D. *et al.* (1989) On the geometry of Dupin cyclides. *The Visual Computer* 5, 277–290

[22] Chandru V., Kochar B.S. (1987) Analytic techniques for geometric intersection problems. In: Farin G.E. (ed.) *Geometric Modeling: Algorithms and New Trends*, SIAM, 305–318

[23] Chen J.J., Ozsoy T.M. (1988) Predictor-corrector type of intersection algorithm for C^2 parametric surfaces. *CAD* 20(6), 347–352

[24] Chen Y.J., Ravani B. (1987) Offset surface generation and contouring in computer-aided design. *Journal of Mechanics, Transmissions and Automation in Design*, Transactions of the ASME 109(3), 133–142

[25] Cheng K.P. (1989) Using plane vector fields to obtain all the intersection curves of two general surfaces. In: Strasser W., Seidel H.P. (ed.) *Theory and Practice in Geometric Modeling*, Springer, New York 187–204

[26] Chionh E.W., Goldman R.N. (1992) Using multivariate resultants to find the implicit equation of a rational surface. *The Visual Computer* 8, 171–180

[27] Choi, B.K., Jerard, R.B. (1998) *Sculptured Surface Machining. Theory and Applications.* Kluwer Academic Publishers, Dordrecht/Boston/London

[28] Choi B.K., Ju S.Y. (1989) Constant-radius blending in surface modeling. *CAD* 21, 213–220

[29] Chou S.Y., Woo T.C. *et al.* (1994) Scallop hull and its offset. *CAD* 26(7), 537–542

[30] Cohen E., Lyche T. *et al.* (1980) Discrete B-splines and subdivision techniques in CAGD and computer graphics. *Computer Graphics and Image Processing* 14, 87–111

[31] DeRose T., Kaas M. *et al.* (1998) Subdivision surfaces in characters animation. *Proceedings of SIGGRAPH'98*, 85–94

[32] Dobkin D.P., Levy S.V.F. *et al.* (1990) Contour tracking by piecewise linear approximations. *ACM Transactions on Graphics* 9, 389–423

[33] Elber G., Cohen E. (1991) Error bounded variable distance offset operator for free form curves and surfaces. *International Journal of Computational Geometry and Applications* 1(1), 67–78

[34] Elber G., Lee I. *et al.* (1997) Comparing offset curve approximation methods. *IEEE Computer Graphics and Applications* 17(3), 62–71

[35] Farin G.E. (1989) Trends in curve and surface design. *CAD* 21(5), 293–296

[36] Farin G.E. (1992) An ISS bibliography. In: Barnhill R.E. (ed.) *Geometric Processing for Design and Manufacturing*, SIAM, 205–207

[37] Farin G.E. (1996) *Curves and Surfaces for Computer Aided Geometric Design*, Fourth Edition. Academic Press, San Diego

[38] Farin G.E. (1999) *NURB Curves and Surfaces: from Projective Geometry to Practical Use*, Second Edition. AK Peters, Wellesley, MA

[39] Farin G.E., Hansford D. (2000) *The Essentials of CAGD*. AK Peters, Wellesley, MA

[40] Farouki R.T. (1986) The approximation of non-degenerate offset surfaces. *CAGD* 3, 15–43

[41] Farouki R.T. (1987) Direct surface section evaluation. In: Farin G.E. (ed.) *Geometric Modeling. Algorithms and New Trends*, SIAM, 319–334

[42] Farouki R.T. (1992) Pythegorean-hodograph curves in practical use. In: Barnhill R.E. (ed.) *Geometric Processing for Design and Manufacturing*, SIAM, 3–33

[43] Farouki, R.T., Neff. C.A. *et al.* (1989) Automatic parsing of degenerate quadric-surface intersections. *ACM Transactions on Graphics* 8(3), 319–334

[44] Faux I.D., Pratt M.J. (1979) *Computational Geometry for Design and Manufacture*. Ellis Horwood, Chichester

[45] Filip D.J. (1989) Blending parametric surfaces. *ACM Transactions on Graphics* 8(3), 164–173

[46] Glaeser G., Groller E. (1998) Efficient volume-generation during the simulation of NC-milling. In: Hege H.C., Polthier K. (eds.) *Mathematical Visualization. Algorithms, Applications and Numerics*. Springer-Verlag, Berlin Heidelberg, 89–106

[47] Goldman R.N. (1983) Subdivision algorithms for Bezier triangles. *CAD* 15, 159–166

[48] Grandine T.A. (2000) Applications of contouring. *SIAM Review* 42, 297–316

[49] Grandine T.A., Klein F.W. (1997) A new approach to the surface intersection problem. *CAGD* 14, 111–134

[50] Gregory J.A., Sarfraz M. (1990) A rational cubic spline with tension. *CAGD* 7, 1–13

[51] Gregory J. A., Sarfraz M. *et al.* (1994) Interactive curve design using C^2 rational splines. *Computers and Graphics* 18(2), 153–159

[52] Griffiths J.G. (1975) A data structure for the elimination of hidden surfaces by patch subdivision. *CAD* 7, 171–178

[53] Hansen A., Arbab F. (1992) An algorithm for generating NC tool paths for arbitrarily shaped pockets with islands. *ACM Transactions on Graphics* 11(2), 152–182

[54] Hartmann E. (1990) Blending of implicit surfaces with functional splines. *CAD* 22, 500–506

[55] Hartmann E. (1995) Blending an implicit with a parametric surface. *CAGD* 12, 825–835

[56] Hartmann E. (1998) Numerical implicitization for intersection and G^n-continuous blending of surfaces. *CAGD* 15(4), 377–397

[57] Held M. (1991) On the Computational Geometry of Pocket Machining. *Lectures Notes in Computer Science*, Springer Verlag, Berlin New York

[58] Heo H.S., Kim M.S. *et al.* (1999) The intersection of two ruled surfaces. *CAD* 31, 33–50

[59] Hermann T. (1998) On the smoothness of offset surfaces. *CAGD* 15, 529–533

[60] Hoffmann C.M. (1989) *Geometric and Solid Modeling*. Morgan Kaufmann Publishers, San Francisco, CA

[61] Hoffmann C.M. (1990) Algebraic and numerical techniques for offsets and blends. In: Micchelli, S., Gasca M. *et al.* (eds.) *Computations of Curves and Surfaces*, Kluwer Academic Publishers, 499–528

[62] Hohmeyer M.E. (1991) A surface intersection algorithm based on loop detection. *International Journal of Computational Geometry and Applications* 1(4), 473–490

[63] Hoffmann C.M., Hopcroft J. (1985) Automatic surface generation in computer aided design. *The Visual Computer* 1, 92–100

[64] Hoschek J., Lasser D. (1993) *Fundamentals of Computer Aided Geometric Design*. A.K. Peters, Wellesley, MA

[65] Hoschek J., Schneider F.J. (1992) Approximate spline conversion for integral and rational Bezier and B-spline surfaces. In: Barnhill R.E. (ed.) *Geometric Processing for Design and Manufacturing*, SIAM, 45–86

[66] Houghton E.G., Emnett R.F. *et al.* (1985) Implementation of a divide-and-conquer method for intersection of parametric surfaces. *CAGD* 2, 173–183

[67] Huang Y., Oliver J.H. (1994) NC milling error assessment and tool path correction. *Proceedings of SIGGRAPH'94, Computer Graphics* 28, 287–294

[68] Hui K.C. (1994) Solid sweeping in image space-application in NC simulation. *The Visual Computer* 10, 306–316

[69] Hui K.C. (1999) Shape blending of curves and surfaces with geometric continuity. *CAD* 31, 819–828

[70] Jerard R.B., Hussaini, S.Z. *et al.* (1989) Approximate methods for simulation and verification on NC machining programs. *The Visual Computer* 5, 329–348

[71] Kalkbrener M. (1990) Implicitization of rational parametric curves and surfaces. *Technical Report*, Kepler Universitat, Linz, Austria

[72] Kapur D., Lakshman Y.N. (1992) Elimination methods. In: Donald B., Kapur D. *et al.* (eds.) *Symbolic and Numerical Computing for Artificial Intelligence*, Academic Press, San Diego, CA

[73] Kawashima Y., Itoh K. *et al.* (1991) A flexible quantitative method for NC machining verification using a space-division based solid model. *The Visual Computer* 7, 149–157

[74] Kay T.L., Kajiya J.T. (1986) Ray tracing complex scenes. *Computer Graphics* 20, 269–278

[75] Kim K.J., Kim M.S. (1998) Torus/sphere intersection based on configuration space approach. *Graphical Models and Image Processing* 60(1), 77–92

[76] Kimura F. (1984) Geomap III: designing solids with free-form surfaces. *IEEE Computer Graphics and Applications* 4, 58–72

[77] Klass R., Kuhn B. (1992) Fillet and surface intersections defined by rolling balls. *CAGD* 9, 185–193

[78] Klass R., Schramm P. (1991) NC milling of CAD surface data. In: Hagen H., Roller D. (eds.) *Geometric Modeling. Methods and Applications*, Springer Verlag, Berlin Heidelberg, 213–226

[79] Koparkar P.A. (1991) Designing parametric blends: surface model and geometric correspondence. *The Visual Computer* 7, 39–58

[80] Kriezis G.A., Patrikalakis N.M. *et al.* (1992) Topological and differential-equation methods for surface reconstructions. *CAD* 24(1), 41–55

[81] Kriezis G.A., Prakash P.V. *et al.* (1990) Method for intersecting algebraic surfaces with rational polynomial patches. *CAD* 22(10), 645–654

[82] Krishnan S., Manocha D. (1997) Efficient surface intersection algorithm based on lower-dimensional formulation, *ACM Transactions on Graphics* 16(1), 74–106

[83] Lasser D. (1986) Intersection of parametric surfaces in the Bernstein-Bezier representation. *CAGD* 3, 186–192

[84] Lee R.B., Fredricks D.A. (1984) Intersection of parametric surfaces and a plane. *IEEE Computer Graphics and Applications* August 1984, 48–51

[85] Li J., Hoscheck J. *et al.* (1990) G^1 functional splines for interpolation and approximation of curves, surfaces and solids. *CAGD* 7, 209–220

[86] Lu W. (1995) Offset-rational parametric plane curves. *CAGD* 12, 601–616

[87] Lu W. (1996) Rational parameterization of quadrics and their offsets. *Computing* 57(2), 135–147

[88] Lukacs, G. (1994) Simple singularities in surface-surface intersections. In: Bowyer A. (ed.) *Computer-Aided Surface Geometry and Design, The Mathematics of Surfaces IV*. Oxford University Press, 213–230

[89] Ma Y., Lee Y.S. (1998) Detection of loops and singularities of surface intersections. *CAD* 30, 1059–1067

[90] Ma Y., Luo R.C. (1995) Topological method for loop detection of surface intersection problems. *CAD* 27(1), 811–820

[91] Maekawa T. (1999) An overview of offset curves and surfaces. *CAD* 31, 165–173

[92] Maekawa T., Cho W. *et al.* (1997) Computation of self-intersections of offsets of Bezier surface patches. *Journal of Mechanical Design, Transactions of ASME* 119(2), 275–283

[93] Manocha D., Canny J.F. (1992) Algorithm for implicitizing rational parametric surfaces. *CAGD* 9, 25–50

[94] Manocha D., Canny J.F. (1992) Implicit representations of rational parametric surfaces. *Journal of Symbolic Computation* 13, 485–510

[95] Middleditch A.E., Sears K.H. (1985) Blend surfaces for set theoretic volume modeling systems. *Proceedings of SIGGRAPH'85, Computer Graphics* 19, 161–170

[96] Miller J., Goldman R.N. (1995) Geometric algorithms for detecting and calculating all conic sections in the intersection of any two natural quadric surfaces. *Graphical Models and Image Processing* 57(1), 55–66

[97] Nasri A.H. (1987) Polyhedral subdivision methods for free-form surfaces. *ACM Transactions on Graphics* 6, 29–73

[98] Nasri A.H. (1991) Surface interpolation on irregular networks with normal conditions. *CAGD* 8, 89–96

[99] Nasri A.H. (1991) Boundary-corner control in recursive-subdivision surfaces. *CAD* 23, 405–410

[100] Owen J.C., Rockwood A.P. (1987) Intersection of general implicit surfaces. In: Farin G.E. (ed.) *Geometric Modeling: Algorithms and New Trends*, SIAM, 335–345

[101] Patrikalakis N.M. (1993) Surface-to-surface intersections. *IEEE Computer Graphics and Applications* 13, 89–95

[102] Patrikalakis N.M., Bardis L. (1989) Offsets of curves on rational B-spline surfaces. *Engineering with Computers* 5, 39–46

[103] Patrikalakis N.M., Maekawa T. (2002) *Shape Interrogation for Computer Aided Design and Manufacturing*. Springer Verlag, New York Berlin Heidelberg

[104] Pegna J., Wilde D.J. (1990) Spherical and circular blending of functional surfaces. *Transactions of ASME, Journal of Offshore Mechanics and Artic Engineering* 112, 134–142

[105] Persson H. (1978) NC machining of arbitrarily shaped pockets. *CAD* 10, 169–174

[106] Petrie G., Kennie T.K.M. (1987) Terrain modeling in surveying and civil engineering. *CAD* 19, 171–187

[107] Pham B. (1992) Offset curves and surfaces: a brief survey. *CAD* 24, 223–229

[108] Piegl L. (1989) Key developments in Computer-Aided Geometric Design. *CAD* 21(5), 262–273

[109] Piegl L., Tiller W. (1997) *The NURBS Book,* Second Edition. Springer Verlag, Berlin Heidelberg

[110] Piegl L., Tiller W. (1999) Computing offsets of NURBS curves and surfaces. *CAD* 31, 147–156

[111] Pottmann H. (1995) Rational curves and surfaces with rational offsets. *CAGD* 12, 175–192

[112] Pottmann H. (1997) General offset surfaces. *Neural, Parallel and Scientific Computations* 5, 55–80

[113] Pottmann H., Glaeser G. *et al.* (1999) Collision-free three-axis milling and selection of cutting tools. *CAD* 31, 225–232

[114] Pottmann H., Lu W. *et al.* (1996) Rational ruled surfaces and their offsets. *Graphical Models and Image Processing* 58(6), 544–552

[115] Pratt M.J., Geisow A.D. (1986) Surface-surface intersection problems. In: Gregory J.A. (ed.) *The Mathematics of Surfaces,* Clarendon Press, Oxford 117–142

[116] Ratschek H., Rokne J. (1993) Test for intersection between plane and box. *CAD* 25(4), 249–250

[117] Ricci A. (1973) A constructive geometry for computer graphics. *The Computer Journal* 16, 157–160

[118] Rockwood A.P. (1984) Introducing sculptured surfaces into a geometric modeler. In: Pickett M.S., Boyse J.W. (eds.) *Solid Modeling by Computers. From Theory to Applications.* Plenum Press, 237–258

[119] Rockwood A.P. (1989) The displacement method for implicit blending of surfaces in solid modeling. *ACM Transactions on Graphics* 8(4), 279–297

[120] Rogers D.F. (2000) *An Introduction to NURBS with Historical Perspective.* Morgan Kaufmann, San Diego, CA

[121] Rohmfeld R.F. (1998) IGB-offset curves – loop removal by scanning the interval sequences. *CAGD* 15(3), 339–375

[122] Rossignac J.R., Requicha A.A.G. (1986) Offsetting operations in solid modeling. *CAGD* 3, 129–148

[123] Saeed S.E.O., de Pennington A. *et al.* (1988) Offsetting in geometric modeling. *CAD* 20, 67–74

[124] Salmon G. (1885) *Lessons Introductory to the Modern Higher Algebra,* G.E. Stechert & Co., New York

[125] Saito T., Takahashi T. (1991) NC machining with G-buffer method. *Proceedings of SIGGRAPH'91, Computer Graphics* 25(4), 207–216

[126] Saitoh T., Hosaka M. (1990) Interpolating curve networks with blending patches. *Proceedings of EURO-GRAPHICS'90,* 137–146

[127] Sarfraz M. (1992) A C^2 rational cubic alternative to the NURBS. *Computers and Graphics* 16(1), 69–77

[128] Sarfraz M. (1992) Interpolatory rational cubic spline with biased, point and interval tension. *Computers and Graphics* 16(4), 427–430

[129] Sarfraz M. (1993) Designing of curves and surfaces using rational cubics. *Computers and Graphics* 17(5), 529–538

[130] Sarfraz M. (1995) Curves and surfaces for CAD using C^2 rational cubic splines. *Engineering with Computers* 11(2), 94–102

[131] Sarma R., Dutta D. (1997) The geometry and generation of NC tool paths. *Journal of Mechanical Design: ASME Transactions* 119, 253–258

[132] Sarraga R.F. (1983) Algebraic methods for intersections of quadric curfaces in GMSOLID. *Computer Vision, Graphics and Image Processing* 22(2), 222–238

[133] Sederberg T.W. (1983) Implicit and parametric curves and surfaces for computer aided geometric design. *Ph.D. thesis*, Purdue University, West Lafayette, IN

[134] Sederberg T.W. (1987) Algebraic geometry for surface and solid modeling. In: Farin G.E. (ed.) *Geometric Modeling: Algorithms and New Trends,* SIAM, 29–42

[135] Sederberg T.W., Anderson D.C. *et al.* (1984) Implicit representation of parametric curves and surfaces. *Computer Vision, Graphics and Image Processing* 28, 72–74

[136] Sederberg T.W. Christiansen H.N. *et al.* (1989) An improved test for closed loops in surface intersections. *CAGD* 21, 505–508

[137] Sederberg T.W., Meyers R.J. (1988) Loop detection in surface patch intersections. *CAGD* 5, 161–171

[138] Shene C.K., Johnstone J. (1994) On the lower degree intersections of two natural quadrics. *ACM Transactions on Graphics* 13(4), 400–424

[139] Sourin A.I., Pasko A.A. (1996) Function representation for sweeping by a moving solid. *IEEE Transactions on Visualization and Computer Graphics* 2(2), 11–18

[140] Suresh K., Yang D.C.H. (1994) Constant scallop-height machining of freeform surfaces. *Journal of Engineering for Industry: Transactions of the ASME* 116, 253–259

[141] Sutcliffe D.C. (1980) Contouring over rectangular and skewed rectangular grids. In: Brodlie K. (ed.) *Mathematical Methods in Computer Graphics and Design*, Academic Press, 39–62

[142] Sweeney M., Bartels R. (1986) Ray tracing free-form B-spline surfaces. *IEEE Computer Graphics and Applications* 6, 41–49

[143] Vafiadou M.E., Patrikalakis N.M. (1991) Interrogation of offsets of polynomial surface patches. *Proceedings of EUROGRAPHICS'91*, 247–259

[144] Van Hook T. (1986) Real time shaded NC milling display. *Proceedings of SIGGRAPH'86, Computer Graphics* 20(4), 15–20

[145] Van Overveld K., Tigges M. *et al.* (1999) Soft shadows for soft objects. *Proceedings of Fourth Eurographics Workshop on Implicit Surfaces*, Bordeaux, France

[146] Varady T., Martin R.R. *et al.* (1989) Topological considerations in blending boundary representations solid models. In: Strasser W., Seidel H.P. (eds.) *Theory and Practice of Geometric Modeling*. Springer, 205–220

[147] Vida J., Martin R. *et al.* (1994) A survey of blending methods that use parametric surfaces. *CAD* 26, 341–365

[148] Vries-Baayens A.E. (1991) Conversion of a Composite Trimmed Bezier Surface into Composite Bezier Surfaces. In: Laurent P.J., Le Mehaute *et al.* (eds.) *Curves and Surfaces in Geometric Design*. Academic Press, Boston, 485–489

[149] Vries-Baayens A.E., Seebregts C.H. (1992) Exact Conversion of a Composite Trimmed Nonrational Bezier Surface into Composite or Basic Nonrational Bezier Surfaces. In: Hagen H. (ed.) *Topics in Surface Modeling*. SIAM, Philadelphia, 115–143

[150] Wang Y. (1996) Intersections of offsets of parametric surfaces. *CAGD* 13, 453–465

[151] Warren J. (1989) Blending algebraic surfaces. *ACM Transactions on Graphics* 8(4), 263–278

[152] Welbourn D.B. (1996) Full three-dimensional CAD/CAM. *CAE Journal* 13, 54–60, 189–192

[153] Wentland K., Dutta D. (1993) Method for offset-curve generation for sheetmetal design. *CAD* 25(9), 662–670

[154] Wilf I., Manor Y. (1993) Quadric-surface intersection curves: shape and structure. *CAD* 25(10), 633–643

[155] Woodwark J.R. (1987) Blends in geometric modeling. In: Martin R.R. (ed.) *The Mathematics of Surfaces II*, Oxford University Press, 255–297

[156] Woodwark J.R. (1990) Blends in geometric modeling. In: Creasy C., Craggs C. (eds.) *Applied Surface Modeling*, Ellis Horwood, 85–103

[157] Wu W.T. (1986) Basic principles of mechanical theorem proving in geometries. *J. of Systems Sciences and Mathematical Sciences* 4, 207–235

[158] Wyvill B., Bloomenthal J. *et al.* (1993) *SIGGRAPH'93*, Course #25, Modeling and animating with implicit surfaces

[159] Wyvill B., McPheeters C. *et al.* (1986) Animating soft objects. *The Visual Computer* 2(4), 235–242

[160] Wyvill B., Van Overveld K. (1996) Polygonization of implicit surfaces with constructive solid geometry. *Journal of Shape Modelling* 2(4), 257–274

[161] Wyvill B., Wyvill G. *et al.* (1987) Solid texturing of soft objects. *IEEE Computer Graphics and Applications* 7(12), 20–26

[162] Xu G., Bajaj C. *et al.* (2001) C^1 modeling with A-patches from rational trivariate functions. *CAGD* 18(3), 221–243

[163] Yan C.G. (1987) On speeding up ray tracing of B-spline surfaces. *CAD* 19, 122–130

[164] Yen J., Spach S. *et al.* (1991) Parallel boxing in B-spline intersection. *IEEE Computer Grapics and Applications* 11, 72–79

6

Weighted Nu Splines: An Alternative to NURBS

Muhammad Sarfraz

Department of Information and Computer Science, King Fahd University of Petroleum and Minerals, Dhahran 31261, Saudi Arabia.

A cubic spline curve method is considered to be a decent approach for designing applications in the area of computer graphics and geometric modeling. However, due to its various limitations like lack of freedom in shape control, a designer may not have much help in using this method. In this study, the weighted v-spline method has been reviewed. This curve design method, in addition to enjoying the good features of cubic splines, possesses interesting shape design features too. It has two families of shape parameters working in such a way that one family of parameters is associated with intervals and the other with points. These parameters provide a variety of shape control, like point and interval tension. This is an interpolatory curve scheme, which utilizes a piecewise cubic function in its description. However, it is desirable to extend this idea to freeform curves, which can enjoy all the ideal properties related to B-spline theory. This work is mainly concerned with developing such a theory. A constructive approach has been adopted to build B-spline like basis for cubic spline curves with the same continuity constraints as those for interpolatory weighted v-splines. These are local basis functions with local support and having the property of being positive everywhere. The design curve, constructed through these functions, possesses all the ideal geometric properties like partition of unity, convex hull, and variation diminishing. This curve method not only provides a variety of very interesting shape control, like point and interval tensions but, as a special case, also recovers the cubic B-spline curve method. In addition, it also provides B-spline like design curves for weighted splines, v-splines and weighted v-splines. The method for evaluating these splines is suggested by a transformation to Bézier form.

Advances in Geometric Modeling. Edited by M. Sarfraz
© 2003 John Wiley & Sons, Ltd   ISBN: 0-470-85937-7

1. Introduction

Designing curves, especially those curves which are robust and easy to control and compute, has been one of the significant problems of computer graphics and geometric modeling. Specific applications including font designing, capturing hand-drawn images on computer screens, data visualization, and computer-supported cartooning are the main motivations towards curve design. In addition, various other applications in CAD/CAM/CAGD also provide a good reason to study this topic. Many authors have worked in this direction. For more in depth information, the reader is referred to [1–20].

Weighted splines [7] were discovered as a cubic spline method. It provides a C^1 computationally simpler alternative to the exponential splineunder-tension [4, 13, 20]. Regarding shape characteristics, it has shape control parameters associated with each interval, which can be used to flatten or tighten the curve locally. Nusplines [11–12] were discovered as another cubic spline method. It provides a GC^2 computationally simpler alternative to the exponential spline-under-tension [4, 13, 20]. Regarding shape characteristics, it has shape control parameters associated with each point, which can be used to tighten the curve both locally and globally. The ideas of weighted splines and Nusplines were married together to formulate another spline called weighted Nuspline [11–12]. This curve design method covers the shape features of both of its counterparts and provides a C^1 computationally economical method.

B-splines are a useful and powerful tool for computer graphics and geometric modeling. They can be found frequently in existing CAD/CAM (Computer Aided Design/Computer Aided Manufacturing) systems. They form a basis for the space of nth degree splines of continuity class C^{n-1}. Each B-spline is a nonnegative nth degree spline that is nonzero only on $n + 1$ intervals. The B-splines form a partition of unity, that is, they sum up to one. Curves generated by summing control points multiplied by the B-splines have some very desirable shape properties, including the local convex hull property and variation diminishing property.

It is desirable to generalize the idea of B-spline like local basis functions for the classes of splines with shape parameters considered in the description of continuity. The first local basis for GC^2 splines was developed by Lewis [10]. In 1981, Barsky [1] generalized B-splines to Beta splines. These splines preserve the geometric smoothness of the design curve whilst allowing the continuity conditions on the spline functions at the knots to be varied by certain parameters, thus giving greater flexibility. Later in 1984, Bartels and Beatty [2] developed local bases for Beta spline curves that are equivalent to Boehm's [3] Gamma splines. Foley [7], in 1987, constructed a B-spline like basis for weighted splines; different weights were built into the basis functions so that the control point curve was a C^1 piecewise cubic with local control of interval tension.

This paper reviews the weighted v-spline method [8] in Section 2 and then constructs, in Section 3, a B-spline like local basis for the weighted v-spline. The design curve, in Section 4, maintains the C^1 continuity of the weighted v-splines. This description of freeform weighted v-spline not only provides a variety of interesting shape control like point, and interval tensions but, as a special case, also recovers the cubic B-spline curve method. In addition, it also provides B-spline like design curves for weighted splines, v-splines and weighted v-splines.

The approach adopted in the construction of local basis for the weighted v-splines is quite different from those adopted for different spline methods in [1–7, 8, 15]. The way for evaluating the weighted v-splines representation of a curve is suggested by a transformation to piecewise defined Bézier form. This form will also expedite a proof of the variation diminishing property for the Bézier representation.

2. Review of Weighted Nu Splines

This section gives a brief review of the weighted v-splines.

2.1. Preliminaries

For a detailed description, the reader is referred to [8]. Assume that we are given $t_1 < t_2 < \ldots < t_n$, $y_1, y_2, \ldots, y_n$, point tension factors v_i for $i = 1, 2, \ldots, n$, and interval weights $w_i > 0$ for $i = 1, 2, \ldots, n$, and interval weights $w_i > 0$ for $i = 1, 2, \ldots, n$. The weighted v-spline interpolant is a C^1 piecewise cubic function $S(t)$ that minimizes:

$$V(f) = \sum_{i=1}^{n-1} w_i \int_{t_i}^{t_{i+1}} [f''(t)]^2 dt + \sum_{i=1}^{n} v_i [f'(t_i)]^2,$$

subject to the interpolation conditions $f(t_i) = y_i$ for $i = 1, 2, \ldots, n$ and one of the following end conditions:

- **Type 1:** First derivative end conditions,
- **Type 2:** Natural end conditions, or
- **Type 3:** Periodic end conditions.

The v_i are termed point tension factors because they 'tighten' a parametric curve at the ith point in the same way that they do for the v-splines in [11–12]. The w_i are termed interval weights because they 'tighten' the curve on the ith interval in the same way that they do for the weighted splines in [14]. If $v_i = 0$ and all $w_i = q$, where q is some constant value, then the weighted v-spline equals the v-spline in [11–12] with tension factors v_i/q. If all $v_i = 0$, then it equals the weighted spline given in [14].

The approach taken in [8] uses piecewise cubic Hermite basis functions to represent the weighted v-splines. Given y_i and m_i for $i = 1, 2, \ldots, n$, a unique C^1 piecewise cubic function $f(t)$ exists that satisfies for $f(t_i) = y_i$ and $f'(t_i) = m_i$ for $i = 1, 2, \ldots, n$. The unknowns are the first derivative values, m_i, $i = 1, 2, \ldots, n$, and once they are computed, the function $f(t)$ can be easily evaluated using the standard piecewise cubic Hermite form. The necessary and sufficient conditions for the function $S(t)$ to be the weighted v-spline interpolant are that its derivatives m_i satisfy:

$$c_{k-1} m_{k-1} + \left(\frac{1}{2} v_k + 2c_{k-1} + 2c_k \right) m_k + c_k m_{k+1}$$
$$= b_k(y_{k+1} - y_k) + b_{k-1}(y_k - y_{k-1}), \tag{6.1}$$

for $k = 1, 2, \ldots, n$, where $c_i = w_i/h_i$, $b_i = 3c_i/h_i$ and $h_i = t_{i+1} - t_i$. The above system of equations provides $(n-2)$ equations for n unknowns, $m_1, \ldots, m_n$, and the additional equations come from the given end conditions. The equations for Type I first derivative end conditions are $m_1 = f'(t_1)$ and $m_n = f'(t_n)$. For Type II natural end conditions they are:

$$\left(\frac{1}{2}v_1 + 2c_1\right)m_1 + c_1 m_2 = b_1(y_2 - y_1),$$

and

$$c_{n-1}m_{n-1} + \left(\frac{1}{2}v_n + 2c_{n-1}\right)m_n = b_{n-1}(y_n - y_{n-1}).$$

For Type 3 periodic end conditions, they are:

$$\left(\frac{1}{2}v_1 + \frac{1}{2}v_n + 2c_1 + 2c_{n-1}\right)m_1 + c_1 m_2 + c_{n-1}m_{n-1}$$
$$= b_1(y_2 - y_1) + b_{n-1}(y_n - y_{n-1}),$$

and $m_1 = m_n$. The linear system of equations that occurs when Type 1 or 2 end conditions are used is tridiagonal and diagonally dominant, thus it can be solved efficiently by using a standard tridiagonal system solver.

2.2. Parametric Representation

For parametric interpolation, suppose that we are given data points $F_i = (x_i, y_i, z_i)$ and $v_i \geq 0$ for $i = 1, \ldots, n$ and $w_i > 0$ for $i = 1, \ldots, n - 1$. If we let $X(t)$ be the weighted v-spline interpolant to the data (t_i, x_i), $Y(t)$ be the weighted v-spline interpolant to the data (t_i, y_i) and $Z(t)$ be the weighted v-spline interpolant to the data (t_i, z_i), then the parametric curve $S(t) = (X(t), Y(t), Z(t))$, where $t_1 \leq t \leq t_n$ is the weighted v-spline interpolant. It is a C^1 piecewise cubic function:

$$S(t) \equiv S_i(t) = F_i(1 - \theta)^3 + 3\theta(1 - \theta)^2 V_i + 3\theta^2(1 - \theta)W_i + F_{i+1}\theta^3, \tag{6.2}$$

where

$$\theta = \frac{t - t_i}{h_i}, \tag{6.3}$$

and

$$V_i = F_i + \frac{h_i M_i}{3}, \quad W_i = F_{i+1} - \frac{h_i M_{i+1}}{3}. \tag{6.4}$$

It is obvious that the piecewise cubic function Equation (6.2), holds the following interpolatory properties:

$$\left.\begin{array}{ll} S(t_i) = F_i, & S(t_{i+1}) = F_{i+1} \\ S^{(1)}(t_i) = M_i, & S^{(1)}(t_{i+1}) = M_{i+1} \end{array}\right\}, \tag{6.5}$$

where $S^{(1)}$ denotes first derivative with respect to t and M_i denotes derivative values given at the knots t_i. This leads the piecewise cubic Equation (6.2) to the piecewise Hermite interpolant $S \in C^1[t_1, t_n]$.

The parametric weighted v-spline can be computed by solving for M_i's. This can be done by generalizing the system of equations in Equation (6.1) as follows:

$$c_{i-1}M_{i-1} + \left(\frac{v_i}{2} + 2c_{i-1} + 2c_i\right)M_i + c_i M_{i+1} = 3c_i \Delta_i + 3c_{i-1}\Delta_{i-1}, \qquad (6.6)$$

where

$$\Delta_i = (F_{i+1} - F_i)/h_i.$$

for $i = 2, \ldots, n - 1$. For given appropriate end conditions (Type 1, Type 2, or Type 3), this system of equations is a tridiagonal linear system. This is also diagonally dominant for the following constraints on the shape parameters:

$$v_i \geq 0, \quad i = 1, 2, \ldots, n, \quad \text{and} \quad w_i > 0, \quad i = 1, 2, \ldots, n - 1, \qquad (6.7)$$

and hence has a unique solution for M_i's. As far as the computation method is concerned, it is much more economical to adopt the LU-decomposition method to solve the tridiagonal system. Therefore, the above discussion can be concluded in the following:

Theorem 1. *For the shape parameter constraints (in Equation (6.7)), the spline solution of the weighted v-spline exists and is unique.*

Remark 1: Each component of the parametric weighted v-spline is a C^1 function in general, but it has second order geometric continuity at t_i if $w_{i-1} = w_i$ and the tangent vector at t_i is nonzero and it is C^2 at t_i if $w_{i-1} = w_i$ and $v_i = 0$.

2.3. Demonstration

Figure 6.1 is the parametric weighted v-spline interpolant to the points denoted by circles using periodic end conditions. In Figure 6.2, interval weight, w_i, of 30 is used in the base interval, while point tension factors, v_i, of 10 are used on the four vertices defining the 'neck'. The rest of the parameters are taken as $w_i = 1$ and $v_i = 0$.

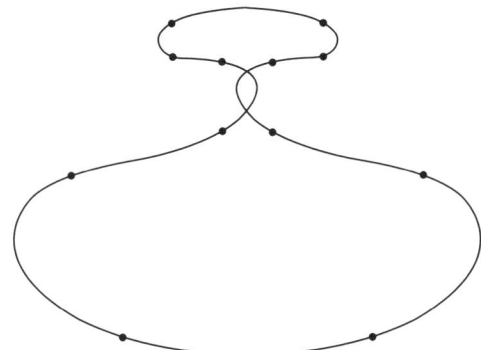

Figure 6.1   The default weighted v-spline with periodic end conditions.

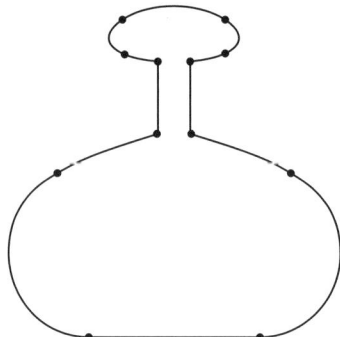

Figure 6.2   The weighted v-spline with periodic end conditions using $w_i = 30$ on the base interval, $w_i = 1$ otherwise, $v_i = 10$ on the four vertices defining the 'neck', and $v_i = 0$ otherwise.

3. Development of Local Support Basis

This section is devoted to constructing the local support basis B_i's to compute the cubic weighted v-spline $P(t)$ satisfying the following constraints:

$$\begin{bmatrix} P(t_{i+}) \\ P^{(1)}(t_{i+}) \\ P^{(2)}(t_{i+}) \end{bmatrix} = \begin{bmatrix} 1 & 0 & 0 \\ 0 & 1 & 0 \\ 0 & \dfrac{v_i}{w_i} & \dfrac{w_{i-1}}{w_i} \end{bmatrix} \begin{bmatrix} P(t_{i-}) \\ P^{(1)}(t_{i-}) \\ P^{(2)}(t_{i-}) \end{bmatrix}. \tag{6.8}$$

For the purpose of the analysis, let additional knots be introduced outside the knot partition $t_i < t_2 < \ldots < t_n$ of the interval $[t_1, t_n]$, defined by:

$$t_{-2} < t_{-1} < t_0 < t_1 \quad \text{and} \quad t_n < t_{n+1} < t_{n+2} < t_{n+3}. \tag{6.9}$$

Let:

$$a_i = 1/c_i, \tag{6.10}$$

and ϕ_i be cubic weighted v-spline.

$$\phi_i^{(t)} = \begin{cases} 0, & t \le t_{i-2}, \\ 1, & t \ge t_{i+1}. \end{cases} \tag{6.11}$$

Imposing weighted v-spline constraints (Equation (6.8)), we have:
At t_{i-2}:

$$\phi_i^{(1)}(t_{i-1}) = \frac{3}{h_{i-2}} \phi_i(t_{i-1}). \tag{6.12}$$

At t_{i-1}:

$$\left[\frac{1}{2} a_{i-1} a_{i-2} v_{i-1} + 2a_{i-1} + 2a_{i-2} \right] \phi_i^{(1)}(t_{i-1}) + a_{i-2} \phi_i^{(1)}(t_i)$$

$$= \frac{3a_{i-2}}{h_{i-1}} [\phi_i(t_i) - \phi_i(t_{i-1})] + \frac{3a_{i-1}}{h_{i-2}} \phi_i(t_{i-1}). \tag{6.13}$$

At t_i:

$$a_i \phi_i^{(1)}(t_{i-1}) + \left[\frac{a_i a_{i-1}}{2} v_i + 2a_{i-1} + 2a_i \right] \phi_i^{(1)}(t_i)$$

$$= \frac{3a_{i-1}}{h_i}[1 - \phi_i(t_i)] + \frac{3a_i}{h_{i-1}}[\phi_i(t_i) - \phi_i(t_{i-1})]. \tag{6.14}$$

At t_{i+1}:

$$\phi_i^{(1)}(t_i) = \frac{3}{h_i}[1 - \phi_i(t_i)]. \tag{6.15}$$

From Equations (6.12) and (6.15):

$$\phi_i(t_{i-1}) = \frac{h_{i-2}}{3}\phi_i^{(1)}(t_{i-1}),$$

and:

$$\phi_i(t_i) = 1 - \frac{h_i}{3}\phi_i^{(1)}(t_i).$$

Substituting these in Equations (6.13) and (6.14), we have:

$$\overline{U}_i \phi_i^{(1)}(t_{i-1}) + \overline{V}_i \phi_i^{(1)}(t_i) = \overline{W}_i, \tag{6.16}$$

$$\overline{X}_i \phi_i^{(1)}(t_{i-1}) + \overline{Y}_i \phi_i^{(1)}(t_i) = \overline{Z}_i, \tag{6.17}$$

where:

$$\overline{U}_i = \frac{a_{i-1} a_{i-2}}{2} v_{i-1} + a_{i-1} + a_{i-2}\left(2 + \frac{h_{i-2}}{h_{i-1}}\right),$$

$$\overline{V}_i = a_{i-2}\left(1 + \frac{h_i}{h_{i-1}}\right), \quad \overline{W}_i = \frac{3a_{i-2}}{h_{i-1}},$$

$$\overline{X}_i = a_i\left(1 + \frac{h_{i-2}}{h_{i-1}}\right), \quad \overline{Y}_i = \frac{a_i a_{i-1}}{2} v_i + a_{i-1} + a_i\left(2 + \frac{h_i}{h_{i-1}}\right),$$

and:

$$\overline{Z}_i = \frac{3a_i}{h_{i-1}}.$$

Equations (6.16) and (6.17) give:

$$\phi_i^{(1)}(t_{i-1}) = \frac{(\overline{W}_i \overline{Y}_i - \overline{V}_i \overline{Z}_i)}{(\overline{U}_i \overline{Y}_i - \overline{X}_i \overline{V}_i)},$$

and:

$$\phi_i^{(1)}(t_i) = \frac{(\overline{U}_i \overline{Z}_i - \overline{W}_i \overline{X}_i)}{(\overline{U}_i \overline{Y}_i - \overline{X}_i \overline{V}_i)},$$

where, if $d_i = \dfrac{1}{2} a_i a_{i-1} v_i + a_{i-1} + a_i$, then:

$$\overline{W}_i \overline{Y}_i - \overline{V}_i \overline{Z}_i = \frac{3 a_{i-2}}{h_{i-1}} d_i,$$

$$\overline{U}_i \overline{Z}_i - \overline{W}_i \overline{X}_i = \frac{3 a_i}{h_{i-1}} d_{i-1},$$

$$\overline{U}_i \overline{Y}_i - \overline{X}_i \overline{V}_i = d_i d_{i-1} + \frac{a_i}{h_{i-1}} (h_{i-1} + h_i) d_{i-1} + \frac{a_{i-2}}{h_{i-1}} (h_{i-1} + h_{i-2}) d_i.$$

Let,

$$D_i = h_{i-1} d_i d_{i-1} + a_i (h_{i-1} + h_i) d_{i-1} + a_{i-2} (h_{i-1} + h_{i-2}) d_i,$$
$$\mu_i = \phi_{i+1}(t_i), \qquad \lambda_i = 1 - \phi_i(t_i),$$
$$\hat{\mu}_i = \phi_{i+1}^{(1)}(t_i), \qquad \hat{\lambda}_i = \phi_i^{(1)}(t_i).$$

Then,

$$\hat{\lambda}_i = \frac{3 a_i d_{i-1}}{D_i}, \qquad \hat{\mu}_i = \frac{3 a_{i-1} d_{i+1}}{D_{i+1}},$$

$$\mu_i = \frac{h_{i-1}}{3} \hat{\mu}_i, \qquad \lambda_i = \frac{h_i}{3} \hat{\lambda}_i,$$

and hence:

$$0 \le \mu_i \le 1, \quad 0 \le \lambda_i \le 1, \quad \text{and} \quad 0 \le \mu_i + \lambda_i \le 1.$$

Now define:

$$B_i(t) = \phi_i(t) - \phi_{i+1}(t).$$

Then B_i has the local support (t_{i-2}, t_{i+2}) and an explicit representation of B_j on any interval (t_i, t_{i+1}) (in particular for $i = j - 2, j - 1, j, j + 1$) can be calculated as:

$$B_j(t) = (1 - \theta)^3 B_j(t_i) + \theta(1 - \theta)^2 \big(3 B_j(t_i) + h_i B_j^{(1)}(t_i)\big)$$
$$+ \theta^2 (1 - \theta)\big(3 B_j(t_{i+1}) - h_i B_j^{(1)}(t_{i+1})\big) + \theta^3 B_j(t_{i+1}), \tag{6.18}$$

where:

$$B_j(t_i) = B_j^{(1)}(t_i) = 0, \qquad \text{for } i \ne j - 1, j, j + 1,$$

and:

$$\left. \begin{aligned} B_j(t_{j-1}) &= \mu_{j-1}, \; B_j^{(1)}(t_{j-1}) = \hat{\mu}_{j-1}, \\ B_j(t_j) &= 1 - \lambda_j - \mu_j, \; B_j^{(1)}(t_j) = \hat{\lambda}_j - \hat{\mu}_j, \\ B_j(t_{j+1}) &= \lambda_{j+1}, \; B_j^{(1)}(t_{j+1}) = -\hat{\lambda}_{j+1}. \end{aligned} \right\} \tag{6.19}$$

Careful examination of the Bézier vertices of $(B_j(t)$ in Equation (6.18) shows these to be nonnegative for v_i, w_i satisfying Equation (6.7) and thus $B_j(t) \ge 0, \forall\, t$. This leads to the following:

Proposition 1. The local support basis functions (Equation (6.18)) are such that the following properties hold:

(i) (Local Support) $B_j(t) = 0 \quad for \quad t \notin (t_{j-2}, t_{j+2})$,
(ii) (Partition of unity) $\sum_{j=-1}^{n+1} B_j(t) = 1 \quad for \quad t \in [t_1, t_n]$,
(iii) (Positivity) $B_j(t) \geq 0$ for all t.

4. Design Curve

Now, we need a convenient method to compute the curve representation. It is desirable to apply the above local basis functions to develop a freeform weighted v-spline curve as follows:

$$P(t) = \sum_{j=-1}^{n+1} B_j(t) P_j, t \in [t_1, t_n], \tag{6.20}$$

where $P_j \in R^N$, $j = 0, 1, \ldots, n+1$, define the control points of the representation. By the local support property,

$$P(t) = \sum_{j=i-1}^{i+2} B_j(t) P_j, t \in [t_i, t_{i+1}), i = 1, \ldots, n-1.$$

Substitution of Equation (6.18), $t \in [t_i, t_{i+1})$, then gives the piecewise defined Bézier representation,

$$P(t) \equiv P_i(t) = F_i(1-\theta)^3 + 3\theta(1-\theta)^2 V_i + 3\theta^2(1-\theta)W_i + F_{i+1}\theta^3, \tag{6.21}$$

where:

$$\left. \begin{array}{l} F_i = \lambda_i P_{i-1} + (1 - \lambda_i - \mu_i)P_i + \mu_i P_{i+1}, \\ V_i = (1 - \alpha_i)P_i + \alpha_i P_{i+1}, \\ W_i = \beta_i P_i + (1 - \beta_i)P_{i+1}, \end{array} \right\} \tag{6.22}$$

with,

$$\alpha_i = \mu_i + h_i \, \hat{\mu}_i / 3 = \frac{\hat{\mu}_i}{3}(h_{i-1} + h_i),$$

$$\beta_i = \lambda_{i+1} + h_i \, \hat{\lambda}_i = \frac{\hat{\lambda}_{i+1}}{3}(h_i + h_{i+1}).$$

This transformation to Bézier form is very convenient for computational purposes and also leads to:

Proposition 2. (Variation Diminishing Property) The weighted v-spline curve $P(t)$, $t \in [t_0, t_n]$, defined by Equation (6.20), crosses any (hyper) plane of dimension $N - 1$ no more times than it crosses the 'control polygon' joining the control points $P_1, P_0, \ldots, P_{n+1}$.

Proof. Following the arguments of positivity in the previous proposition, it is straight forward that $0 \leq \alpha_i \leq 1, 0 \leq \beta_i \leq 1$, and $0 \leq \alpha_i + \beta_i \leq 1$. Thus, V_i and W_i lie on the line segment joining P_i and P_{i+1}, where V_i is before W_i. It can also be simply noted that:

$$F_i = (1 - \gamma_i)W_{i-1} + \gamma_i V_i, \tag{6.23}$$

where:

$$0 < \gamma_i = \frac{h_{i-1}}{h_{i-1} + h_i} < 1.$$

Thus, the control polygon of the piecewise defined Bézier representation is obtained by corner cutting of the weighted v-spline control polygon. Since the piecewise defined Bézier representation is variation diminishing, it follows that weighted v-spline representation is variation diminishing.

4.1. Shape Control

The shape parameters, defined in Equation (6.7), can be used to control the local or global shape of the design curve. To analyze such behaviors, the explicit form on (t_i, t_{i+1}) of the weighted v-spline design curve (Equation (6.20)) can be expressed as:

$$P(t) = l_i(t) + e_i(t), \tag{6.24}$$

where:

$$l_i(t) = (1 - \theta)F_i + \theta F_{i+1}, \tag{6.25}$$
$$e_i(t) = \theta(1 - \theta)\{[(F_{i+1} - F_i) - h_i P^{(1)}(t_i)](\theta - 1)$$
$$+ [(F_{i+1} - F_i) - h_i P^{(1)}(t_{i+1})]\theta\}. \tag{6.26}$$

Proposition 3. Let $w_i = w \geq 1$, and $v_i = 0$, $\forall i$ are all bounded then the weighted v-spline design curve is straightaway the standard cubic spline.

Proof. It follows from the last constraint of relation (Equation (6.8)).

Proposition 4. (Global tension) Let $w_i \geq 1$, $\forall i$, be bounded and $v_i \geq v$ then the weighted v-spline curve (Equation (6.20)) converges uniformly to the control polygon $P_0, \ldots, P_n$ as $v \to \infty$.

Proof. Let $v_i = v$, $\forall i$ then from Equation (6.1)

$$\lim_{v \to \infty} P^{(1)}(t_i) = 0. \tag{6.27}$$

Moreover:

$$\lim_{v \to \infty} \hat{\mu}_i = 0 = \lim_{v \to \infty} \hat{\lambda}_i, \forall i.$$

This implies the following:

$$\lim_{v \to \infty} F_i = P_i, \forall i. \tag{6.28}$$

More generally, for $v_i \geq v \geq 0$, it can be shown that:

$$\max_i |\hat{\lambda}_i| \leq r(v),$$

and,

$$\max_i |\hat{\mu}_i| \leq s(v),$$

where,

$$\lim_{v \to \infty} r(v) = 0 = \lim_{v \to \infty} s(v),$$

and again Equations (6.27) and (6.28) hold. Hence the result.

Proposition 5. (Local Tension) Consider an interval $[t_k, t_{k+1}]$ for a fixed k. Then on $[t_k, t_{k+1}]$ weighted v-spline curve converges uniformly to a line segment of the line $P_k P_{k+1}$ as $w_k \to \infty$ where w_{k-1} and v_k are bounded.

Proof. Careful examination shows:

$$\lim_{w_k \to \infty} \mu_k = \frac{h_{k-1}}{(3h_k + h_{k-1} + h_{k+1})} = \hat{\alpha}_k \text{ (say)}$$

$$\lim_{w_k \to \infty} \mu_{k+1} = 0$$

$$\lim_{w_k \to \infty} \lambda_k = 0$$

$$\lim_{w_k \to \infty} \lambda_{k+1} = \frac{h_{k+1}}{(3h_k + h_{k-1} + h_{k+1})} = \hat{\beta}_k \text{ (say)}$$

This implies the following:

$$\lim_{w_k \to \infty} F_k = (1 - \hat{\alpha}_k)P_k + \hat{\alpha}_k P_{k+1} = \hat{F}_k \text{ (say)}$$

and:

$$\lim_{w_k \to \infty} F_{k+1} = \hat{\beta}_k P_k + (1 - \hat{\beta}_k)P_{k+1} = \hat{F}_{k+1} \text{ (say)}$$

Obviously $\hat{F}_k$ and $\hat{F}_{k+1}$ lie on $\overline{P_k P_{k+1}}$ and $\hat{F}_k$ is before $\hat{F}_{k+1}$ as $\hat{\alpha}_k < (1 - \hat{\beta}_k)$. Also:

$$\lim_{w_k \to \infty} (F_{k+1} - F_k) = \lim_{w_k \to \infty} h_k P^{(1)}(t_k)$$

$$= \lim_{w_k \to \infty} h_k P^{(1)}(t_{k+1}) = \frac{3h_k(P_{k+1} - P_k)}{(3h_k + h_{k-1} + h_{k+1})}$$

Hence from Equations (6.24), (6.25), (6.26) if $P(t) = P_k(t)$ for $t \in (t_k, t_{k+1})$, then:

$$\lim_{w_k \to \infty} P_k(t) = (1 - \theta)\hat{F}_k + \theta \hat{\hat{F}}_k.$$

Proposition 6. (Local Tension) Consider an interval like that in Proposition 5. Then on $[t_k, t_{k+1}]$, weighted v-spline curve converges uniformly to the linear interpolant $l_i(t)$ as both $v_k, v_{k+1} \to \infty$, where w_{k-1}, w_k, w_{k+1} are bounded.

Proof. It can be noted that:

$$\lim \mu_k = \lim \mu_{k+1} = 0,$$
$$\lim \lambda_k = \lim \lambda_{k+1} = 0,$$

and:

$$\lim P^{(1)}(t_k) = \lim P^{(1)}(t_{k+1}) = 0.$$

This gives the desired result.

4.2. Demonstration

The tension behaviour of the weighted Nu spline is illustrated by the following simple examples for data set in R^2. Unless otherwise stated, in all the figures, the parameter v_i will be assumed as zero $\forall i$ and the parameters w_i as 1 for all i.

Figure 6.3 is the default curve, which is a cubic spline for $v_i = 0$, and $w_i = 1$, for all i. The control polygon, together with the control points, is shown in the figure.

Figure 6.4 shows the effect of interval tension with $w = 100$ in the base of the figure. The effect of the high tension parameter is clearly seen in the corresponding interval in the base of the figure.

Figure 6.5 shows the effect of point tension with $v = 100$ at two opposite points in the figure.

Figures 6.6–6.8 illustrate the effect of progressively increasing the values of the point tension parameters v_i's $= 0, 5, 100$, respectively, at all the points of the figure. This is the global tension effect due to progressive increase.

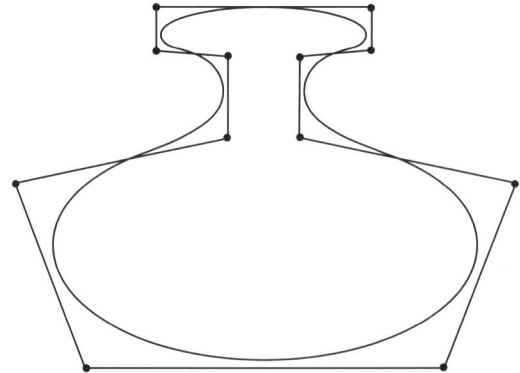

Figure 6.3   The default weighted Nu spline.

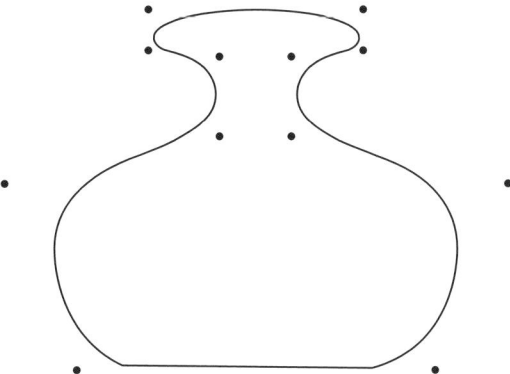

Figure 6.4   The weighted Nu spline with interval tension at the base.

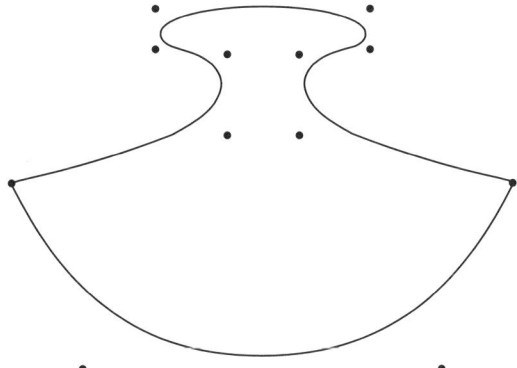

Figure 6.5   The weighted Nu spline with corner tension at two points.

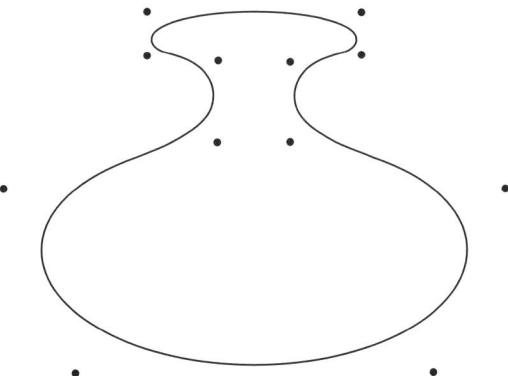

Figure 6.6   The weighted Nu spline with global tension $v = 1$ (the default curve).

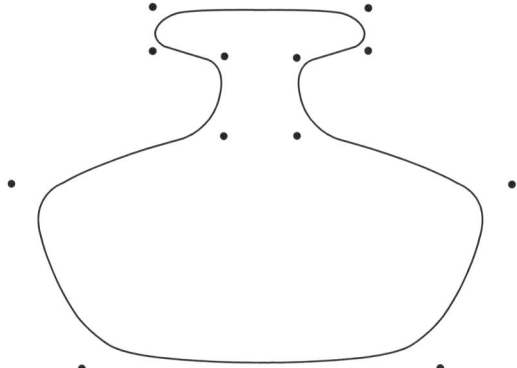

Figure 6.7   The weighted Nu spline with global tension $v = 5$.

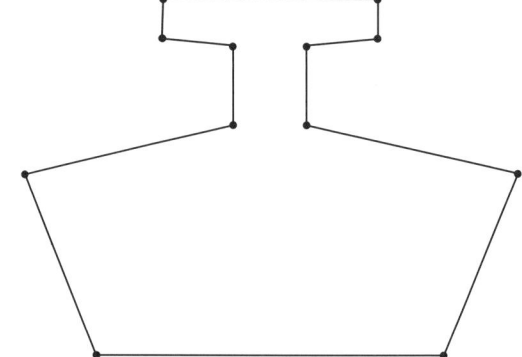

Figure 6.8   The weighted Nu spline with global tension $v = 100$.

5. Summary

A freeform C^1 weighted Nu spline curve design has been developed through the construction of local support B-spline like basis functions. This cubic spline method has been developed with a view to its application in computer graphics, geometric modeling, and CAGD. It is quite reasonable to construct a freeform cubic spline method, which involves two families of shape parameters in a similar way to the interpolatory weighted v-spline. These parameters provide a variety of local and global shape controls like interval and point shape effects. The visual smoothness of the proposed method is also C^1, which is the same as the smoothness of interpolatory weighted v-spline. The freeform C^1 weighted Nu spline method can be applied to tensor product surfaces but unfortunately, in the context of interactive surface design, this tensor product surface is not that useful because any one of the tension parameters controls an entire corresponding interval strip of the surface. Thus, as an application of C^1 spline for the surfaces, a method similar to Nielson's [12] spline blended methods may be attempted. This will produce local shape control, which is quite useful regarding computer graphics and geometric modeling applications.

Acknowledgments

This work has been supported by the King Fahd University of Petroleum and Minerals under Project No. FT/2001-18.

References

[1] Barsky, B. A. (1981), The Beta-spline: A local representation based on shape parameters and fundamental geometric measure, *Ph.D. Thesis*, University of Utah.

[2] Bartels, R. and Beatty, J. (1984), Beta-splines with a difference, Technical Report cs-83-40, Computer Science Department, University of Waterloo, Waterloo, Canada.

[3] Boehm, W. (1985), Curvature Continuous Curves and Surfaces, *Comp. Aided Geom. Design* 2(2), 313–323.

[4] Cline

[5] Dierckx, P. and Tytgat, B. (1989), Generating the Bézier Points of β-spline Curve, *Comp. Aided Geom. Design* 6, 279–291.

[6] Farin, G. E. (1996), *Curves and Surfaces for CAGD*, Academic Press, New York.

[7] Foley, T. A. (1987), Local Control of Interval Tension using Weighted Splines, *Comp. Aided Geom. Design* 3, 281–294.

[8] Foley, T. A. (1987), Interpolation with Interval and Point Tension Controls using Cubic Weighted v-splines, *ACM Trans. Math. Software* 13, 68–96.

[9] Goodman, T. N. T. and Unsworth, K. (1985), Generation of Beta Spline Curves using a Recursive Relation, In *Fundamental Algorithms for Computer Graphics*. R. E. Earnshaw (Ed.), Springer, Berlin, 326–357.

[10] Lewis, J. (1975), 'B-spline' bases for splines under tension, Nu-splines, and fractional order splines, Presented at the SIAM-SIGNUM-meeting, San Francisco, USA.

[11] Nielson, G. M. (1974), Some Piecewise Polynomial Alternatives to Ssplines under Tension, In *Comp. Aided Geom. Design*, R. F. Barnhill (Ed.), Academic Press. New York.

[12] Nielson, G. M. (1986), Rectangular v-splines, *IEEE Comp. Graph*, Appl. 6, 35–40.

[13] Pruess, S. (1979), Alternatives to the Exponential Spline in Tension, *Math. Comp.* 33, 1273–1281.

[14] Salkauskas, K. (1984), C^1 Splines for Interpolation of Rapidly Varying Data, Rocky Mtn. *J. Math.* 14, 239–250.

[15] Sarfraz, M. (1992), A C^2 Rational Cubic Spline Alternative to the NURBS, *Comp. & Graph.* 16(l), 69–78.

[16] Sarfraz, M. (1995), Curves and surfaces for CAD using C2 rational cubic splines, *International Journal of Engineering with Computers*, Springer-Verlag, Vol. 11(2), 94–102.

[17] Sarfraz, M. (1994), Freeform Rational Bicubic Spline Surfaces with Tension Control, FACTA UNIVERSITATIS (NIS), Ser. *Mathematics and Informatics*, Vol. 9, 83–93.

[18] Sarfraz, M. (1994), Cubic Spline Curves with Shape Control, *International Journal of Computers & Graphics*, Elsevier Science, Vol. 18(5), 707–713.

[19] Schoenburg, I. J. (1946), Contributions to the Problem of Approximation of Equidistant Data by Analytic Functions, *Appl. Math* 4, 45–99.

[20] Schweikert, D. G. (1966), An Interpolation Curve using a Spline in Tension, *J. Math. Phys.* 45, 312–317.

7

Generation of Parting Surfaces Using Subdivision Technique

C. L. Li

Department of Manufacturing Engineering and Engineering Management,
City University of Hong Kong, Tat Chee Avenue, Kowloon, Hong Kong.

Automation in the various tasks of the plastic injection mould design process has attracted a lot of attention in the past decades. Techniques for the automatic selection of a parting direction, determination of complex 3D parting line and the generation of a parting surface have been reported. A common strategy in parting surface generation is the method of extrusion. Our investigation reveals that this method fails under certain conditions. A new method based on the subdivision method of surface generation is proposed. A variation of the Catmull-Clark subdivision scheme is employed which ensures that the resulting surface interpolates the main parting line of the part. The initial control mesh used in the subdivision process is obtained by 'projecting' the 2D medial axis of an approximating polygon to 3D. An experimental program has been implemented to verify the feasibility of the method.

1. Introduction

CAD/CAM systems are widely used in the injection mould design and manufacturing process. Given a plastic part in the form of a CAD model, a mould design engineer uses a CAD/CAM system to construct the detailed design of the entire mould structure and other relevant components (such as electrodes for electric-discharge machining) that are used in the design analysis and manufacturing of the mould. The CAD data of the mould are then used to generate manufacturing information (such as CNC machining toolpath, CMM inspection instructions, machine setup instructions etc.). Due to the widespread application, stand-alone or add-on software packages that are specific to plastic injection mould design, such as IMOLD (developed in Singapore), MoldWizard (used in Unigraphics II) and PiMould (being developed in the University of Hong Kong and due to be released soon), are commercially available. These software

Advances in Geometric Modeling. Edited by M. Sarfraz
© 2003 John Wiley & Sons, Ltd   ISBN: 0-470-85937-7

packages provide automatic and semi-automatic tools to assist the human mould designer in the various design tasks. The algorithms employed by these software packages in automating some of the design tasks are the result of recent research efforts, which can be found in the literature. In this chapter, we report our investigation on the automation of one of these design tasks, namely, the automation of the parting surface generation. While automatic methods for parting surface generation have been reported in the literature, almost all existing methods use extrusion. In addition to this method of extrusion, we propose the use of a variation of the Catmull-Clark subdivision surface method to solve the limitations inherent to extrusion.

This chapter is organized as follows. In the next section, related research work on automation in the injection mould design process is summarized. Section 3 discusses the commonly used method of extrusion and explains its two limitations. Section 4 discusses the application of the Catmull-Clark subdivision surface method in parting surface generation. Section 5 briefly explains the implementation and illustrates a design example. The chapter is concluded in Section 6.

2. Related Work

Research in the automatic determination of parting direction, parting line and parting surface, and the detection of an undercut feature have been reported. Hui and Tan [1] reported a method that automates the task of parting direction determination. They devised a heuristic function called blocking factor that measures the extent of blockage on moulding open due to undercut. This function guides a searching process to identify a parting direction that gives minimum undercut. Visibility analysis was used by Chen *et al.* [2, 3] to determine parting direction. They utilized the fact that the optimal parting direction should give the maximum visibility of the part along the parting direction in order to minimize undercut. Weinstein [4] studied the use of heuristic rules to select the optimum parting line location and draw direction. Mouldability analysis based on external and internal undercut features was reported by Hui [5]. Wong *et al.* [6] reported a method that determines the parting line on a part with a free-form shape by an adaptive slicing algorithm. The complex parting line is obtained by tracing the extreme points on adjacent slices. A comprehensive classification and recognition of an undercut feature is reported by Fu *et al.* [7]. Other methods that determine the parting line and parting surfaces were also reported [8–10]. Other tasks in the mould design process that have been investigated include the determination of gating locations [11], ejector types and ejecting locations [12], and the extraction of sharp corner uncut that requires electric-discharge machining [13]. Recently, the author has developed a feature-based approach to the cooling system design [14–15]. This work is different from all existing work in CAE cooling analysis. While CAE tools only aim at the analysis or optimization of a given cooling system design, the author's work focuses on the automatic design synthesis of the initial cooling system design.

The reported work that is most relevant to this project includes the methods reported by Tan *et al.* [8], Fu *et al.* [16] and Kong *et al.* [10]. One common strategy employed by these methods in the generation of the parting surface is the method of extrusion.

3. The Method of Extrusion

The parting surface is the surface that separates the mould insert into the core and cavity halves. Given a plastic part and the main parting line, a common approach to generate the parting surface

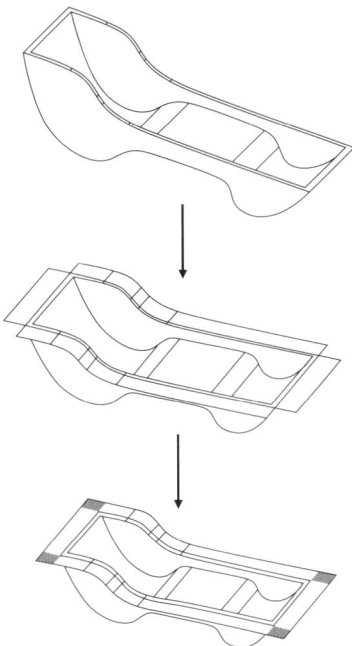

Figure 7.1   Generating parting surface by the method of extrusion.
Source: CLLi, "Application of Catmull-Clark Subdivision Method in Plastic Injection Mould Parting Surface Design", the proceedings of the Sixth International Conference on Information Visualisation, 10–12 July 2002, London, England, pp 477–482, (2002 IEEE).

is illustrated in Figure 7.1. For each segment on the parting line, a surface patch is generated by extruding the parting line segment along an extrusion direction. This extrusion direction is perpendicular to one of the boundary walls of the mould insert, and is also perpendicular to the parting direction. The extrusion surface is then trimmed by the boundary wall of the mould insert. After extruding all the parting line segments, a set of surface patches is obtained. Then, corner patches are added at the junction between two adjacent parting line segments where their extrusion directions are different. The set of extrusion surfaces and the corner patches then form the parting surface. However, there are two major limitations of the method, namely, inter-locking between the mould halves and intersection between extrusion surface and part surface.

3.1. Inter-Locking Between Mould Halves

Figure 7.2 illustrates the problem of inter-locking. The part is obtained by adding an opening feature in one of the walls in the part shown in Figure 7.1. If the method of extrusion is used to generate the parting surface, the core and cavity halves contain an inter-lock as shown in the figure. This inter-lock will prevent the normal mould opening operation. Notice that the opening feature itself is not an undercut feature of the plastic part. That is, the opening feature will not prevent mould opening or part ejection, and does not require any special design, such as a slide mechanism, for mould ejection. The inter-locking of the mould halves is exclusively due to the limitation of the extrusion method in generating the parting surface.

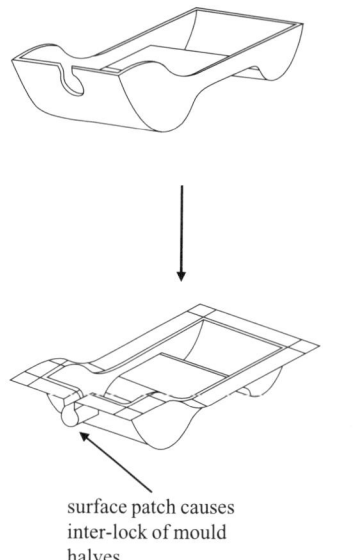

surface patch causes
inter-lock of mould
halves

Figure 7.2   The method of extrusion causes inter-lock between mould halves.
Source: CLLi, "Application of Catmull-Clark Subdivision Method in Plastic Injection
Mould Parting Surface Design", the proceedings of the Sixth International Conference on
Information Visualisation, 10–12 July 2002, London, England, pp 477–482, (2002 IEEE).

3.2. Intersecting the Part Surface

Figure 7.3 illustrates an example of this problem. The opening feature of the part in Figure 7.2
is further extended to the shape shown in Figure 7.3. The figure shows that the extrusion
surfaces associated with the opening feature intersect a face of the mould cavity. Obviously,
such extrusion surfaces cannot be used to construct the parting surface.

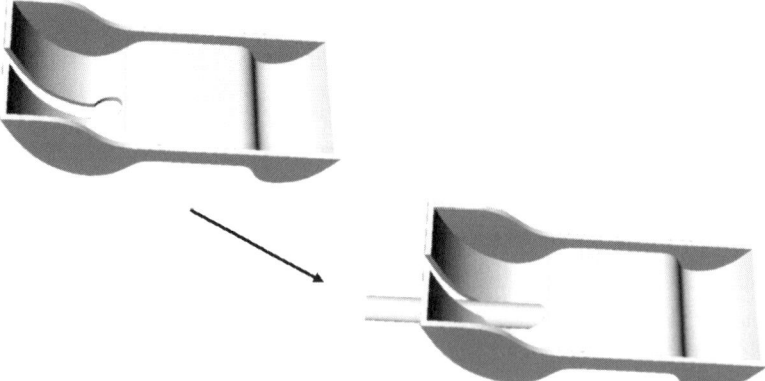

Figure 7.3   The method of extrusion causes intersection between extruded surface and part surface.
Source: CLLi, "Application of Catmull-Clark Subdivision Method in Plastic Injection
Mould Parting Surface Design", the proceedings of the Sixth International Conference on
Information Visualisation, 10–12 July 2002, London, England, pp 477–482, (2002 IEEE).

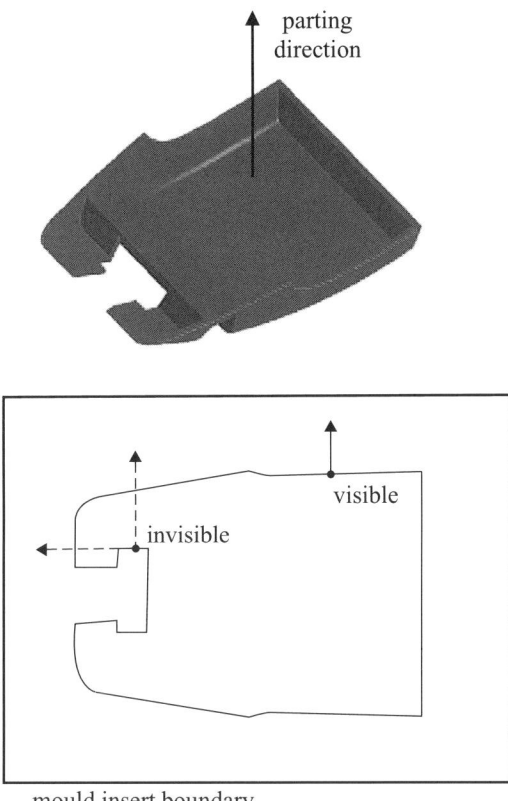

Figure 7.4   Use of 2D visibility test to determine validity of the extrusion method.

3.3. Validity Test

Due to the two problems mentioned earlier, it is necessary to perform a test in order to determine if the extrusion method can create a valid parting surface. Instead of testing for inter-locking between the extrusion surfaces and testing for intersection between the extrusion surfaces and the part surface, which are obviously computationally intensive and thus not efficient, we propose the use of a visibility test [17]. It can be shown that a simple 2D visibility analysis can serve the purpose. Figure 7.4 illustrates the basic idea of the method. Given a plastic part P, the main parting line PL of P that consists of a set of parting line segments $\{PLSi\}$ and parting direction d_p, the visibility test can be summarized into the following steps:

1. A 2D projection of the parting line PL' is obtained by projecting the 3D parting line PL along the parting direction d_p onto a plane perpendicular to d_p.
2. For each segment $PLSi'$ of the projected parting line PL', its visibility along the directions perpendicular to the mould wall boundary is checked. A segment is considered visible along a direction if, for any point on the segment, the semi-infinite line originates at the point along that direction and does not intersect PL'.
3. If all segments are visible in any one of the directions perpendicular to the mould inset boundary, the entire parting line is visible and the method of extrusion can be used.

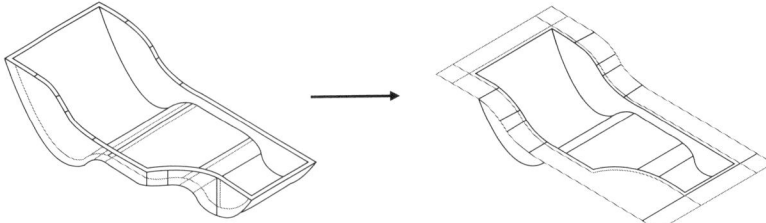

Figure 7.5   The method of extrusion applied to a part where the project image of its parting line is
non-convex.
Source: CLLi, "Application of Catmull-Clark Subdivision Method in Plastic Injection
Mould Parting Surface Design", the proceedings of the Sixth International Conference on
Information Visualisation, 10–12 July 2002, London, England, pp 477–482, (2002 IEEE).

4. If not all segments are visible, then the method of extrusion can only be used on the visible
   segments. For the invisible segments, a new method based on the Catmull-Clark subdivision
   method is used.

All the parting line segments of the part shown in Figure 7.1 are visible in at least one direction
that is perpendicular to the mould wall insert. The method of extrusion can thus generate a
valid parting surface. For the part shown in Figures 7.2 and 7.3, there are obviously segments
of the parting line with their projected image not visible in any directions that are perpendicular
to walls of the mould insert. These segments do not pass the validity test. To fill the region
associated with these segments, the method of subdivision has to be used.

   It is obvious that if the projected parting line *PL'* is convex, then *PL'* always passes the
validity test and thus the method of extrusion can be used. However, it should be noted that
the convexity of *PL'* is a sufficient condition and not a necessary condition. Figure 7.5 shows
a part where *PL'* is non-convex but visible, and thus the parting surface can be generated by
the method of extrusion.

4. Applying the Catmull-Clark Method

The subdivision methods in surface generation are becoming very popular in computer graphics
applications. Subdivision curve was first introduced by Chaikin [18], and the extension of the
method to surface generation was pioneered by Doo [19] and Catmull-Clark [20]. The basic
approach to surface generation by subdivision can be explained as follows. Given an initial
mesh, a new mesh is obtained by adding new mesh elements along the edges and at the corner
of the previous mesh. The addition of the new mesh elements can be considered as a 'corner
cutting' process whereby the sharp edges and corners are replaced by new mesh elements.
When this process is repeated indefinitely, Doo and Sabin [21] proved that in the limit the
mesh becomes a smooth surface.

   The first step in the generation of a subdivision surface is to construct the initial control
mesh. In our current application, the initial mesh can be considered as the skeleton of the
final surface. To determine this 3D skeleton, a 2D skeleton is first determined, which can be
obtained by computing the medial axis [22] of an approximating polygon. Given a 2D polygon,
the medial axis of the polygon is defined as the set of points such that for each point P_m in

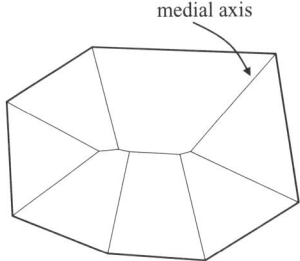

medial axis

Figure 7.6   The medial axis of a polygon.

the set, there are at least two points in the boundary of the polygon that are closest to P_m. Figure 7.6 illustrates the medial axis of a 2D polygon. From the 2D medial axis, the 3D mesh is then obtained by 'projecting' the 2D skeleton to 3D.

Given a set of curves that defines the boundary of a region to be filled by a subdivision surface, the procedure that generates the initial mesh for subdivision can be summarized into the following major steps:

1. The set of boundary curves C is projected along the parting direction d_p onto a plane perpendicular to the d_p to obtain a 2D closed curve C'. To simplify the discussion, assume that the plane lies on the x-y plane and the z axis is thus parallel to the parting direction d_p.
2. A polygon P' is constructed to approximate C'.
3. The medial axis MA of P' is computed.
4. As P' may be non-convex, the medial axis MA may contain parabolic segments. MA is simplified by replacing these curved segments by line segments.
5. For each boundary vertex of MA (i.e. those vertices lying on P'), its z-coordinate is obtained from a corresponding point on the 3D boundary C.
6. For each internal vertices V of MA (i.e. those vertices that have two or three closest points on the polygon P'), the closest points P_i on the boundary of P' is determined. The z-coordinate of P_i is obtained from a corresponding point on the 3D boundary C. The z-coordinate of the internal vertex V is obtained by a weighted average of the z-coordinates of P_i, given by the following formula.

$$z_v = \frac{\sum_{i=0}^{n-1} \frac{1}{2}(z_i + z_{(i+1) \bmod n}) \times DIST(p_i, p_{(i+1) \bmod n})}{\sum_{i=0}^{n-1} DIST(p_i, p_{(i+1) \bmod n})}$$

where n is the number of points that are at the same distance from V, and $DIST(p_i, p_j)$ is the Euclidean distance between 3D points p_i and p_j. This formula ensures that the z-coordinate of V will not be biased towards the z-coordinates of two points that are close together. (i.e. P_1 and P_2 in the example shown in Figure 7.7).

After the initial mesh is computed, a subdivision scheme is invoked to generate the surface. Various subdivision schemes have been developed and they differ by the way the new mesh elements are determined from the previous mesh. In this research, a variation of the Catmull-Clark method developed by Levin [23] is used. Using this method, the subdivision surface can be made to interpolate a set of curves that defines the boundary of the surface. This is

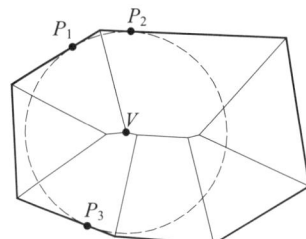

Figure 7.7   Computation of the 3D skeleton.

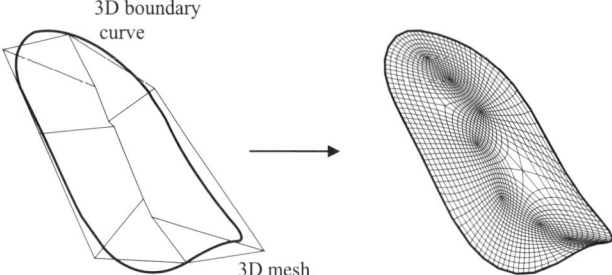

Figure 7.8   Applying subdivision technique to generate a surface that interpolates a given boundary curve.

an important property as the surface patch of the parting surface must attach to the parting line segment. Figure 7.8 shows the 3D mesh obtained by applying the above forumula to the polygon given in Figure 7.7, and the resulting subdvision surface generated from the 3D mesh.

5. Implementation and Design Example

An experimental program has been implemented to verify the feasibility of the proposed method. The program is written in C++ and run on a PC. The program accepts as input a set of NURBS curves that define the main parting line of a plastic part. The output of the program is a set of surface patches that define the parting surface. The patches are generated by either the method of extrusion or the subdivision method. Figure 7.9 illustrates a design example generated by the program. Figure 7.9a shows the plastic part. Figure 7.9b shows that extrusion surface patches are generated from two segments of the parting line, and the region to be filled by a subdivision surface is shown. Figure 7.9c shows the polygonal approximation of the region, and the control mesh for subdivision. Figure 7.9d shows the subdivision surface after 4 iterations. The extrusion surface patches together with the subdivision surface patch are shown in Figure 7.9(e).

6. Summary

There are two problems in most existing algorithms for automatic generation of parting surfaces for injection mould design. These algorithms usually use the method of extrusion to generate a

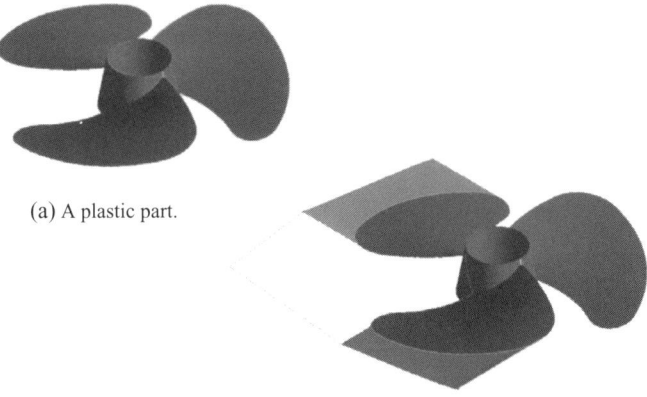

(a) A plastic part.

(b) Extrusion surface patches and a region to be filled by a subdivision surface.

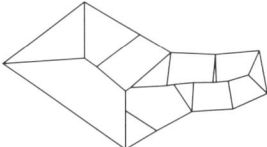

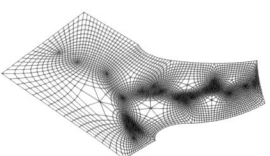

(c) A polygonal approximation of the region boundary and the initial control mesh for subdivision.

(d) The subdivision surface after 4 iterations.

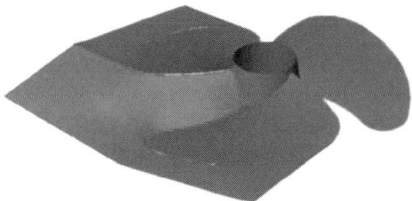

(e) Subdivision surface patch and extrusion surface patches forming part of the parting surface.

Figure 7.9 A design example that shows the use of the subdivision method in constructing the parting surface.
Source: CLLi, "Application of Catmull-Clark Subdivision Method in Plastic Injection Mould Parting Surface Design", the proceedings of the Sixth International Conference on Information Visualisation, 10–12 July 2002, London, England, pp 477–482, (2002 IEEE).

surface patch from a parting line segment to form the parting surface. Our investigation reveals that the extrusion surface may intersect with the part surface, or may result in interlocking of mould halves and thus affects mould opening. A new method for automatic parting surface generation is thus developed to solve the problems. This new method employs a variation of the Catmull-Clark method to generate subdivision surface patches for parting line segments where the method of extrusion fails. An experimental program has been developed to implement the new method. The program has been tested with real parts to verify the feasibility of the method.

References

[1] Hui K C and Tan S T (1992) Mould design with sweep operations. A heuristic search approach. *Computer-Aided Design*, vol 24 n 2, 81–91.

[2] Chen L L, Chou S Y and Woo T C (1993) Parting directions for mould and die design. *Computer-Aided Design*, Vol 25, 762–768.

[3] Chen L L, Chou S Y and Woo T C (1995) Partial visibility for selecting a parting direction in mould and die design. *Journal of Manufacturing Systems*. Vol 14, 319–330.

[4] Weinstein M and Manoochehri S (1996) Geometric influence of a molded part on the draw direction range and parting line locations. *Transactions of ASME, Journal of Mechanical Design*. Vol. 118, pp 29–39.

[5] Hui K C (1997) Geometric aspects of the mouldability of parts. *Computer-Aided Design*. Vol 29, 197–208.

[6] Wong T, Tan S T and Sze W S (1998) Parting line formation by slicing a 3D CAD model. *Engineering with Computers*. Vol 14, 330–342.

[7] Fu M W, Fuh J Y H, A Y C Nee (1999) Undercut feature recognition in an injection mould design system. *Computer-Aided Design*. Vol 31, 777–790.

[8] Tan S T, Yuen M F, Sze W S and Kwong K W (1990) Parting lines and parting surfaces of injection moulded parts. *Proc I. Mech. E Part B: Journal of Engineering Manufacture*. Vol 204, 211–221.

[9] Nee A Y C, Fu M W, Fuh J Y H, Lee K S and Zhang Y F (1998) Automatic determination of 3-D parting lines and surfaces in plastic injection mould design. *Annual of CIRP*. 47/1, 95–98.

[10] Kong L, Fuh J Y H and Lee K S (2001) Auto-generation of patch surfaces for injection mould design. *Proc I. Mech. E, Part B*. Vol 215, 105–110.

[11] Saxean M and Irani R K (1992) Automated gating plan synthesis for injection molds. *Proceedings of the ASME Computers in Engineering Conference*, CA, 381–389.

[12] Wang Z, Lee K S, Fuh J Y H, Li Z, Zhang Y F, Nee A Y C and Yang D C H (1996) Optimum ejector system design for plastic injection mould. *Int. J. of Materials and Product Technology*. Vol 11, No 5/6, 371–385.

[13] Ding X M, Fuh J Y H, Lee K S, Zhang Y F and Nee A Y C (2000) A computer-aided EDM electrode design system for mold manufacturing. *Int. J. Prod. Res*. Vol 38 no. 13, 3079–3092.

[14] C L Li (2000) A feature-based approach to injection mould cooling system design. *Computer-Aided Design*. Vol 33 no. 14, 1073–1090.

[15] C L Li (2001) Automatic synthesis of cooling system design for plastic injection mould. ASME Design Engineering Technical Conferences and Computers and Information in Engineering Conference, Design Automation Conference, Pittsburgh, Pennsylvania.

[16] Fu M W, Fuh J Y H and Nee A Y C (2001) Core and cavity generation method in injection mould design. *Int. J. Prod. Res*. Vol 39 No 1, 121–138.

[17] Yin Z P, Ding H and Xiong Y L (2000) Visibility theory and algorithms with application to manufacturing processes. *Int. J. Prod Res*. Vol 38 No 13, 2891–2909.

[18] Chaikin G (1974) An algorithm for high speed curve generation. *Computer Graphics and Image Processing*. Vol 3, 346–349.

[19] Doo D (1978) A subdivision algorithm for smoothing down irregularly shaped polyhedrons. *Int. Conf. Interactive Techniques in Computer-Aided Design*, Bologna, Italy. 157–165.

[20] Catmull E and Clark J (1978) Recursively generated B-spline surfaces on arbitrary topological meshes. *Computer-Aided Design*. Vol 10, 350–355.

[21] Doo D and Sabin M (1978) Behaviour of recursive division surfaces near extraordinary points. *Computer-Aided Design*. Vol 10, 356–360.

[22] Preparata F P (1977) The medial axis of a simple polygon. *Proc. 6th Symp. Math. Foundations of Comput. Sci.*, 443–450.

[23] Levin A (1999) Interpolating nets of curves by smooth subdivision surfaces. *Siggraph 99*, California, USA, pp 57–64.

8

Triadic Subdivision of Non Uniform Powell-Sabin splines

Evelyne Vanraes, Paul Dierckx, Adhemar Bultheel

Department of Computer Science, Katholieke Universiteit Leuven,
Celestijnenlaan 200A, 3001 Heverlee, Belgium.

This chapter examines the recent advances in the research on Powell-Sabin splines. We already had at our disposal a very useful normalized B-spline representation in which the basis functions form a convex partition of unity. This representation has been intensively used in different applications for uniform Powell-Sabin splines, that is on triangulations with all equilateral triangles, but not for Powell-Sabin splines on general triangulations. Recently we developed a subdivision scheme for the non uniform case. It is a triadic scheme; every edge is split into three new edges and every original triangle is split into nine new triangles. The scheme can be used in, and leads to, many other applications. We mention multiresolution analysis and decomposition (wavelets), local editing and visualization.

1. Introduction

Powell-Sabin splines are functions in the space $S_2^1(\Delta^*)$ of C^1 continuous piecewise quadratic functions on a Powell-Sabin refinement. Such a refinement Δ^* can be obtained from an arbitrary triangulation Δ by splitting each triangle into six subtriangles with a common interior point [5]. Working with triangles makes it possible to design surfaces with an arbitrary number of edges, which is not possible with the widely used tensor product B-spline representation that is restricted to rectangular domains. In contrast to Bézier triangles [2], where imposing smoothness conditions between the patches requires a great number of nontrivial relations between the coefficients to be satisfied, the C^1 continuity of a PS-spline is guaranteed for any choice of the coefficients.

A first attempt to describe PS-splines was by Shi *et al.* [6], but their method has some serious drawbacks from the numerical point of view. This was solved by the improved algorithm of

Advances in Geometric Modeling. Edited by M. Sarfraz
© 2003 John Wiley & Sons, Ltd   ISBN: 0-470-85937-7

Dierckx [1] to construct a normalized B-spline basis. This representation has the advantage that the basis functions form a convex partition of unity which is a useful property in CAGD applications. Furthermore it leads to a nice geometric interpretation with control triangles that are tangent to the surface.

The last advance in the area of Powell-Sabin splines is the development of a subdivision scheme [9]. Given a surface on a certain triangulation, we can now calculate a B-spline representation of the surface on a refinement of that triangulation. The result is a denser set of control points. The availability of a subdivision scheme opens the door to many other applications. Among them are multiresolution analysis, graphical display and the design of wavelets.

In Section 2 we recall some basic concepts of polynomials on triangles and the definition of Powell-Sabin splines. Section 3 gives the construction of a normalized B-spline basis and how to choose the remaining parameters. Then, in Section 4, we discuss the subdivision algorithm. Finally we give an overview of possible applications in Section 5.

2. Powell-Sabin splines

2.1. Bézier Polynomials

Let $\lambda = (\lambda_1, \lambda_2, \lambda_3)$, $|\lambda| = \lambda_1 + \lambda_2 + \lambda_3 = d$, $\lambda_i \ \square \ \{0, 1, \ldots, d\}$ using standard multi-index notation. Consider a non degenerate triangle T (T_1, T_2, T_3) in a plane with its vertices having Cartesian coordinates T_i (x_i, y_i), $i = 1, 2, 3$. Any point $P(x, y)$ in that plane can be expressed in terms of barycentric coordinates $\tau = (\tau_1, \tau_2, \tau_3)$ with respect to T : $P = \sum_{i=1}^{3} \tau_i T_i$, where $|\tau| = 1$.

A Bézier polynomial [2] of degree d over the triangle T is defined by:

$$b_T^d(P) = b_T^d(\tau) = \sum_{|\lambda|=d} b_\lambda B_\lambda^d(\tau), \qquad (8.1)$$

in which b_λ are called Bézier ordinates, and:

$$B_\lambda^d(\tau) = \frac{d!}{\lambda_1! \lambda_2! \lambda_3!} \tau_1^{\lambda_1} \tau_2^{\lambda_2} \tau_3^{\lambda_3}, \qquad (8.2)$$

are the Bernstein-Bézier polynomials on the triangle.

The domain point ξ_λ associated with the Bézier ordinate b_λ is the point in the (x, y) plane with barycentric coordinates $\left(\dfrac{\lambda_1}{d}, \dfrac{\lambda_2}{d}, \dfrac{\lambda_3}{d} \right)$. The points (ξ_λ, b_λ) are the control points for the surface $z = b_T^d(\tau)$ and the piecewise linear interpolant to these points is the Bézier net or control net. This is displayed schematically in Figure 8.1 for the case $d = 2$. The domain points ξ_λ are marked with dots. The control net mimics the shape of the surface and is tangent to the polynomial surface at the three vertices of the triangle.

Continuity conditions between triangles can be expressed as relations between the Bézier ordinates. Let $b_T^d(\tau)$ be a polynomial with Bézier ordinates b_{ijk} on the triangle T (T_1, T_2, T_3), and $c_T^d(\tau)$ be a polynomial with Bézier ordinates c_{ijk} on the triangle T* (T_1^*, T_2, T_3), where T_1^* has barycentric coordinates λ with respect to T (T_1, T_2, T_3). A necessary and sufficient condition for $b_T^d(\tau)$ and $c_T^d(\tau)$ to be C^1 continuous across the common boundary is:

$$C^0 : c_{0jk} = b_{0jk}, \qquad (8.3)$$

$$C^1 : c_{1jk} = \lambda_1 b_{1jk} + \lambda_2 b_{0(j+1)k} + \lambda_3 b_{0j(k+1)}. \qquad (8.4)$$

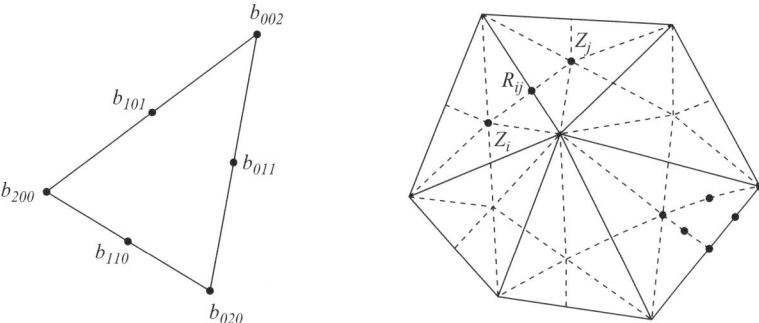

Figure 8.1   Left: Positions of Bézier ordinates for $d = 2$. Right: PS-refinement. Each triangle ρ_j is split into six smaller triangles with a common vertex Z_j.

Representing complex shapes, however, requires the use of patch complexes with a great number of Bézier triangles. Keeping up continuity conditions between all the neighboring patches results generally in nontrivial relations between their Bézier ordinates. The use of split triangles can overcome this problem.

2.2. Powell-Sabin split and the Space $S_2^1(\Delta^*)$

Consider a simply connected subset $\Omega \square \mathrm{R}^2$ with polygonal boundary $\delta\Omega$. Suppose we have a conforming triangulation Δ of Ω, being constituted of triangles ρ_j, $j = 1, \ldots, t$, and having vertices V_k with Cartesian coordinates (x_k, y_k), $k = 1, \ldots, n$. Let Δ^* be a Powell-Sabin refinement of Δ, which divides each triangle ρ_j into six smaller triangles with a common vertex Z_j as follows (Figure 8.1)

1. Choose an interior point Z_j in each triangle ρ_j, so that if two triangles ρ_i and ρ_j have a common edge, then the line joining these interior points Z_i and Z_j intersects the common edge at a point R_{ij} between its vertices. Choosing Z_j as the incentre of each triangle ρ_j ensures the existence of the points R_{ij}. Other choices may be more appropriate from the practical point of view.
2. Join each point Z_j to the vertices of ρ_j.
3. For each edge of the triangle ρ_j
   - which belongs to the boundary $\delta\Omega$, join Z_j to an arbitrary point of the edge.
   - which is common to a triangle ρ_i, join Z_i to R_{ij}.

Now we consider the space $S_2^1(\Delta^*)$ of piecewise C^1 continuous quadratic polynomials on Δ^*, the Powell-Sabin splines. Each of the $6t$ triangles resulting from the PS-refinement becomes the domain triangle of a quadratic Bernstein-Bézier polynomial, i.e. we choose $d = 2$ in Equations (8.1) and (8.2), as indicated for one subtriangle in Figure 8.1. Powell and Sabin [5] proved that the dimension of the space $S_2^1(\Delta^*)$ equals $3n$: there exists a unique solution $s(x,y) \square S_2^1(\Delta^*)$ for the interpolation problem,

$$s(V_k) = f_k, \quad \frac{\partial s}{\partial x}(V_k) = f_{x,k}, \quad \frac{\partial s}{\partial y}(V_k) = f_{y,k}, \quad k = 1, \ldots, n. \tag{8.5}$$

So given the function and derivative values at each vertex V_k, the Bézier ordinates on the domain subtriangles are uniquely defined and the continuity conditions between subtriangles are automatically fulfilled.

3. A Normalized B-spline Representation

3.1. Convex Partition of Unity

Dierckx [1] showed that each piecewise polynomial $s(x, y) \in S_2^1(\Delta^*)$ has a unique representation:

$$s(x, y) = \sum_{i=1}^{n} \sum_{j=1}^{3} c_{ij} B_i^j(x, y), \quad (x, y) \in \Omega, \tag{8.6}$$

where the basis functions satisfy:

$$B_i^j(x, y) \geq 0, \tag{8.7}$$

$$\sum_{i=1}^{n} \sum_{j=1}^{3} B_i^j(x, y) \equiv 1, \tag{8.8}$$

and have local support: $B_i^j(x, y)$ is nonzero only on the so-called molecule M_i of V_i, being the set of triangles ρ_l that have V_i as a vertex. The number of triangles in M_i is called the molecule number m_i.

The basis functions are constructed as follows:

1. For each vertex $V_i \in \Delta$, identify its PS-points. This is a number of particular surrounding Bézier domain points and the vertex V_i itself. Figure 8.2 shows the PS-points S, $\tilde{S}$, S' and V_1 for the vertex V_1 in the triangle $\rho(V_1, V_2, V_3)$.

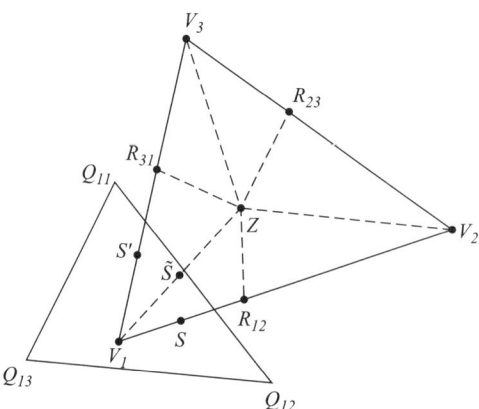

Figure 8.2   PS-points and PS-triangle.

2. For each vertex V_i, find a triangle $t_i(Q_{i1}, Q_{i2}, Q_{i3})$ containing all the PS-points of V_i (from all the triangles ρ_l in the molecule M_i). Denote its vertices $Q_{ij}(X_{ij}, Y_{ij})$. The triangles t_i, $i = 1, \ldots, n$ are called PS-triangles. Figure 8.2 also shows such a PS-triangle t_1. We denote the barycentric coordinates of the PS-points S_i, $\tilde{S}_i$, and S_i' with respect to the triangle $t_i(Q_{i1}, Q_{i2}, Q_{i3})$ with (L_{i1}, L_{i2}, L_{i3}), $(\tilde{L}_{i1}, \tilde{L}_{i2}, \tilde{L}_{i3})$ and $(L_{i1}', L_{i2}', L_{i3}')$.

3. Given the PS-triangle t_i of a vertex V_i, three linearly independent triplets of real numbers can be found as follows:

$$\alpha_i = (\alpha_{i1}, \alpha_{i2}, \alpha_{i3}),$$

are the barycentric coordinates of V_i with respect to t_i,

$$\beta_i = (\beta_{i1}, \beta_{i2}, \beta_{i3}) = \left(\frac{Y_{i2} - Y_{i3}}{e}, \frac{Y_{i3} - Y_{i1}}{e}, \frac{Y_{i1} - Y_{i2}}{e} \right), \tag{8.9}$$

$$\gamma_i = (\gamma_{i1}, \gamma_{i2}, \gamma_{i3}) = \left(\frac{X_{i3} - X_{i2}}{e}, \frac{X_{i1} - X_{i3}}{e}, \frac{X_{i2} - X_{i1}}{e} \right), \tag{8.10}$$

where $e = \begin{vmatrix} X_{i1} & Y_{i1} & 1 \\ X_{i2} & Y_{i2} & 1 \\ X_{i3} & Y_{i3} & 1 \end{vmatrix}$. $\tag{8.11}$

We have $|\alpha_i| = 1$ and $|\beta_i| = |\gamma_i| = 0$.

4. The basis function $B_i^j(x,y)$ is the unique solution of the interpolation problem Equation (8.5) with all $(f_k, f_{x,k}, f_{y,k}) = (0, 0, 0)$ except for $(f_i, f_{x,i}, f_{y,i}) = (\alpha_{ij}, \beta_{ij}, \gamma_{ij})$.

3.2. PS-Triangles and Control Triangles

We define the control points as:

$$C_{ij} = (Q_{ij}, c_{ij}) = (X_{ij}, Y_{ij}, c_{ij}), \tag{8.12}$$

and the control triangles as:

$$T_i(C_{i1}, C_{i2}, C_{i3}). \tag{8.13}$$

The projection of the control triangles T_i in the (x,y) plane are the PS-triangles t_i. We can prove that the control triangle T_i is tangent to the surface $z = s(x,y)$ at the point $(x_i, y_i, s(V_i))$ and that the surface lies in the convex hull of the control points.

The fact that the PS-triangle t_i contains the PS-points of the vertex V_i guarantees property (see Equation (8.7)). Apart from this requirement there are no restrictions on the choice of the PS-triangle. We know, however, that the larger the PS-triangle, the more linearly dependent the basis functions are. Furthermore we prefer the control points to be close to the surface for design purposes. Therefore we always choose a PS-triangle with a small area. In [1] the PS-triangle with the smallest area is computed, but this requires the solution of a quadratic programming problem for each vertex.

The outline of a practical algorithm that avoids the optimization problem is described in [7]. It is based on the observation that for a molecule with three triangles a trivial solution exists. The six PS-points form a triangle themselves and that is of course the smallest triangle that contains the PS-points. Other configurations with more than three triangles in the molecule are changed step by step into the trivial case.

4. Subdivision

The goal of subdivision is to calculate the B-spline representation (Equation (8.6)) of a PS-spline on a refinement Δ^1 of the given triangulation Δ^0. We first choose the refinement and then compute control points for the new vertices.

4.1. Choosing a Suitable Refinement Δ^1 of Δ^0

4.1.1 Dyadic Subdivision

The most obvious possibility is dyadic subdivision. In this scheme a new vertex is inserted on every edge between two old vertices and every original triangle is split into four new triangles. Because we want to represent exactly the same surface, the lines of the old PS-refinement Δ^{0*} must be included in the new PS-refinement Δ^{1*}. To ensure this, the new vertices have to be placed on the points R_{ij}. This is illustrated in Figure 8.3.

The dyadic scheme was used by Windmolders and Dierckx for uniform Powell-Sabin splines, this is on a triangulation with all equilateral triangles. The subdivision rules for this special case can be found in [10, 11]. In the general case, the idea of dyadic subdivision can only be used under certain conditions. For example, the point Z_{ijk} of the PS-refinement of a triangle, must lie inside the middle triangle $(V_{ij} V_{jk} V_{ki})$. This leads to conditions on the initial triangulation Δ^0 and its PS-refinement Δ^{0*}, i.e. on the placement of the interior points Z_{ijk} and the resulting positions of the R_{ij}: the dyadic scheme is not generally applicable.

4.1.2 Triadic Subdivision

Another possibility, shown in Figure 8.4, is triadic subdivision. In this scheme every edge is split in three instead of two and every original triangle is split into nine new triangles. One

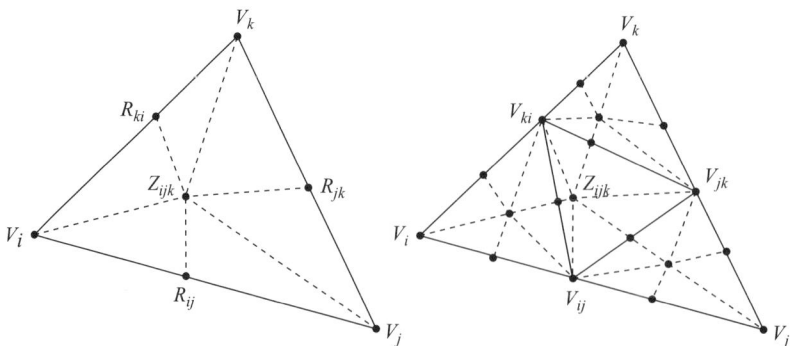

Figure 8.3 Principle of dyadic subdivision.

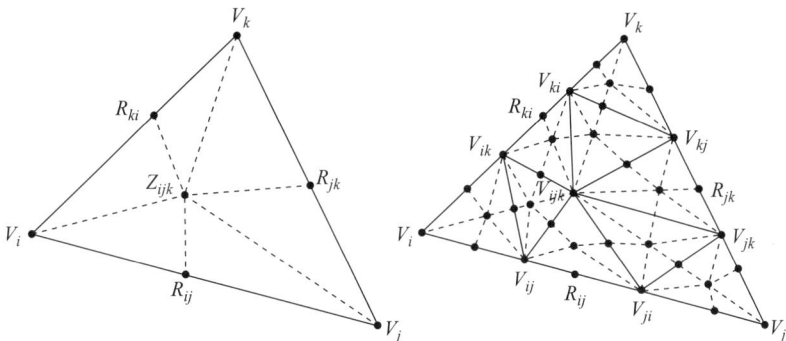

Figure 8.4   Principle of triadic subdivision.

new vertex is added inside the triangle on the place of the interior point Z_{ijk}, and on the edges two new vertices are added each at one side of the points R_{ij}:

$$
\begin{aligned}
V_{ij} &= \omega_{ij} V_i + (1 - \omega_{ij}) R_{ij}, \\
V_{ji} &= \omega_{ji} V_j + (1 - \omega_{ji}) R_{ij}, \\
V_{ijk} &= Z_{ijk}.
\end{aligned}
\tag{8.14}
$$

In these formulas ω_{ij} and ω_{ji} have a value between 0 and 1:

$$
0 < \omega_{ij}, \omega_{ji} < 1.
\tag{8.15}
$$

For the resulting refinement to exist, the interior point Z_{ijk} has to lie inside the hexagon formed by the new vertices ($V_{ij}, V_{ji}, V_{jk}, V_{kj}, V_{ki}, V_{ik}$). It is always possible to place these new vertices, i.e. choose a value for the ω in Equation (8.14), such that this condition is fulfilled: there are no conditions on the initial triangulation Δ^0 or its PS-refinement Δ^{0*}.

4.1.3 $\sqrt{3}$-Subdivision

Triadic subdivision can also be seen as two steps of a $\sqrt{3}$-subdivision scheme. This kind of scheme was first introduced by Kobbelt [3] and Labsik [4] and used for uniform Powell-Sabin splines by Vanraes [8].

The new triangulation $\Delta^{\sqrt{3}}$ is constructed by inserting a new vertex V_{ijk} at the position of the interior point Z_{ijk} of each triangle. Except at the boundaries, the old edges are not preserved in the new triangulation. Instead new edges are introduced connecting every new vertex V_{ijk} with the vertices of the old triangle it lies in, and connecting every two new vertices that lie in neighboring old triangles. The upper part of Figure 8.5 shows (on the right) the result of $\sqrt{3}$-subdivision on the triangulation Δ^0 on the left. In this figure the PS-refinement is not shown, but notice that the new edges in $\Delta^{\sqrt{3}}$ coincide with the lines of the PS-refinement Δ^{0*} and that the original edges of Δ^0 are now part of the new PS-refinement $\Delta^{\sqrt{3}*}$.

Applying the $\sqrt{3}$-subdivision operator a second time again results in new vertices that co-incide with the interior points that, in this case, lie on the edges of the initial triangulation Δ^0. As can be seen in Figure 8.5, this causes a refinement with tri-section of every original edge

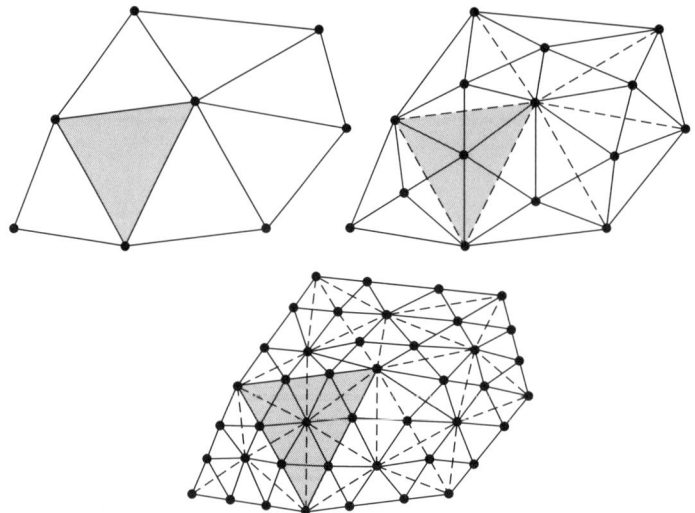

Figure 8.5 Principle of $\sqrt{3}$-subdivision. Applying $\sqrt{3}$-subdivision twice results in triadic subdivision. The PS-refinements are not shown.

and splitting of each original triangle into nine subtriangles. Hence one refinement step of this scheme can be seen as the square root of one step of the triadic scheme.

The boundaries of the domain require a special treatment. Indeed, to end up with a tri-section of the original boundary edges, two new vertices have to be added in the second $\sqrt{3}$-subdivision step in the triangles in question.

4.2. The Subdivision Rules

We prefer to use the triadic or $\sqrt{3}$-scheme for non uniform Powell-Sabin splines because, in contrast to the dyadic scheme, there are no conditions on the initial triangulation and PS-refinement. There is an easy solution for the control points of the vertex added inside each triangle during $\sqrt{3}$-subdivision. We will however always do two steps at once to avoid boundary problems and therefore we call our scheme triadic [9].

To illustrate the subdivision rules we use a triangle of Δ^0 with vertices (V_i, V_j, V_k) as in Figure 8.4. The PS-refinement is also shown, with:

$$
\begin{aligned}
R_{ij} &= \lambda_{ij} V_i + (1 - \lambda_{ij}) V_j, \\
R_{jk} &= \lambda_{jk} V_j + (1 - \lambda_{jk}) V_k, \\
R_{ki} &= \lambda_{ki} V_k + (1 - \lambda_{ki}) V_i, \\
Z_{ijk} &= a_{ijk} V_i + b_{ijk} V_j + c_{ijk} V_k.
\end{aligned}
\tag{8.16}
$$

4.2.1 New Control Points

The new PS-triangles are shown in Figure 8.6. In order not to overload the picture the PS-triangles in the old vertices are not plotted.

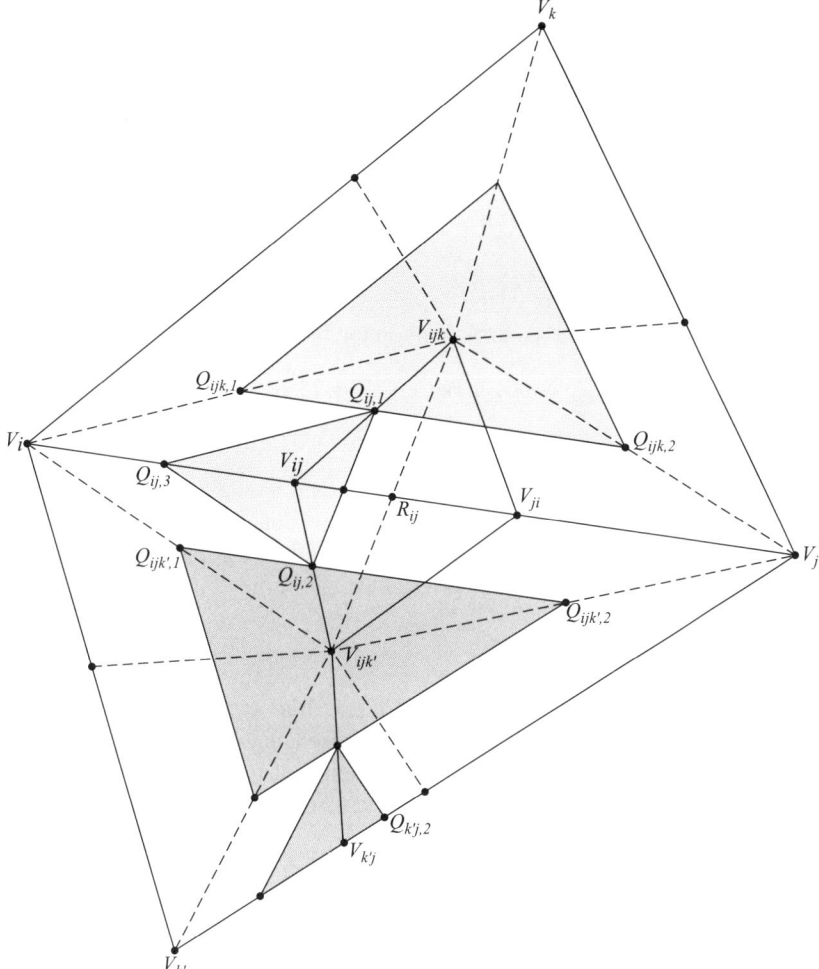

Figure 8.6 PS-triangles for the new vertices.

The vertex V_{ijk} can be considered as added in a first step of $\sqrt{3}$-subdivision. An easy solution for the new control triangle is given by:

$$
\begin{aligned}
C_{ijk,1} &= \tilde{L}_{i1}C_{i1} + \tilde{L}_{i2}C_{i2} + \tilde{L}_{i3}C_{i3}, \\
C_{ijk,2} &= \tilde{L}_{j1}C_{j1} + \tilde{L}_{j2}C_{j2} + \tilde{L}_{j3}C_{j3}, \\
C_{ijk,3} &= \tilde{L}_{k1}C_{k1} + \tilde{L}_{k2}C_{k2} + \tilde{L}_{k3}C_{k3},
\end{aligned}
\tag{8.17}
$$

with $\tilde{L}_i$ the barycentric coordinates of $Q_{ijk,1}$ with respect to the old PS-triangle t_i of V_i, and analogous for the others. We can easily prove that this triangle is tangent to the surface in $(V_{ijk}, s(V_{ijk}))$. The corners of the corresponding PS-triangle coincide with PS-points of the old vertices, and the PS-triangle contains the appropriate PS-points at this intermediate level.

For the new vertex V_{ij} on the original edge, we do a second $\sqrt{3}$-subdivision step and use the same idea for the new control triangle. This leads to:

$$
\begin{aligned}
C_{ij,1} &= (\omega_{ij} + \lambda_{ij} - \omega_{ij}\lambda_{ij})C_{ijk,1} + (1 - \omega_{ij} - \lambda_{ij} + \omega_{ij}\lambda_{ij})C_{ijk,2}, \\
C_{ij,2} &= (\omega_{ij} + \lambda_{ij} - \omega_{ij}\lambda_{ij})C_{ijk',1} + (1 - \omega_{ij} - \lambda_{ij} + \omega_{ij}\lambda_{ij})C_{ijk',2}, \qquad (8.18) \\
C_{ij,3} &= (L_{i1} + \omega_{ij}\alpha_{i1} - \omega_{ij}L_{i1})C_{i1} + (L_{i1} + \omega_{ij}\alpha_{i1} - \omega_{ij}L_{i2})C_{i2} \\
&\quad + (L_{i2} + \omega_{ij}\alpha_{i3} - \omega_{ij}L_{i3})C_{i3},
\end{aligned}
$$

with L_i the barycentric coordinates of $Q_{ij,3}$ with respect to the old PS-triangle t_i of V_i. If the new vertex lies on the boundary of the domain, for example $V_{k'j}$, there is no neighboring triangle and we choose the second control point on the boundary,

$$
\begin{aligned}
C_{k'j,2} &= (\omega_{k'j} + \lambda_{k'j} - \omega_{k'j}\lambda_{k'j})(L_{k'1}C_{k'1} + L_{k'2}C_{k'2} + L_{k'3}C_{k'3}) \\
&\quad + (1 - \omega_{k'j} - \lambda_{k'j} + \omega_{k'j}\lambda_{k'j})(L_{j1}C_{j1} + L_{j2}C_{j2} + L_{j3}C_{j3}).
\end{aligned}
\qquad (8.19)
$$

For the old vertices we keep the control points. This is a valid choice because the control triangle is still tangent to the surface and the PS-triangle contains the new PS-points. In all these formulas we only use convex combinations of the control points on the previous level, which means that this subdivision scheme is a stable algorithm.

4.2.2 Optimization

In the subdivision scheme as we have described it up till now, the control points in the old vertices do not change when going to a finer level. Indeed, a control triangle of the previous level remains valid because the corresponding PS-triangle also contains the PS-points on the refined level. However, these PS-points are now closer to the vertex than before, so it is possible to choose a smaller PS-triangle.

To find a better suited PS-triangle, we shrink the old PS-triangle until it hits a PS-point, but without changing the shape. We can do the same for the interior vertex V_{ijk} because the given PS-triangle is not optimal anymore after the second $\sqrt{3}$-step. This generally does not lead to the smallest control triangle possible, but the overhead of the optimization problem to find the optimal triangle cannot be justified.

5. Applications

With the development of the subdivision scheme many other applications become possible. We give a short description of some of them here.

5.1. Multiresolution Analysis

If we collect the B-spline coefficients on a certain level j in a column vector $\mathbf{c}^j$ and the corresponding B-spline basis functions in a row vector $\mathbf{\Phi}^j$, we can write the representation (Equation (8.6)) on different levels as:

$$
\begin{aligned}
s(x, y) &= \mathbf{\Phi}^j \mathbf{c}^j \\
&= \mathbf{\Phi}^{j+1} \mathbf{c}^{j+1}.
\end{aligned}
\qquad (8.20)
$$

If we also write the subdivision scheme in block matrix form:

$$\mathbf{C}^{j+1} = \mathbf{P}^j \mathbf{C}^j, \tag{8.21}$$

with the matrix $\mathbf{P}^j$ the subdivision matrix, and combine this with Equation (8.20), we obtain the relation between the basis functions on different levels:

$$\mathbf{\Phi}^j = \mathbf{\Phi}^{j\oplus 1} \mathbf{P}^j. \tag{8.22}$$

This equation establishes refinability since it states that each of the functions in $\mathbf{\Phi}^j$ can be written as a linear combination of the functions in $\mathbf{\Phi}^{j+1}$. A strictly increasing sequence of subspaces $V^j = S_2^1(\Delta^{j*})$ is associated with the base triangulation Δ^0:

$$V^0 \subset V^1 \subset V^2 \ldots \tag{8.23}$$

This is called a multiresolution analysis (MRA).

5.2. Multiresolution Editing

The support of a basis function is the molecule of the vertex. After subdivision the molecules are smaller, and so is the support. This gives the designer more local control when manipulating surfaces. It is now possible to make local changes on different resolutions. A change on a finer level will influence a smaller neighborhood of the involved vertex than a change on a coarser level.

5.3. Wavelets

Each space V^{j+1} of the multiresolution analysis contains splines on a finer triangulation than the previous coarser space V^j and therefore can describe more detail of a surface. These details are captured in the algebraic complement W^j such that:

$$V^j \oplus W^j = V^{j+1} \tag{8.24}$$

where $\oplus$ denotes the inner sum of disjoint spaces. The complement space W^j is not necessarily orthogonal to V^j and we refer to the basis functions $\mathbf{\Psi}^j$ of W^j as wavelets.

A wavelet transform is a change of basis from the B-spline basis functions on a fine level V^{j+1} to B-splines on a coarser level V^j completed with wavelet functions in W^j. To calculate the control points and the wavelet coefficients after the change of basis we use the lifting scheme (Figure 8.7).

First the control points $\mathbf{C}^{j+1}$ are split into two sequences. The first sequence, $\mathbf{C}_o^{j+1}$, contains the control points that correspond to vertices in Δ^j, and the second sequence, $\mathbf{C}_n^{j+1}$, contains control points that correspond to vertices that are in Δ^{j+1}, but not in Δ^j.

Then we treat $\mathbf{C}_o^{j+1}$ as the control points of a surface defined on Δ^j and apply the subdivision algorithm $\mathbf{P}^j$ that leads to the control points for the new vertices. The result is used as a prediction for $\mathbf{C}_n^{j+1}$ and substracted from this sequence. This yields the wavelet coefficients $\mathbf{W}^j$ on the lower branch in the picture. We call this step the prediction step.

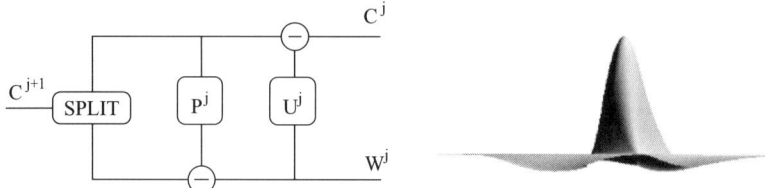

Figure 8.7   Left: Subdivision can be used as the prediction in the lifting scheme. Right: A wavelet
function.

Often an additional update step is used to give the wavelet functions properties such as
vanishing moments. $\mathbf{C}_o^{j+1}$ is updated with a linear combination α^j of wavelet coefficients $\mathbf{W}^j$.
This yields the control points $\mathbf{C}^j$ on the upper branch in the picture. We call this step the update
step.

With the development of Powell-Sabin wavelets come other typical wavelet applications
such as compression, noise reduction, progressive display and transmission, level of detail
control.

5.4. Visualization

The control triangles are tangent to the surface and in case of repeated subdivision, the linear
interpolant of the tangent points converges to the surface itself. Therefore subdivision is a
common technique for displaying surfaces graphically. Note that in this case the additional
shrinkage of the PS-triangles is not needed because in the end we only need the tangent points
and the corresponding shape of the basis functions is irrelevant.

6. Summary

We recalled the definition of Powell-Sabin splines, piecewise C^1 continuous quadratic functions on Powell-Sabin splits, and how to construct a normalized B-spline basis for a certain
triangulation. This representation leads to the notion of control triangles that are tangent to the
surface. We also know that the surface lies in the convex hull of the control points.

For a triangulation with a certain PS-refinement, there are different possibilities for the
PS-triangles. The only requirement is that a PS-triangle must contain all its PS-points. We
prefer the area of the PS-triangle to be small but, as we mentioned, it is not necessary to use
the very smallest one: other more practical choices are also possible.

We described an algorithm for computing the B-spline representation of a Powell-Sabin
spline surface on a refinement of a given triangulation. We used a triadic subdivision scheme
because there are no restrictions on the initial triangulation, as opposed to the dyadic scheme.
$\sqrt{3}$-subdivision can be seen as an intermediate level for the triadic scheme and this idea is used
in the development of the subdivision rules. Since the algorithm uses only convex combinations
it is numerically stable.

The availablity of a subdivision scheme makes other multiresolution based applications
possible for non uniform Powell-Sabin splines. We described how a multiresolution analysis
can be associated with a base triangulation and how this can be used for multiresolution editing

and the construction of wavelets. Subdivision is also a commonly used tool for the visualization of surfaces.

References

[1] P. Dierckx. On calculating normalized Powell-Sabin splines. *CAGD*, 15(3):61–78, 1997.

[2] G. Farin. Triangular Bernstein-Bézier patches. *CAGD*, 3(2):83–128, 1986.

[3] L. Kobbelt. $\sqrt{3}$-subdivision. In *Computer Graphics Proceedings*, Annual Conference Series. *ACM SIGGRAPH*, 2000.

[4] U. Labsik and G. Greiner. Interpolatory $\sqrt{3}$-subdivision. In Sabine Coquilart and Jr. Duke, David, editors, *Proceedings of the 21st. European Conference On Computer Graphics* (*Eurographics-00*) (volume 19, 3 of *Computer Graphics Forum*, pp. 131–138), Cambridge, August 21–25 2000. Blackwell Publishers.

[5] M. J. D. Powell and M. A. Sabin. Piecewise quadratic approximations on triangles. *ACM Transactions on Mathematical Software*, 3:316–325, 1977.

[6] X. Shi, S. Wang, W. Wang and R. H. Wang. The C^1 quadratic spline space on triangulations. Report 86004, Departement of Mathematics, Jilin University, Changchun, 1986.

[7] E. Vanraes, P. Dierckx, and A. Bultheel. On the choice of the PS-triangles. *TW Report 353*, Department of Computer Science, Katholieke Universiteit Leuven, Belgium, February 2003.

[8] E. Vanraes, J. Windmolders, A. Bultheel and P. Dierckx. Dyadic and $\sqrt{3}$-subdivision for uniform Powell-Sabin splines. In *Information Visualisation Proceedings*. IEEE Computer Society, 2002.

[9] E. Vanraes, J. Windmolders, A. Bultheel and P. Dierckx. Subdivision for Powell-Sabin spline surfaces. *TW Report 345*, Department of Computer Science, Katholieke Universiteit Leuven, Belgium, September 2002.

[10] J. Windmolders and P. Dierckx. Subdivision of uniform Powell-Sabin splines. *CAGD*, 16:301–315, 1999.

[11] J. Windmolders. Powell-Sabin splines for Computer Aided Geometric Design. *P.h.d. Thesis*, Department of Computer Science, Katholieke Universiteit Leuven, Belgium, February 2003.

9

Surface Interpolation Scheme by Distance Blending over Convex Sets

Lizhuang Ma

Department of Computer Science and Engineering, Shanghai Jiaotong University,
Huashan Road 1954, Shanghai 200030, China

Qiang Wang

State Key Lab. Of CAD and CG, Zhejiang University
Hangzhou 310027, China

Tony Chan K Y

Center for Advanced Media Technology, Nanyang Technological University,
Singapore 639798

Interpolation and approximation of different types of data points are very important in the field of CAD/CAM, virtual reality and computer graphics. Besides interpolating sample data, we also need to interpolate sample pieces of surfaces, textures or functions located over different areas or sets. Here we actually intend to interpolate an infinite number of data points of any function or pattern. In this chapter, we introduce the concept of distance from a point to a set or area. The square of distance function is C^1 smooth when the set is convex. The distance function is then used to construct a blending basis for interpolating functions or surfaces defined on related convex sets. An explicit set-splines basis is then derived, which has many useful properties similar to the traditional B-spline basis. These set-splines can be used to not only interpolate sample data, but sample pieces of functions as well. The potential of this new method is great, for instance, blending or smoothing surfaces, interpolating textures and images, and interpolating volume data.

Advances in Geometric Modeling. Edited by M. Sarfraz
© 2003 John Wiley & Sons, Ltd   ISBN: 0-470-85937-7

1. Introduction

Interpolation of data points is a fundamental problem in the field of CAD/CAM, geometric modeling and computer graphics and so on [1, 11–19]. Sampled or given data points are usually interpolated by using B-splines in many applications [1, 2, 3, 11–16, 19]. However, we also need to interpolate a set of infinitive data points, for instance, pieces of surfaces, textures, tiling and continuous functions defined on a subset of Euclidean space. Here we interpolate some surfaces as a whole instead of some sampled data points. The problem of finding fillet surfaces for a set of given surfaces can be considered as an example. We aim to construct a smooth surface in the whole domain, which interpolates these given surfaces defined on some sub-domains. In fact, some special cases have already been studied with different methods and view points. Given four boundary curves and corresponding derivatives along the input curves, the Coons patches are bicubically blended with the given curve data which consists of infinite data points [1, 5]. The C^0 Barnhill, Birkhoff and Gordon approach for triangular Coons patches can be explained as follows. Suppose we are given three boundary curves. We seek a surface that interpolates to all three of them, namely, a transfinite triangular interpolant [1, 6]. More Coons type interpolants are proposed over triangles, and a general interpolant called N-sided surface patches is proposed by J. Gregory [7]. However, they only consider a special case that the function values are given on some curves or line segments.

In this chapter we consider a much more general problem. Suppose that we are given some functions or patterns, $F_1, F_2 \ldots, F_n$, defined respectively on some convex sets, $A_1, A_2, \ldots, A_n$, we seek a surface that interpolates all these given functions on the interpolation sets, $A_1, A_2, \ldots, A_n$. The interpolation sets that we often encounter are points, rectangles, discs, polygons and polyhedrons. The distance from a point P to a set A, $d(p, A)$, can be used to measure how the set A and point p are related to each other. Thus, it can be used to blend the functions defined on different sets or points. We have shown that the square of distance function, $d^2(p, A)$ is C^1 smooth when the set A is convex. The distance function is then used to construct blending basis for interpolating functions or surfaces defined on related convex sets. Some interpolation formula are discussed when the interpolation sets are rectangles, polyhedrons and discs. We also study the C^k smooth interpolation problem. An explicit set-splines basis of smoothness order k is explicitly derived, which has many useful properties similar to the traditional B-spline basis. Our method has great potential, for instance, blending or smoothing surfaces, interpolating textures and images, and interpolating volume data. Experimental results show that our method works well.

2. Distance Functions over Convex Sets

Let A be a convex subset of two-dimensional Euclidean space, $A \subset R^2$, and $d(p, q)$ denote the Euclidean distance between two points, p and q. Namely, $d(p, q) = \sqrt{(x_p - x_q)^2 + (y_p - y_q)^2}$, if $p = (x_p, y_p)$, $q = (x_q, y_q)$. Then the distance from a point p to a set A, $d(p, A)$, is given by:

$$d(p, A) = \min_{q \in A} d(p, A). \tag{9.1}$$

We use the notations int (A) and $bd(A)$ to denote the interior and boundary points of the set A respectively. It is clear that $f(p) = d(p, A)$ is a continuous positive function defined on R^2. We define a function f to be a convex function on a subset S of R^2 if the epigraph of f is

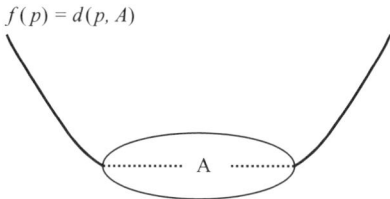

$f(p) = d(p, A)$

A

Figure 9.1 The section line of the distance function $f(p) = d(p, A)$.

convex as a subset of R^3. It is obvious that $f(p) = d(p, A)$ is also a convex function [see [8], p. 23–p. 28]. By analyzing the partial derivatives of this convex function, we can further show that $g(p) = d^2(p, A)$ is a continuously differentiable function, namely C^1 on R^2 (refer to the Appendix of [9]).

Suppose that A_i, $i = 1, 2, \ldots, n$ be convex sets, and $\lambda_i = d^2(p, A_i)$. Let F_i be given continuous functions defined on A_i, $i = 1, 2, \ldots, n$. To find a function $F(p)$ that interpolates a given set of functions F_i, we set:

$$F(p) = \sum_{i=1}^{n} w_i(p) F_i(p), \tag{9.2}$$

where

$$w_i(p) = \left(\prod_{\substack{j=1 \\ j \neq i}}^{n} \lambda_j \right) \bigg/ \left(\sum_{k=1}^{n} \prod_{\substack{j=1 \\ j \neq i}}^{n} \lambda_j \right). \tag{9.3}$$

$F(p)$ is actually a weighted summary of the given functions $F_i(p)$, $i = 1, 2, \ldots, n$. Since λ_i is C^1 and so is w_i, we conclude that:

Theorem 1. Let A_i, $i = 1, 2, \ldots, n$ be convex subsets of R^2, $A_i \bigcap_{j \neq i} A_j = \phi$ (the empty set), $i, j = 1, 2, \ldots, n$, and F_i be given C^1 continuous functions defined on A_i, $i = 1, 2, \ldots, n$, namely, the domain of F_i is A_i. Set w_i and F(p) as in Equations (9.2) and (9.3), then w_i, $i = 1, 2, \ldots, n$ form a basis of blending functions, which have the following properties:

(1) $0 < w_i(p) \leq 1$, $\sum_{i=1}^{n} w_i(p) = 1$,
(2) $w_i(p) = 1$ if and only if $p \in A_i$, $i = 1, 2, \ldots n$.
(3) $F(p)$ is C^1 continuous and interpolates all given functions, namely, $F(p) = F_i(p)$, $p \in A_i$, $i = 1, 2, \ldots, n$.
(4) If the dimension of set A_i is 2, then F coherently interpolates the first order derivatives of F_i on A_i.

The first three properties can be verified directly. For a convex subset S of R^2, the dimension of S is 2 if and only if three different points exist that are not collinear in S. Suppose the dimension of set A_i is 2 and p_0 is an arbitrary member. If p_0 is an interior point of the set, then it is clear that the partial derivatives of F coincides with that of F_i since F and F_i coincide with each other on A_i. Now, suppose $p_0 \in bd(A_i)$, then for every neighborhood of p_0, $N(p_0) = \{p | d(p, p_0) \leq r\}$, there exists infinitive number of points in $N(p_0)$ because A_i is convex and so is $N(p_0) \cap A_i$. We can thus find a sequence of points in int (A_i), $q_1, q_2, q_3, \ldots, q_k, \ldots$, which converges to p_0. Since F is C^1 continuous, $\displaystyle \lim_{k \to \infty} \frac{\partial F_i(q_k)}{\partial x_i} = \frac{\partial F_i(p_0)}{\partial x_i}$, $\displaystyle \lim_{k \to \infty} \frac{\partial F_i(q_k)}{\partial y_i} = \frac{\partial F_i(p_0)}{\partial y_i}$. The desired result follows. When the dimension of set A_i is less than 2, the conclusion in Equation (9.4) may be false (see Section 5).

From the above Theorem, we know that the interpolated function $F(p)$ has the well-known convex hull property [1, 3, 10]. It is not only useful for us to estimate and control the approximation error, but also important in modeling complex objects. The so-called Barnhill, Birkhoff and Gordon's BBG operator for infinite interpolation over triangles can be considered as a special case of the above formula when set $A_i, i = 1, 2, 3$, are three edges of the given triangle.

3. Interpolating Scheme for Some Typical Convex Sets

Let us consider the simple and useful cases when the sample sets are rectangles, discs and convex polygons. Here we use the notation of truncated polynomials for representing B-splines as follows, where m is an integer and a is a real number:

$$(x - a)_+^m = \begin{cases} (x - a)^m, & \text{if } x \geq a, \\ 0, & \text{otherwise.} \end{cases} \tag{9.4}$$

(a) If $A = [a, b; c, d]$, a rectangular set, and $p = (x, y)$. Then it is obvious that:

$$d^2(p, A) = (a - x)_+^2 + (c - y)_+^2 + (x - b)_+^2 + (y - d)_+^2. \tag{9.5}$$

Therefore, the distance function is actually a quadric B-spline and is continuously differentiable. The distance function has a simpler polynomial form over the 9 regions determined by the rectangle A (see Figure 9.2). It equals $(a - x)^2 + (y - d)^2$, $(x - b)^2 + (y - d)^2$, $(x - b)^2 + (c - y)^2$, $(a - x)^2 + (c - y)^2$, $(y - d)^2$, $(x - b)^2$, $(c - y)^2$, $(a - x)^2$ and 0 over religions 1, 2, 3, 4, 5, 6, 7, 8 and A respectively. If there are n functions F_i defined over n disjoint rectangles $A_i = [a_i, b_i; c_i, d_i], i = 1, 2, \ldots, n$. Then:

$$\lambda_i = d^2(p, A_i)$$
$$= (a_i - x)_+^2 + (c_i - y)_+^2 + (x - b_i)_+^2 + (y - d_i)_+^2. \tag{9.6}$$

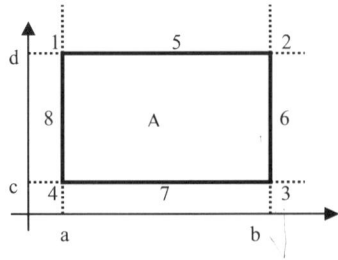

Figure 9.2 The 9 regions determined by a rectangle A.

The function $F(p)$ given by Equation (9.2) interpolates $F_i, i = 1, 2, \ldots, n$. Notice that the relative locations of these rectangles can be arbitrary without intersections.

(b) If A is a circular disc, $A = \{p = (x, y) | (x - x_0)^2 + (y - y_0)^2 \le r_0^2\}$, then:

$$\lambda(p) = \left(\sqrt{(x - x_0)^2 + (y - y_0)^2} - r_0\right)_+^2 \tag{9.7}$$
$$= (r - r_0)_+^2.$$

In the second part of the above formula, the point p is represented by polar coordinates.

(c) If A is a convex polygonal area with vertices $p_i, i = 1, 2, \ldots, m$. Suppose that:

$$L^i : a_i x + b_i y + c_i = 0,$$

is the line equation on which segment $p_i p_{i+1}$ lies.

We can choose appropriate signs of L^i such that the interior of A is the intersection of negative half planes $L^i_- : a_i x + b_i y + c_i \le 0$, namely, $A = \bigcap_{i=1}^m L^i_-$. Let d_i denote the distance from p to segment $p_i p_{i+1}$.

In Figure 9.3, draw two perpendicular rays through the end points p_i and p_{i+1}. The half plane is subdivided into three regions, I, II and III, which are related with the computation of distances. It is clear that:

$$d_i^2 = \begin{cases} (p - p_i)^2, & \text{if } (p - p_i) \cdot (p_{i+1} - p_i) \le 0, (p \in I) \\ (p - p_{i+1})^2, & \text{if } (p - p_{i+1}) \cdot (p_i - p_{i+1}) \le 0, (p \in III) \\ (a_i x + b_i y + c)_+^2, & \text{otherwise}, (p \in II) \end{cases} \tag{9.8}$$

Hence,

$$d^2(p, A) = \min_i d_i.$$

The above formula can then be used to compute the distance from p to a given convex set A.

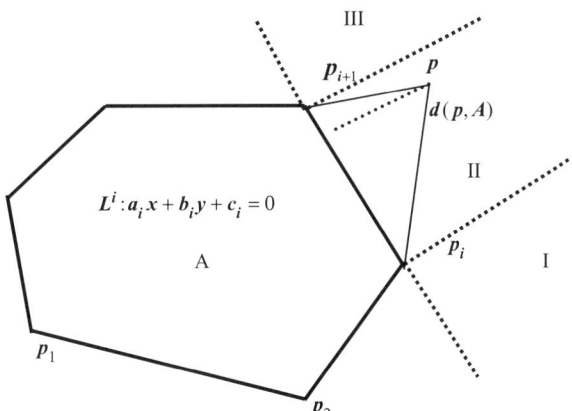

Figure 9.3 The polygonal set and the computation of distance.

A more efficient but complicated algorithm can be implemented if the convex properties of the set are carefully considered.

4. Set-splines

Although the method for set interpolation can be easily implemented, it has similar disadvantage to the Lagrangian interpolating method, for example, it has a high order when interpolating many functions over sets and computation may be unstable. We try to find a set-spline basis that is a kind of extension to the conventional B-spline functions. We will first derive the set-spline basis in a one-dimensional case and then extend the basis to a two-dimensional case using tensor product method. For a set-spline basis of smoothness order k, Ψ_i, $i = 1, 2, \ldots, n$ it must satisfy:

$$\begin{cases} \sum_{i=1}^{n} \Psi_i^k(x) \equiv 1, \\ 1 \geq \Psi_i \geq 0, \\ \Psi_i^k(x) = 1, x \in A_i = [a_i, b_i]. \end{cases} \tag{9.9}$$

Then we can assume that Ψ_i^k, $i = 1, 2, \ldots, n$, is an order $2k + 1$ spline statisfying the following equations,

$$\begin{cases} \Psi_i^k(x) = 0, & if\ x \notin (b_{i-1}, a_{i+1}), \\ \dfrac{\partial^j \Psi_i^k(x)}{\partial x^j} = 0, j = 1, \ldots, k, \\ if\ x = b_{i-1}, a_i, b_i, a_{i+1} \\ \Psi_i^k(x) = 1, & if\ x \in A_i = [a_i, b_i]. \end{cases} \tag{9.10}$$

It is clear that $\Psi_i^k(x)$ is an interpolating spline of Herimite type on interval $[b_{i-1}, a_i]$, and $[b_i, a_{i+1}]$. Therefore, we can derive:

$$\Psi_i^k(x) = \begin{cases} \int_x^{a_i+1}(t - b_i)^k(t - a_{i+1})^k\ dt / \int_{b_i}^{a_i+1}(t - b_i)^k(t - a_{i+1})^k\ dt, x \in [b_i, a_{i+1}] \\ 1, & x \in [a_i, b_i] \\ 0, & x \notin [b_{i-1}, a_{i+1}] \\ \int_{b_{i-1}}^{x}(x - b_{i-1})^k(x - a_i)^k dt / \int_{b_{i-1}}^{a_i}(x - b_{i-1})^k(x - a_i)^k dt, x \in [b_{i-1}, a_i] \end{cases} \tag{9.11}$$

We can obtain the set-spline without integration signs. Let $\delta_i = (a_{i+1} + b_i)/2$, $h_i = (a_{i+1} - b_i)/2$. Then,

$$(t - a_{i+1})^k(t - b_i)^k = ((t - \delta_i)^2 - (h_i)^2)^k$$

$$= \sum_{j=0}^{k}(-1)^j \frac{k!(t - \delta_i)^{2k-2j}(h_i)^{2j}}{j!(k - j)!}. \tag{9.12}$$

Hence, the set-splines $\psi_i^k(x)$ can be represented as follows:

$$\Psi_i^k(x) = \begin{cases} \dfrac{\sum_{j=0}^{k} \dfrac{(-1)^j (h_i)^{2j}((h_i)^{2k+!-2j} - (x - \delta_i)^{2k+1-2j})}{j!(k-j)!(2k+1-j)}}{\sum_{j=0}^{k}(-1)^j \dfrac{2}{j!(k-j)!(2k+1-j)}(h_i)^{2k+1}}, \\ \qquad x \in [b_i, a_{i+1}] \\ 1, \qquad x \in [a_i, b_i] \\ 0, \qquad x \notin [b_{i-1}, a_{i+1}] \\ \dfrac{\sum_{j=0}^{k} \dfrac{(-1)^j (h_{i-1})^{2j}((x - \delta_{i-1})^{2k+1-2j} + (h_{i-1})^{2k+!-2j})}{j!(k-j)!(2k+1-j)}}{\sum_{j=0}^{k}(-1)^j \dfrac{2}{j!(k-j)!(2k+1-j)}(h_{i-1})^{2k+1}}, \\ \qquad x \in [b_{i-1}, a_i] \end{cases} \tag{9.13}$$

One important case is the cubic spline when $k = 1$. The cubic set-spline $\Psi_i^3(x)$ can be represented as follows,

$$\Psi_i^3(x) = \begin{cases} \dfrac{2x^3 - 3(a_{i+1} + b_i)x^2 + 6a_{i+1}b_i x - (-a_{i+1} + 3b_i)a_{i+1}^2}{(a_{i+1} - b_i)^3}, \\ \qquad x \in [b_i, a_{i+1}] \\ 1, \qquad x \in [a_i, b_i] \\ 0, \qquad x \notin [b_{i-1}, a_{i+1}] \\ \dfrac{-2x^3 + 3(a_i + b_{i-1})x^2 - 6a_i b_{i-1}x + (-b_{i-1} + 3a_i)b_{i-1}^2}{(a_i - b_{i-1})^3}, \\ \qquad x \in [b_{i-1}, a_i] \end{cases} \tag{9.14}$$

Notice that:

$$\Psi_i^k(x) + \Psi_{i+1}^k(x) \equiv 1. \tag{9.15}$$

It is clear that $\Psi_i^k(x)$ is a piecewise polynomial with order at most $2k + 1$, and is a very special type of B-spline. We can verify that the above set-splines satisfy the desired constraints. The shape of $\Psi_i^k(x)$ is similar to a hat. For the two-dimensional sets $[a_i, b_i] \times [c_j, d_j], i = 1, 2, \ldots, m; j = 1, 2, \ldots, n$ we can construct a tensor-product spline basis, $\Psi_i^{k_1}(x)\Psi_j^{k_2}(y)$ of order $k_i \times k_2$. For the given mn functions F_{ij} defined on $[a_i, b_i] \times [c_j, d_j], i = 1, 2, \ldots, m; j = 1, 2, \ldots, n$, the interpolating function F can be represented as follows,

$$F = \sum_{i=1}^{m} \sum_{j=1}^{n} F_{ij} \Psi_i^{k_1}(x)\Psi_j^{k_2}(y) \tag{9.16}$$

It is clear that F is a smooth surfaces interpolating given function F_{ij} on $[a_i, b_i] \times [c_j, d_j], i = 1, 2, \ldots, m; j = 1, 2, \ldots, n$.

5. Set Interpolation for Unusual Cases

For the set-splines, the domain sets of the given functions are supposed to be convex, and there is no intersection for any pair of these sets. Some unusual cases may occur, for instance; (1) some convex sets intersect with each other; (2) the domain sets degenerate to points; (3) some sets have lower dimensions than usual. Let int(A), bd(A) denote the set of interior points and set of boundary points of set A respectively. We discuss the above unusual cases accordingly.

(1) Suppose some domain sets intersect with each other. We can still use our method to construct a C^1 continuous surface F interpolating the given C^1 functions $F_i, i = 1, 2, \ldots, n$, if these given functions meet with C^1 along their intersections of these domain sets. Notice that the denominator in Equation (9.3) may vanish if the point p lies in the intersection of some sets. However, we can modify the interpolation formula (9.2) by defining the value of F coincides with the interpolated functions at the domain sets, $A_i, i = 1, 2, \ldots, A_n$. Namely,

$$F(p) = \begin{cases} F_i, & \text{if } p \in A_i \\ \sum_{i=1}^{n} w_i(p) F_i(p), & \text{if } p \notin \bigcup_{i=1}^{n} A_i, \end{cases} \tag{9.17}$$

where

$$w_i(p) = \left(\prod_{\substack{j=1 \\ j \neq i}}^{n} \lambda_j \right) \Big/ \left(\sum_{k=1}^{n} \prod_{\substack{j=1 \\ j \neq k}}^{n} \lambda_j \right). \tag{9.18}$$

Since $F_i, i = 1, 2, \ldots, n$ meet with C^1 along respective common boundary (intersection of domain set) and $w_i(p)$ is C^1 too, the above formula is well defined and the interpolation function F is also smooth.

For example, let $A_1 = [-1, 1; -1, 1]$, $A_2 = [1, 3; -1, 1]$, and:

$$\begin{cases} F_1(p) = (x - 1)^2 + y^2/3.0, & (x, y) \in A_1 \\ F_2(p) = (x - 1)^3/30.0 + y^2/3.0, & (x, y) \in A_2 \end{cases},$$

Then it is clear that these two given functions meet C^1 along common boundary curve $F_1(p) = F_2(p) = y^2, x = 1, 0 \leq y \leq 1$. Now,

$$\lambda_1 = (-1 - x)_+^2 + (x - 1)_+^2 + (-1 - y)_+^2 + (y - 1)_+^2,$$
$$\lambda_2 = (1 - x)_+^2 + (x - 3)_+^2 + (-1 - y)_+^2 + (y - 1)_+^2.$$

It can be directly verified that the interpolation function F derived from Equation (9.17) is C^1 (see Figure 9.6). However, if the given two functions do not meet with C^1 along the common boundary themselves, the interpolated result is not C^1 either.

(2) If the domain sets degenerate to points, then the domain sets have only one boundary point and no interior point. Let $A_i = (x_i, y_i), i = 1, 2, \ldots, n$, then $\lambda_i = (x - x_i)^2 + (y - y_i)^2$. The blending functions $w_i, i = 1, 2, \ldots, n$ turn out to be rational polynomials.

(3) If the domain sets are degenerate, then the interpolated function F may not interpolate the derivatives of the given function. One simple case is that if every set has only one point, then the function F only interpolates the function value of every function at the corresponding point. However, the derivative of F_i at $p_i = A_i$ is determined not only by the function value at the point p_i, but also its function values at the neighborhood set of P_i.

Let A_1, A_2, A_3 be the three edges of a triangle, and $\lambda_i = d(p, A_i), i = 1, 2, 3$. Then, the interpolation function F given by Equations (9.2) and (9.3) is a C^0 transfinite triangular interpolant, which is similar to the so-called BBG operator (see [1]).

6. Summary

We implemented the proposed algorithm on a Pentium II 233 computer and it ran very well. One important character of our method is that our algorithm can deal with different types of interpolating sets and functions. In Figure 9.4, the function F interpolates three different types of functions defined on a triangle, circle and rectangular sets respectively. Here, A_1 is a triangle with vertices, [2, 2], [2, 4] and [4, 2], and $F_1 = \sin(\pi x)\sin(\pi y)$; A_2 is a circular set defined by $A_1 = \{(x, y)|(x + 1)^2 + (y - 1)^2 \le 1\}$, $F_2 = x^2/2 - y^2/2$; A_3 is a rectangle, $A_3 = [-5, -3; -5, -3]$, and $F_3 = 0.5\sin(\pi x)\sin(\pi y)$. In Figure 9.5, tensor product set-spline interpolation basis is used. The sets are defined by $[a_i, b_i] \times [c_i, d_i], i = 1, 2$ where $[a_1, b_1] = [-5, -2], [a_2, b_2] = [2, 5], [c_1, d_1] = [-5, -2]$ and $[c_2, d_2] = [1, 5]$, $F_{11} = x^2/2 - y^2/4$, where, $F_{11} = x^2/2 - y^2/4$, $F_{12} = \sin(\pi x)\sin(\pi y)$, $F_{21} = \sin(0.75\pi x)\cos(\pi y)$, $F_{22} = x^2/3 - y^2/3$.

In Figure 9.6, unusual cases are considered when the given sets have a common boundary where $F_1 = (x - 1)^2 + y^2/3.0$ and $F_2 = (x - 1)^3/300 + y^2/3.0$ are defined on $[-1, 1; -1, 1]$ and $[1, 3; -1, 1]$ respectively. The implementation of the proposed algorithm is simple. We

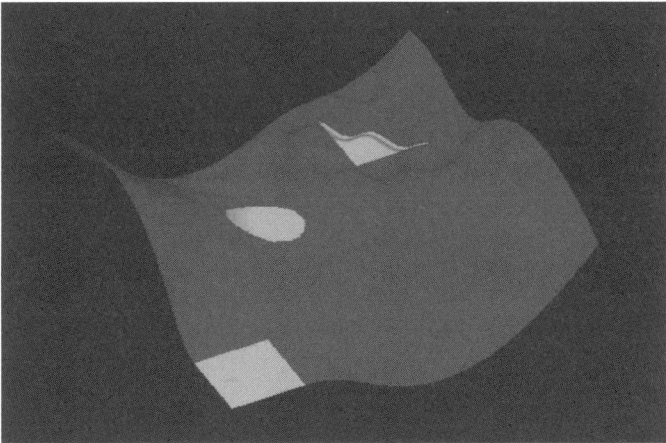

Figure 9.4   Different types of functions are interpolated. The three interpolating sets are respectively, triangle, circle and rectangle.

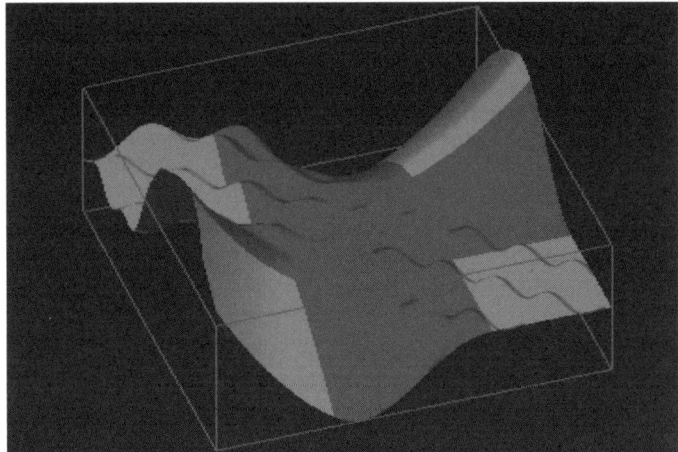

Figure 9.5   Interpolation result of four given functions. Red surface patches show the given functions
$F_{ij}, i, j = 1, 2$, and the blue one is the transition surface.

design a basic function to calculate the distance from a point to a convex polygon, which is
classified into four cases, triangle, rectangle, disc and general ones. The experimental result
shows that different types of functions, for instance, polynomials and trigonometric functions,
can be easily blended using our method. When scattered functions or functions over regular sets
are given, we will use the general interpolation formula (9.2) or setspline interpolation formula
(9.17) respectively. Note that set-spline basis has the local property that the function defined
on set A_{ij} only affects the interpolated function over the 9 neighboring sets $A_{i+k,j+l}, k, l =$
$-1, 0, 1$. However, general interpolation formula (9.2) is a global method, and the greater the
distance $d(p, A_i)$, the lower the function F_i effect on the interpolated function F. The proposed

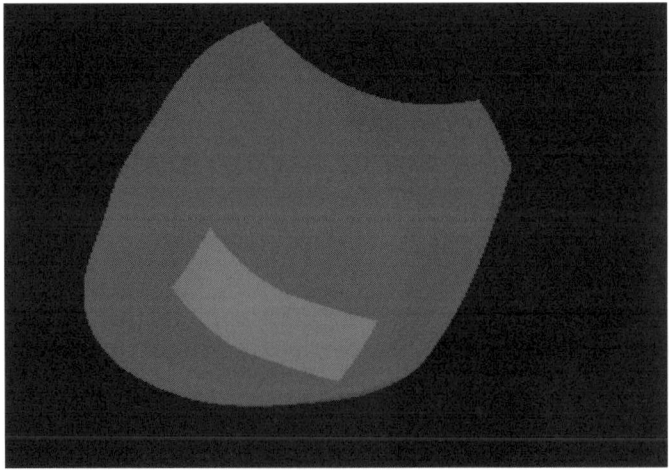

Figure 9.6   The interpolation result is smooth when the given two functions meet with C^1 along the
common boundary.

method can be applied in many applications, for example, blending or smoothing surfaces, interpolating textures and images, and interpolating volume data. We are now trying to extend our method for interpolating volume data and find a more efficient algorithm for computing the distance function from a point to a convex polyhedron or volume.

Acknowledgement

This work was supported by NSF China No. 60173035 and 69973043.

References

[1] Farin, G. (1993), *Curves and Surfaces for Computer Aided Geometric Design – A Practical Guide*. Academic Press, INC, 2nd edition.
[2] de Boor, C. (1978), *A Practical Guide to Splines*. Springer-Verlag.
[3] Hoschek, J. and Lasser, D. (1993), *Computer Aided Geometric Design*. English translation by L. L. Schumaker. A. K. Peters, Wellesley, Massachusetts.
[4] Piegl, L. and Tiller, W. (1995), *The NURBS Book*. Springer-Verlag Berlin.
[5] Coons, S. (1964), Surfaces for computer aided design. Technical Report, MIT. Available as AD 663 504 from the National Technical Information service, Springfield, VA 22161.
[6] Barinhill, R., Birkhoff, G. and Gordon, W. (1973), Smooth interpolation in triangles. *J. Approx. Theory*, 8(2), 114–128.
[7] Gregory, J. (1986), N-sided surface patches, In J. Gregory, editor, *The* Mathematics of Surfaces, 217–232, Clarendon Press.
[8] Rockafellar, R. T. (1972), *Convex Analysis*. Princeton University Press.
[9] Lizhuang, M., Youdong, L. and Qunsheng, P. (1992), Equidistant smoothing of polyhedron with arbitrary topologies *Computer Graphics Forum*, 11 (3), 405–414.
[10] Lizhuang, M. and Rongliang, W. (1996), *Techniques for Computer Aided Geometric Modeling and Its Applications*. Science Press, Beijing (in Chinese).
[11] Sarfraz, M. (2003), Optimal curve fitting to digital data, *International Journal of WSCG*, Vol 11(1).
[12] Sarfraz, M. (2003), Curve fitting for large data using rational cubic splines, *International Journal of Computers and Their Applications*, Vol 10(3).
[13] Sarfraz, M. and Razzak, M. F. A. (2003), A web based system to capture outlines of arabic fonts, *International Journal of Information Sciences*, Elsevier Science Inc., Vol. 150(3-4), 177–193.
[14] Sarfraz, M. (2002), Visualization of positive and convex data by a rational cubic spline, *International Journal of Information Sciences*, Elsevier Science Inc., Vol. 146(1-4), 239–254.
[15] Sarfraz, M. and Razzak, M. F. A. (2002), An algorithm for automatic capturing of font outlines, *International Journal of Computers & Graphics*, Elsevier Science, Vol. 26(5), 795–804.
[16] Sarfraz, M. (2002), Fitting curves to planar digital data. *Proceedings of IEEE International Conference on Information Visualization IV'02-UK*: IEEE Computer Society Press, USA, 633–638.
[17] Sarfraz, M. and Raza, A. (2002), Visualization of Data using Genetic Algorithm, *Soft Computing and Industry: Recent Applications*, Eds.: R. Roy, M. Koppen, S. Ovaska, T. Furuhashi, and F. Hoffmann, ISBN: 1-85233-539-4, Springer, 535–544.
[18] Sarfraz, M. and Raza, A. (2002), Towards automatic recognition of fonts using genetic approach, *Recent Advances in Computers, Computing, and Communications*, Eds.: N. Mastorakis and V. Mladenov, ISBN: 960-8052-629, WSEAS Press, 290–295.
[19] Sarfraz, M. (1995), Curves and surfaces for CAD using C^2 rational cubic splines, *Engineering with Computers*, 11(2), 94–102.

10

Family of G^2 Spiral Transition Between Two Circles

Zulfiqar Habib
Manabu Sakai

Department of Mathematics and Computer Science, Graduate School of Science and Engineering, Kagoshima University, Japan.

A family of fair G^2 cubic and Pythagorean Hodograph (PH) quintic transition curves connecting two circles has been obtained. We have discussed both C- and S-shaped curves. It is shown that transition curve is a pair of two spirals. A S-shaped curve has no curvature extremum and a C-shaped curve has a single curvature extremum. We simplified and completed the analysis of Meek and Walton. Our scheme can generate a family of transition curves with less restrictive constraints which are more flexible and hence most reasonable for practical use. The results for the cubic curve are extended to PH quintic transition curves. A quintic is the lowest degree PH curve that may have an inflection point with attractive properties that's arc-length is a polynomial of its parameter, and its offset is rational. S- and C-shaped transition curves are suitable in CAD applications like highway design, rounding corners, or for smooth transition between two curves, e.g. two circular arcs.

1. Introduction

Parametric cubic curve segments are widely used in Computer Aided Geometric Design (CAGD) and Computer Aided Design (CAD) applications because their flexibility makes them suitable for use in the interactive design of curves and surfaces. Also they are the lowest degree polynomial curves that allow inflection points (where curvature is zero), so they are suitable for the composition of G^2 blending curves. The Bezier form of a parametric cubic curve is usually used in CAD and CAGD applications because of its geometric and numerical properties. Many authors have advocated their use in different applications like data fitting and font designing.

Advances in Geometric Modeling. Edited by M. Sarfraz
© 2003 John Wiley & Sons, Ltd   ISBN: 0-470-85937-7

The importance of fair curves in the design process is well documented in the literature [2, 5, 8, 9, 11, 12–18]. Consumer products such as pingpong paddles can be designed by blending circles. To be visually pleasing it is also desirable that the blend be fair. For applications such as the design of highways or railways, it is also desirable that transitions be fair. In the discussion about geometric design standards in AASHO (American Association of State Highway officials), Hickerson [7, p. 17] states that "Sudden changes between curves of widely different radii or between long tangents and sharp curves should be avoided by the use of curves of *gradually increasing or decreasing radii* without at the same time introducing an appearance of forced alignment". The importance of this design feature is highlighted in [3] which links vehicle accidents to inconsistency in highway geometric design.

Cubic curves, although smoother, are not always helpful since they may have unwanted inflection points and singularities [4, 9, 10]. A cubic segment has the following undesirable features:

- Its arc-length is the integral part of the square root of a polynomial of its parameter.
- Its offset is neither a polynomial, nor a rational algebraic function of its parameter.
- It may have more curavture extrema than necessary.

PH curves do not suffer from the first two of the aforementioned undesirable features. A quintic is the lowest degree PH curve that may have an inflection point, as required for an *S*-shaped transition curve. For application such as highway design, it is desirable that transitions G^2 are with a small number of curvature extrema. Spirals have several advantages of containing neither inflection points and singularities nor curvature extrema [19]. Such curves are also very useful for transition between two circles. The clothoid or Cornu spiral has been used in highway design for many years [1, 6]. A major drawback in using this spiral is the fact that the highway spiral currently used is neither polynomial nor rational. It is thus not easily incorporated in CAD/CAM/CAGD packages that are manly based on NURBS (Non Uniform Rational B Splines). Walton and Meek [18, 20] considered planar G^2 cubic and PH Quintic transition between two circles with a fair Bezier curve. They showed there is no curvature extremum in the case of an *S*-shaped transition, and there is a curvature extremum in the case of *C*-transition. The constraints used by Walton and Meek are more restrictive than necessary. We not only simplified and completed the analysis of Meek and Walton but also offered constraints which are less restrictive, more reasonable and comfortable for practical applications. Also we found constraints for PH quintic transition are more flexible than constraints for cubic transition.

The objectives of this chapter are:

- To simplify and complete the analysis of Walton and Meek [18, 20].
- To obtain a family of fair G^2 cubic and PH quintic transition curves between two non-enclosing circles.
- To achieve more flexible and less restrictive constraints.
- To discuss and prove all the shape features of transition curve.
- To find the locus of the center of smaller circle.
- To compare PH quintic transition curve with cubic transition.

Our compact scheme also guarantees the absence of interior curvature extremum for an *S*-shaped transition curve and one curvature extremum for an *C*-shaped transition curve. The problem of finding a fair parametric transition curve between two circles Ω_0, Ω_1 with centers C_0, C_1 and radii r_0, r_1 respectively may be solved in a Hermite-like manner, where $r = \| C_1 - C_0 \|$. We consider the following problems:

- For $r_0 + r_1 < r$ (the circles Ω_i, $i = 0, 1$ do not intersect), find an S-shaped transition curve from Ω_0 to Ω_1.
- For $r_0 - r_1 < r$ (the smaller circles Ω_1 is not enclosed in the larger circle Ω_0), find a C-shaped transition curve from Ω_0 to Ω_1.

In this chapter, '$\times$' stands for the two-dimensional cross product, $(x_0, y_0) \times (x_1, y_1) = x_0 y_1 - x_1 y_0$ and $\| \; \|$ means the Euclidean norm. Then, for the planar curve $z(t)$ ($0 \le t \le 1$), its signed curvature $k(t)$ and $k'(t)$ are given by:

$$k(t) = (z'(t) \times z''(t))/\|z'(t)\|^3 (= \phi(t)/\|z'(t)\|^3), \tag{10.1}$$

$$w(t)(= k'(t)\|z'(t)\|^5) = -3\phi(t)\{z'(t) \cdot z''(t)\} + \phi'(t)\|z'(t)\|^2. \tag{10.2}$$

A spiral is a curve whose curvature does not change sign and whose curvature is monotone. G^2 (Geometric continuity of second order) means continuity in position, in unit tangent, and in signed curvature. A curve is said to match G^2 Hermite data if it passes from one given point to another given point, if its unit tangent matches given unit tangents at the two given points, and its signed curvature matches given signed curvatures at the two given points. The organization of this chapter is as follows. In each case, this chapter treats the curve whose initial curvature is positive. Without loss of generality, a shift and rotation enables us to assume that $p_0 = (0, 0)$ is the first point of transition curve lying on larger circle with center $C_0 = (0, r_0)$; refer to Figures 10.1–10.4. Two cases of an S-shaped and a C-shaped transition curves for both cubic and PH quintic transition are now considered in the following sections. Sections 2 and 3 give a description of a method for cubic and PH quintic transition curves respectively. Critical analysis and illustrative examples are then presented followed by the summary of this chapter.

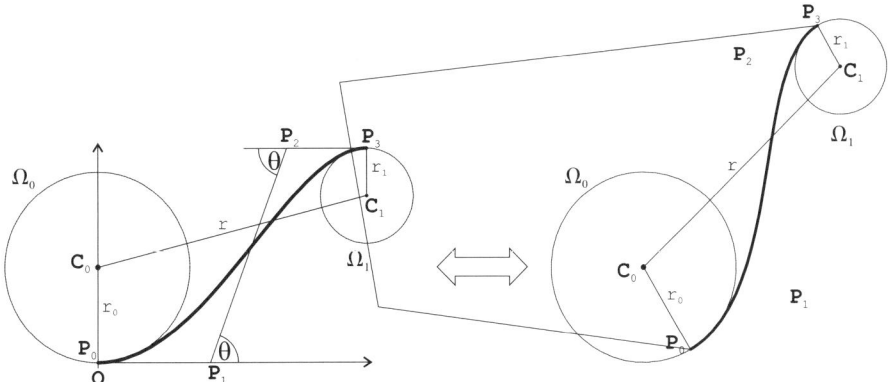

Figure 10.1 An S-shaped cubic Bezier transition curve.

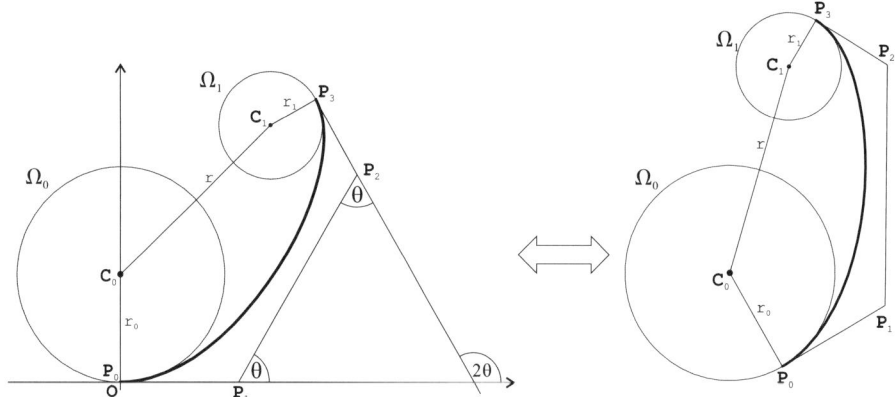

Figure 10.2   A C-shaped cubic Bezier transition curve.

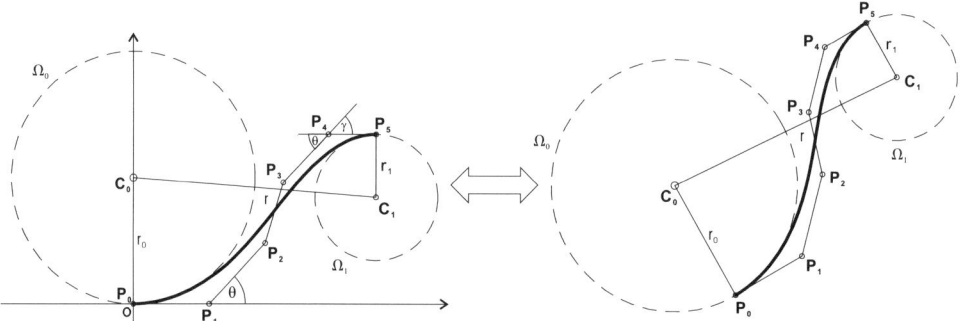

Figure 10.3   An S-shaped PH quintic Bezier transition curve.

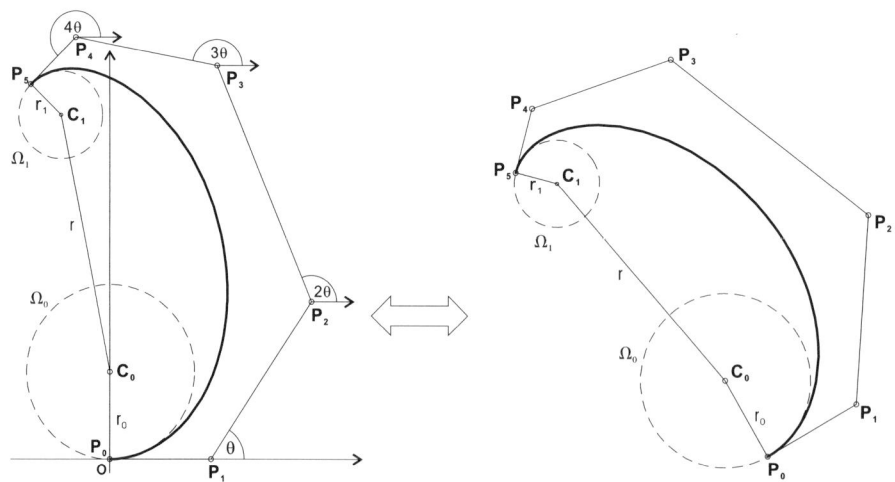

Figure 10.4   A C-shaped PH quintic Bezier transition curve.

2. G^2 **Cubic Transition**

With $r_1 = \lambda^2 r_0, 0 < \lambda \leq 1$, we use the cubic curve $z(t)$ $(= (x(t), y(t)))$, $0 \leq t \leq 1$ of the form:

$$z(t) = \sum_{i=0}^{3} \binom{3}{i} p_i (1-t)^{3-i} t^i, \tag{10.3}$$

with its derivative:

$$z'(t) = z_0(1-t)^2 + 2z_1 t(1-t) + z_2 t^2, \tag{10.4}$$

where:

$$z_i = 3(p_{i+1} - p_i), i = 0, 1, 2. \tag{10.5}$$

Letting $T_i = p_{i+1} - p_i$, define θ to be the angle from T_0 to T_1 and γ to be the angle from T_1 to T_2. As in [18], $\gamma = -\theta$ and $\gamma = \theta$ are to be taken for the S- and C-shaped curves, respectively. The two cases of the transition curves are now considered separately.

2.1. S-shaped Cubic Transition Curve

Here we consider an S-shaped transition curve $z(t)$ of the Equation 10.3. G^2 transition requires:

Lemma 1: With $u_0 = 2mr_0 \tan \theta$ $(0 < \theta < \pi/2)$,

$$z_0 = u_0(1, 0), z_1 = mu_0(1, \tan \theta), z_2 = \lambda u_0(1, 0). \tag{10.6}$$

Then we obtain the following theorem to guarantee a single cubic transition curve with monotone decreasing curvature.

Theorem 1 $(r > r_0 + r_1)$: *Assume that* $1/6 \leq \lambda \leq 1$. *Each value of* m $(\geq 2/3)$ *determines a* G^2 *cubic S-shaped transition curve of the form* (10.3) *with* (10.6) *between the two circles with no curvature extremum. The curve with monotone decreasing curvature of positive and negative signs is free of loops and cusps and has a single inflection point, i.e., it is a pair of two spirals.*

Proof: Note that the center C_1 of the smaller circle Ω_1 is given by:

$$C_1 = p_3 - r_1(1, 0) = \left(u_0(1 + m + \lambda)/3, u_0^2/(6r_0) - r_0\lambda^2\right). \tag{10.7}$$

Since $\|C_1 - C_0\| = r$ gives the determining equation $f(\rho) = 0$ in $\rho(= u_0^2/(4r_0^2))$ as:

$$\begin{aligned} f(\rho) = \rho^2 + \{m^2 + 2(1+\lambda)m - 2(1 - \lambda + \lambda^2)\}\rho \\ -9\left\{r^2 - (1+\lambda^2)^2 r_0^2\right\}/(4r_0^2). \end{aligned} \tag{10.8}$$

The above quadratic equation has a unique positive root since its constant term is negative.

Now, we note with $p, q > 0$, $p \neq q$,

$$\{y(1/(1+p)) - y(1/(1+q))\}/(q-p)$$
$$= 6r_0 u_0^2 (1+p)^3 (1+q)^3 \{p^2 + q^2 + pq(3p+3q+10) + 3(p+q)\}. \tag{10.9}$$

Since the right hand side of above equation is positive even when $p = q$, the curve is free of loops and cusps. To show that the transition curve is a pair of spirals for $m \geq 2/3$, note the relations (1) and (2). A symbolic manipulator *Mathematica* gives with $t = 1/(1+s)$, $0 \leq s < \infty$ (from now on, we use this substitution).

$$r_0(1+\rho)^2 \phi(t) = u_0^3(s^2 - \lambda),$$
$$r_0^3(1+s)^5 w(t) = -u_0^5 \{su_0^5 \mu(s) + 2r_0^2 \eta(s)(s^2 + 2ms + \lambda)\}. \tag{10.10}$$

where for $1/6 \leq \lambda \leq 1$ and $m \geq 2/3$,

$$\mu(s) = 3s^3 - s^2 - \lambda s + 3\lambda \geq 3s^3 - s^2 - s + 1/2 > 0, \tag{10.11}$$
$$\eta(s)(= m\mu(s) - 2(s^3 - 2\lambda s^2 - 2\lambda s + \lambda^2)), \tag{10.12}$$
$$\geq (2/3)\{(6\lambda - 1)s^2 + 5\lambda s + 3\lambda(1-\lambda)\} > 0.$$

Hence, $k'(t) < 0$ for $1/6 \leq \lambda \leq 1$, i.e., the transition is a curve with monotone decreasing curvature and has a single inflection point. This completes the proof of Theorem 1.

Remark 1: Note (Equation 5) and refer to Figure 10.1 (the definition of θ) to get:

$$h(= \|\boldsymbol{p}_1 - \boldsymbol{p}_0\|) = u_0/3 = (2mr_0 \tan\theta)/3. \tag{10.13}$$

Hence, a value of $h = (4r_0 \tan\theta)/9$ by Walton and Meek [18] is equivalent to $m = 2/3$. The ratio of the larger to the smaller radii of the given circular arcs is constrained to be less than 36 while the ratio must be less than 9 in [18]. In addition, we note that the coefficient: $2r_0^2(3m-2)$ of s^2 (the highest term in the brackets of (Equation 1.10)) must be nonnegative for the spiral transition curve. Therefore $m \geq 2/3$ is necessary and $m = 2/3$ means $k'(0) = 0$ since the denominator is quintic in s.

2.2. C-shaped Cubic Transition Curve

Here we consider a C-shaped transition curve $z(t)$ of the form (Equation (10.3)). Then G^2 transition requires:

Lemma 2 With $u_0 = 2mr_0 \tan\theta$ $(0 < \theta < \pi/2)$,

$$z_0 = u_0(1, 0), z_1 = mu_0(1, \tan\theta), z_2 = \lambda u_0(\cos 2\theta, \sin 2\theta). \tag{10.14}$$

Then we obtain the following theorem.

Theorem 2 $(r > r_0 - r_1)$: *Each value of $m(\geq (1 + \sqrt{1+3\lambda})/3(= m(\lambda)))$ determines a G^2 cubic C-shaped transition curve of the form (10.3) with (10.14), between the two circles is a cubic curve which is free of inflections, loops and cusps and has a single interior curvature* *extremum. It is a pair of two spirals with monotone decreasing curvature and monotone increasing curvature respectively.*

Proof: Since $C_1 = p_3 - r_1 (\sin 2\theta, -\cos 2\theta) (= (c, d))$, a symbolic manipulator gives:

$$c = -r_0\lambda^2 \sin 2\theta + (2mr_0/3)(m + 1 + \lambda \cos 2\theta) \tan \theta,$$
$$d = r_0\lambda^2 \cos 2\theta + (2mr_0/3)(m + \lambda + \lambda \cos 2\theta) \tan^2 \theta.$$
(10.15)

Conditions $\|C_1 - C_0\| = r$ gives the determining equation $f(\rho) = 0$ in $\rho (= \tan^2 \theta)$ as:

$$
\begin{aligned}
f(\rho) = {} & 4m^4 r_0^2 \rho^3 + 8m^2 r_0^2\{(m + \lambda)(m - \lambda) + (1 + \lambda)(m - 1)\}\rho^2 \\
& + \left[r_0^2\{4m^4 + 8(1 + \lambda)m^3 - 8(1 - \lambda + \lambda^2)m^2 - 24\lambda(1 + \lambda)m \right. \\
& \left. + 9(1 + \lambda^2)^2\} - 9r^2 \right]\rho - 9\{r^2 - r_0^2(1 - \lambda^2)^2\}.
\end{aligned}
$$
(10.16)

Note that the constant term of $f(\rho)$ is negative since the smaller circle Ω_1 is not contained in the larger one Ω_0. For $m \geq m(\lambda)$, Descartes' rule of signs shows the *unique* of the positive zero of the right hand side of $f(\rho)$ since the signs of the coefficients of ρ^i, $i = 3, 2, 1, 0$ are $(+, + \text{ or } 0, ?, -)$. The unique positive ρ determines $\theta \in (0, \pi/2)$.

We examine the shape of the transition for $m \geq m(\lambda) (> 2/3)$. First, consider the following system of equations: $pz'(0) + qz'(1) = \Delta z (= z(1) - z(0))$. Then:

$$p = \frac{m + 2\cos^2 \theta}{6\cos^2 \theta}(> 1/3), \quad q = \frac{m + 2\lambda \cos^2 \theta}{6\lambda \cos^2 \theta}(> 1/3),$$
(10.17)

from which the cubic curve is free of inflection points, loops and cusps [11]. Next, note Equation (10.2) to obtain:

$$(1 + \rho)w(0) = -64r_0^4 m^4\{m(3m - 2)\rho + 3m^2 - 2m - \lambda\} \tan^5 \theta(< 0),$$
$$(1 + \rho)w(1) = 64r_0^4 m^4 \lambda^2\{m(3m - 2\lambda)\rho + 3m^2 - 2m\lambda - \lambda\} \tan^5 \theta(> 0).$$
(10.18)

In addition,

$$w'(t) = -\phi(t)\{3\|z''(t)\|^2 + 4z'(t) \cdot z^{(3)}(t)\}(= -\phi(t)\psi(t)).$$
(10.19)

Use $\rho = \tan^2 \theta$ to obtain:

$$(1 + \rho)(1 + s)^2\phi(t) = 8r_0^2 m^2\{2s\lambda + m(s^2 + \lambda)(1 + \rho)\} \tan^3 \theta > 0,$$
(10.20)
$$(1 + \rho)\psi(t) = 16r_0^2 m^2 \rho(at^2 + bt + c),$$
(10.21)

where

$$
\begin{aligned}
a = {} & 5\lfloor\{2(1 + \rho)m - (1 + \lambda)\}^2 + (1 - \lambda)^2\rho\rfloor \geq 0, \\
b = {} & -10\{2(1 + \rho)^2 m^2 - m(3 + \lambda)(1 + \rho) + 1 + \rho + \lambda(1 - \rho)\}, \\
c = {} & 3(1 + \rho)^2 m^2 - 10(1 + \rho)m + 5(1 + \rho) + 2\lambda(1 - \rho).
\end{aligned}
$$
(10.22)

Since $w(0) < 0$ and $w(1) > 0$, $w(t)$ has a zero in $I(= (0, 1))$ if $\psi(t)$ has none or one zero there. If $\psi(t)$ has two zeroes in I, $a > 0$ means $v'(0) < 0$. Then, w also has a zero in I. This completes the proof of Theorem 2.

Remark 2: Walton and Meek proposed a value of $h = (2r_0 \tan \theta)/3$ [18] which is equivalent to $m = 1$ since $h = u_0/3 = (2mr_0 \tan \theta)/3$.

3. G^2 PH Quintic Transition

With $r_1 = \lambda^3 r_0$, $0 < \lambda \leq 1$, we consider the PH quintic curve $z(t)$ $(= (x(t), y(t)))$, $0 \leq t \leq 1$ of the form:

$$z(t) = \sum_{i=0}^{5} \binom{5}{i} p_i (1-t)^{5-i} t^i, \qquad (10.23)$$

derivative of which is defined as:

$$z'(t) = (u(t)^2 - v(t)^2, 2u(t)v(t)), \qquad (10.24)$$

where

$$u(t) = u_0(1-t)^2 + 2u_1 t(1-t) + u_2 t^2,$$
$$v(t) = v_0(1-t)^2 + 2v_1 t(1-t) + v_2 t^2. \qquad (10.25)$$

For Bezier points p_i, $0 \leq i \leq 5$, the readers are referred to [20, p. 111]. Then, $k'(t)$ is given by:

$$\{u^2(t) + v^2(t)\}^3 k'(t) = 2\{u(t)v''(t) - u''(t)v(t)\}\{u^2(t) + v^2(t)\}$$
$$- 8\{u(t)v'(t) - u'(t)v(t)\}\{u(t)u'(t) + v(t)v'(t)\}(= 2w(t)). \qquad (10.26)$$

Letting $T_i = p_{i+1} - p_i$, define θ to be the angle from T_0 to T_1 and γ to be the angle from T_3 to T_4. As in [20], $T_0 \| T_4$ (note $\gamma = -\theta$; refer to Lemma 3) and $\gamma = \theta$ are to be taken for the S- and C-curves, respectively. The two cases of the transition curves are now considered separately.

3.1. S-shaped PH Quintic Transition Curve

Here we consider an S-shaped transition curve $z(t)$ of the form (10.23). Then G^2 transition requires.

Lemma 3 For $z_i = (u_i, v_i)$, $0 \leq i \leq 2$, the coefficients of $u(t)$ and $v(t)$ by (Equation (10.25))

$$z_0 = u_0(1, 0), \quad z_1 = u_0(m, u_0^2/(4r_0)), \quad z_2 = u_0\lambda(1, 0), \qquad (10.27)$$

where $v_0 = v_2 = 0$ gives $\theta = -\gamma$.

Then we obtain the following theorem to guarantee a unique PH quintic transition curve with monotone decreasing curvature.

Theorem 3 $(r > r_0 + r_1)$: *Assume that $3/10 \leq \lambda \leq 1$. Each value of m $(\geq 3/4)$ determines a G^2 quintic S-shaped transition curve of the form (10.23) with (10.29) between the two circles with no interior curvature extremum. It is free of loops and cusps and has a single inflection* point, i.e., it is a pair of two spirals with monotone decreasing curvature which changes its sign from positive to negative.

Proof: First, note $z(1) = (p, q)$,

$$p = u_0^2 \left\{ -\frac{u_0^4}{r_0^2} + 8(2m^2 + 3m(1 + \lambda) + 3 + \lambda + 3\lambda^2) \right\} /120,$$
$$q = u_0^4(4m + 3 + 3\lambda)/(60r_0). \tag{10.28}$$

With $\rho = u_0^4/(25r_0^2)$, the center $C_1 (= (c, d)) (= z(1) - r_1 (0, 1))$ is given by:

$$c = r_0\sqrt{\rho}\{16m^2 + 24(\lambda + 1)m + 24\lambda^2 + 8\lambda + 24 - 25\rho\}/24,$$
$$d = r_0(20\rho m + 15\rho\lambda + 15\rho - 12\lambda^3)/12. \tag{10.29}$$

$\|C_0 - C_1\| = r$ gives the cubic determining equation $f(\rho) = 0$ where:

$$f(\rho) = \sum_{i=0}^{3} a_i \rho^i, \tag{10.30}$$

with

$$a_3 = 625r_0^2, \quad a_2 = 100r_0^2\{8m^2 + 12(1 + \lambda)m - 3\lambda^2 + 14\lambda - 3\},$$
$$a_1 = 32r_0^2\{8m^4 + 24(1 + \lambda)m^3 + (42 + 44\lambda + 42\lambda^2)m^2$$
$$-24(1 - 2\lambda - 2\lambda^2 + \lambda^3)m - 27\lambda^4 - 33\lambda^3 + 38\lambda^2 - 33\lambda - 27\}, \tag{10.31}$$
$$a_0 = -576\{r^2 - r_0^2(1 + \lambda^3)^2\}.$$

Since the signs of coefficients (a_3, a_2, a_1, a_0) are $(+, +, ?, -)$, combine Descartes' rule of sign and intermediate value theorem to ensure that the above cubic equation has a unique positive root.

Now we examine the shape of the transition curve. First, the second component $y(t)$ of $z(t)(= (x(t), y(t))$ satisfies:

$$r_0(1 + s)^4 y'(t) = su_0^4(s^2 + 2ms + \lambda) (> 0), \tag{10.32}$$

which implies that the curve is free of loops and cusps since 'loop' means $z(\alpha) = z(\beta)$, $0 \le \alpha < \beta \le 1$, i.e., $y'(t)$ has at least one zero and 'cusp' means $z'(t) = \mathbf{O}$ for some $t \in (0, 1)$, i.e., $y'(t)$ has at least one zero. Now we show that the transition curve has monotone decreasing curvature. A symbolic manipulator *Mathematica* gives:

$$4(1 + s)^5 w(t) = -125r_0^2\rho^{3/2}[4\eta(s)(s^2 + 2ms + \lambda) + 25\rho s\mu(s)], \tag{10.33}$$

where for $3/10 \le \lambda \le 1$ and $m \ge 3/4$:

$$\mu(s)(= 2s^3 - s^2 - \lambda s + 2\lambda) \ge 2s^3 - s^2 - s + \frac{3}{5} > 0, \tag{10.34}$$

$$\eta(s)(= 2m\mu(s) - (3s^3 - 5\lambda s^2 - 5\lambda s + 3\lambda^2)),$$
$$= (10\lambda - 3)s^2 + 7\lambda s + s + 6\lambda(1 - \lambda) > 0 \tag{10.35}$$

Hence, $k'(t) < 0$ for $3/10 \le \lambda \le 1$, i.e., the transition curve is a spiral whose curvature is monotone decreasing and has a single inflection point. This completes the proof of Theorem 3.

Here we note that the coefficient $4(4m - 3)$ of s^5 (the highest term in the brackets of (Equation 10.33)) must be nonnegative for the spiral transition curve. Therefore, $m \ge 3/4$ is necessary and $m = 3/4$ means $k'(0) = 0$ since the numerator is quartic and the denominator is quintic in s.

3.2. C-shaped PH Quintic Transition Curve

Now we consider a C-shaped transition curve $z(t)$ of the form (10.23). Note that the angles between $p_i - p_{i-1}$ and $p_{i+1} - p_i$, $1 \le i \le 4$ are all θ; refer to Figure 10.4. Letting $\rho = \tan^2 \theta$, G^2 transition requires:

Lemma 4:

$$z_0 = 2\sqrt{mr_0}\rho^{1/4}(1, 0), \; z_1 = 2m\sqrt{mr_0}\rho^{1/4}(1, \sqrt{\rho}),$$

$$z_2 = \frac{2\sqrt{mr_0}\rho^{1/4}\lambda}{1 + \rho}(1 - \rho, 2\sqrt{\rho}).$$

$$(10.36)$$

Proof: Since the angle from T_0 to T_1 is θ, we obtain $v_1 = u_1 \tan\theta$. Slopes of T_3 and T_4 are given by $(u_2v_1 + u_1v_2)/(u_1u_2 - v_1v_2) (= m_3)$ and $2u_2v_2/(u_2^2 - v_2^2) (= m_4)$, respectively. Therefore, note the angle from T_3 to T_4 is θ to get $(m_4 - m_3)/(1 + m_3m_4) = \tan\theta$, i.e.,

$$u_1v_2 - u_2v_1 = (u_1u_2 + v_1v_2)\tan\theta,$$

$$(10.37)$$

from which $v_1 = u_1 \tan\theta$ and $v_2 = u_2 \tan 2\theta$. Next, note $k(0) = 1/r_0$ and $k(1) = 1/r_1$ to obtain $u_1 = u_0^3 \tan\theta/(4r_0)$, $u_2 = u_0\lambda \cos 2\theta$.

Then we obtain the following theorem to guarantee a unique PH quintic transition curve.

Theorem 4 $(r > r_0 - r_1)$: *Assume $r \le 15.37r_0$. refer to Figures 10.5–10.6. Then, each value of $m (\in [1, 3.22])$ determines a G^2 quintic C-shaped transition curve of the form (10.23) with (10.36) between the two circles with a single inteior curvature extremum. The curve is free of*

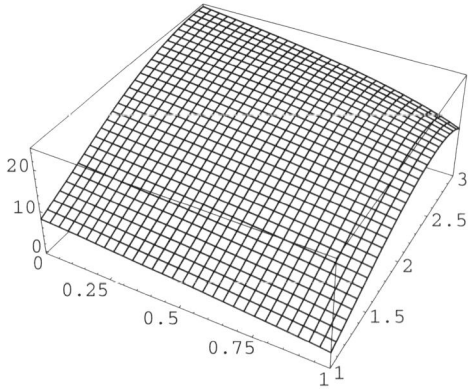

Figure 10.5   Graph of $\psi(m, \lambda)$ for $0 < \lambda \le 1, 1 \le m \le 3$.

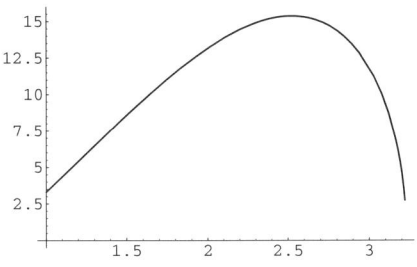 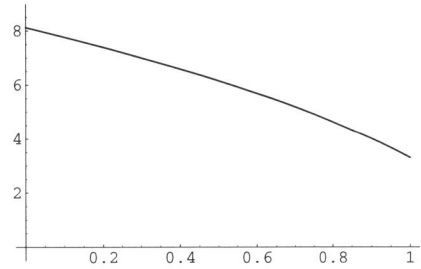

Figure 10.6   Graphs of $\psi(m, 1)$ for $1 \le m \le 3.2$ (left) and $\psi(1, \lambda)$ for $0 < \lambda \le 1$(right).

inflections, loops and cusps, i.e., it is a pair of two spirals with starting monotone decreasing curvature and ending monotone increasing curvature.

Proof: Note:

$$C_1 = p_5 + r_1 \left(\frac{4(-1 + \rho)\sqrt{\rho}}{(1 + \rho)^2}, \frac{1 - 6\rho + \rho^2}{(1 + \rho)^2} \right)$$
$$(= p_5 + r_1(-\sin 4\theta, \cos 4\theta)). \tag{10.38}$$

A symbolic manipulator with $C_1 = (c, d)$ gives:

$$
\begin{aligned}
15(1 + \rho)^2 c &= \lfloor 2(1 - p)(1 - \rho)^2 m^3 + 3(1 + \rho)\{1 + \rho + \lambda(1 - 3\rho)\}m^2 \\
&\quad + \{3(1 + \rho)^2 + \lambda(1 - \rho^2) + 3\lambda^2(1 - 6\rho + \rho^2)\}m - 15\lambda^3(1 - \rho)]4r_0\sqrt{\rho}, \\
15(1 + \rho)^2 d &= r_0[16(1 + \rho)^2 m^3 + 12\rho(1 + \rho)\{1 + \rho + \lambda(3 - \rho)\}m^2 \\
&\quad + 8\lambda\rho\{1 + \rho + 6\lambda(1 - \rho)\}m + 15\lambda^3(1 - 6\rho + 6\rho^2)].
\end{aligned} \tag{10.39}
$$

Condition $\|C_1 - C_0\| = r$ gives the determining equation $f(\rho) = 0$ in $\rho (= u_0^4/(16r_0^2 m^2))$ as:

$$f(\rho) = \{r_0^2/(225(1 + \rho)^2)\} \sum_{i=0}^{5} a_i \rho^i - r^2, \tag{10.40}$$

where the coefficients are given by:

$$
\begin{aligned}
a_5 &= 64m^6, \ a_4 = 16m^4\{16m^2 + 12(1 + \lambda)m - 3\lambda^2 - 14\lambda - 3\}, \\
a_3 &= 8m^2\{48m^4 + 72(1 + \lambda)m^3 + 6(5 - 2\lambda + 5\lambda^2)m^2 \\
&\quad - 24(1 + 4\lambda + 4\lambda^2 + \lambda^3)m - 27 + 33\lambda + 38\lambda^2 + 33\lambda^3 - 27\lambda^4\}, \\
a_2 &= 256m^6 + 576(1 + \lambda)m^5 + 48(13 + 10\lambda + 13\lambda^2)m^4 \\
&\quad - 348(1 + \lambda + \lambda^2 + \lambda^3)m^3 - 16(27 + 45\lambda + 106\lambda^2 + 45\lambda^3 \\
&\quad + 27\lambda^4)m^2 - 240\lambda(1 - 6\lambda - 6\lambda^2 + \lambda^3)m + 225(-1 + \lambda^3)^2, \\
a_1 &= 64m^6 + 192(1 + \lambda)m^5 + 16(21 + 22\lambda + 21\lambda^2)m^4 \\
&\quad - 192(1 - 2\lambda - 2\lambda^2 + \lambda^3)m^3 - 8(27 + 123\lambda - 38\lambda^2 + 123\lambda^3 \\
&\quad + 27\lambda^4)m^2 - 240\lambda(1 + 6\lambda + 6\lambda^2 + \lambda^3)m + 450(1 + 6\lambda^3 + \lambda^6), \\
a_0 &= 225(1 - \lambda^3)^2.
\end{aligned} \tag{10.41}
$$

The intermediate value theorem assures the existence of the positive root since $f(0) (= (r_0 - r_1)^2 - r^2) < 0$, $f(\infty) = \infty$. To show the uniqueness of the positive root, note:

$$225(1 + \rho)^3 f'(\rho) = 8r_0^2 \sum_{i=0}^{5} b_i \rho^i, \tag{10.42}$$

where using a nonnegative $u(= m - 1)$ to show $b_i > 0$ $(1 \le i \le 5)$. If $b_0 > 0$, then the positive root of $f(\rho) = 0$ is unique. If $b_0 < 0$, then $f'(\rho)$ has a single positive zero where $f'(\rho)$ changes its sign from $-$ to $+$. Therefore, $f(0) < 0$ and $f(\infty) = \infty$ mean the unique positive root of $f(\rho) = 0$.

Now we examine the shape of the transition curve. First,

$$u(t)v'(t) - u'(t)v(t) = \frac{16mr_0\rho\{2\lambda s + m(s^2 + \lambda)(1 + \rho)\}}{(1 + s)^2(1 + \rho)} (> 0), \tag{10.43}$$

from which the curve is free of inflection points. Next, 'cusps' require $z'(\alpha) = 0, 0 < \alpha < 1$, i.e., $u(\alpha) = v(\alpha) = 0$. On the other hand:

$$(1 + s)^2(1 + \rho)v(t) = 4\sqrt{mr_0}\rho^{3/4}\{\lambda + ms(1 + \rho)\}(> 0), \tag{10.44}$$

from which the curve is free of cusps. Thirdly, for no loops, note:

$$15(1 + s)^5(1 + \rho)^2 y(t) = 4mr_0\rho \sum_{i=0}^{3} a_i s^i (> 0), \tag{10.45}$$

where $a_i > 0$ $(0 \le i \le 3)$ as:

$$\begin{aligned}
a_3 &= 30m(1 + \rho)^2, \quad a_2 = 10(1 + \rho)\{2\lambda + m(3 + 4m)(1 + \rho)\}, \\
a_1 &= 5(1 + \rho)[4m^2 + 3m + 9m\lambda + 2\lambda + m\{4m + 3(1 - \lambda)\}\rho], \\
a_0 &= 4m^2 + 3m + 9m\lambda + 2\lambda + 12\lambda^2 + 2(4m^2 + 3m + 3m\lambda \\
&\quad + \lambda - 6\lambda^2)\rho + m\{4m + 3(1 - \lambda)\}\rho^2.
\end{aligned} \tag{10.46}$$

If the curve had loops, at least two α and β $(0 < \alpha \ne \beta < 1)$ would exist such that $z(\alpha) = z(\beta)$, i.e.,

$$x(\alpha)/y(\alpha) = x(\beta)/y(\beta). \tag{10.47}$$

Therefore, for no loops, it suffices to show that $x(t)/y(t)$ is monotone decreasing or $x'(t)y(t) - x(t)y'(t) < 0$. A symbolic manipulator gives:

$$x'(t)y(t) - x(t)y'(t) = -\frac{16m^2 r_0^2 \rho^{3/2}}{15(1 + s)^8(1 + \rho)^2} \sum_{i=0}^{6} b_i s^i (< 0), \tag{10.48}$$

where coefficients b_i, $0 \le i \le 6$ are easily shown to be positive for $m \ge 1$.

Finally, we show that the curve has a single curvature extremum. A symbolic manipulator *Mathematica* gives:

$$(1 + \rho)^2 (1 + s)^5 w(t) = -64 m^2 r_0^2 \rho^{3/2} \sum_{i=0}^{5} c_i s^i, \tag{10.49}$$

where a nonnegative $u(= m - 1)$ shows $c_i > 0$, $i = 5, 4, 3, 0$.

Descartes' rule of signs implies that $c_2 \leq 0$ is a sufficient condition for the curvature to have a single extremum, i.e., a local minimum where:

$$c_2 = 4\lambda \{ 2m^3 - 12m^2 + \lambda(13m - 3)$$
$$+ (6m^3 - 24m^2 + 14m\lambda + 3\lambda)\rho + m(6m^2 - 12m + \lambda)\rho^2 + 2m^3\rho^3 \}. \tag{10.50}$$

From now on, we assume that $1 \leq m \leq 3.22$, where 3.22 is a necessary condition for the following ρ_0 to be positive for all $\lambda \in (0, 1)$. Note that $2m^3 - 12m^2 + \lambda(13m - 3) \leq 2m^3 - 12m^2 + 13m - 3 \leq 0$ to obtain a sufficient condition for $c_2 \leq 0$:

$$0 < \rho \leq (-6m^2 + 12m - \lambda + \phi(m, \lambda))/(4m^2)(= \rho_0), \tag{10.51}$$

with:

$$\phi(m, \lambda) = \sqrt{-12m^4 + 48m^3 + 4(36 - 25\lambda)m^2 - 48\lambda m + \lambda^2}. \tag{10.52}$$

Hence, $f(\rho_0) \geq 0$ gives a bound $\psi(m, \lambda)$ for r/r_0 where *Mathematica* would give its explicit form. This completes the proof of Theorem 4.

Figures 10.5 and 10.6 give the graphs of numerically determined upper bound $\psi(m, \lambda)$ for r/r_0.

4. Numerical Examples and Critical Analysis

This section gives some numerical examples to assure our theoretical analysis. The figures with P_0 fixed on the larger circles show the effect of parameter m for $r_0 = 1$ where $(r_1, r) = r_0$ $(0.5, 2)$. In Figures 10.7–10.14, case $m = 1$ is shown with thick curves. Figures 10.7, 10.9 show cubic S-shaped curves ($m = 2/3, 1$) and C-shaped curves ($m = 1, 3/2$) respectively. Similarly,

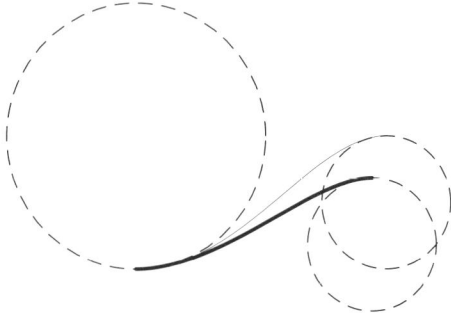

Figure 10.7   Family of cubic S-shaped transition ($z(t)$, $0 \leq t \leq 1$).

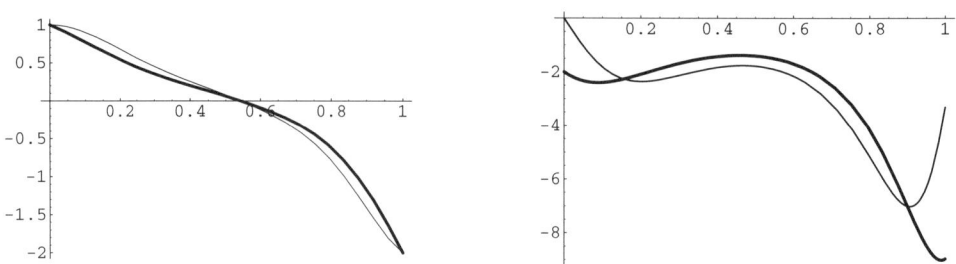

Figure 10.8   Curvature plot (left) and its derivative plot (right) of $z(t)$ in Figure 10.7.

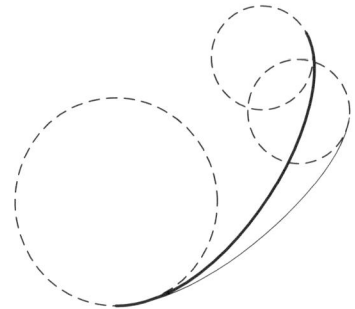

Figure 10.9   Family of cubic C-shaped transition $(z(t), 0 \leq t \leq 1)$.

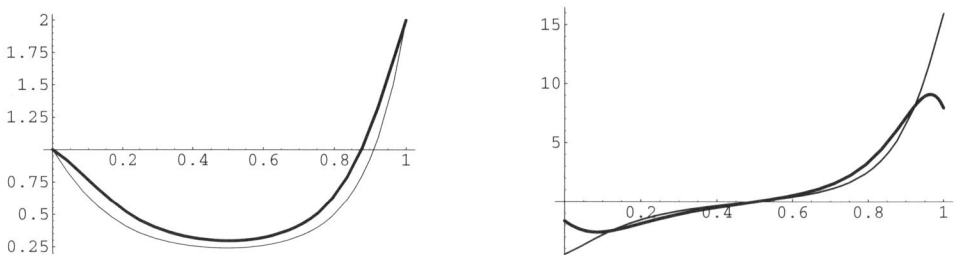

Figure 10.10   Curvature plot (left) and its derivative plot (right) of $z(t)$ in Figure 10.9.

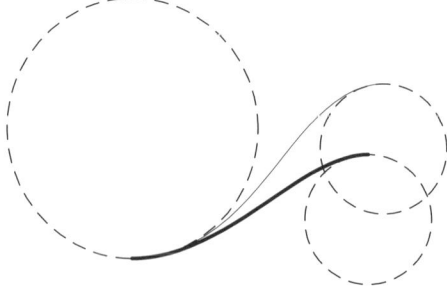

Figure 10.11   Family of PH quintic S-shaped transition $(z(t), 0 \leq t \leq 1)$.

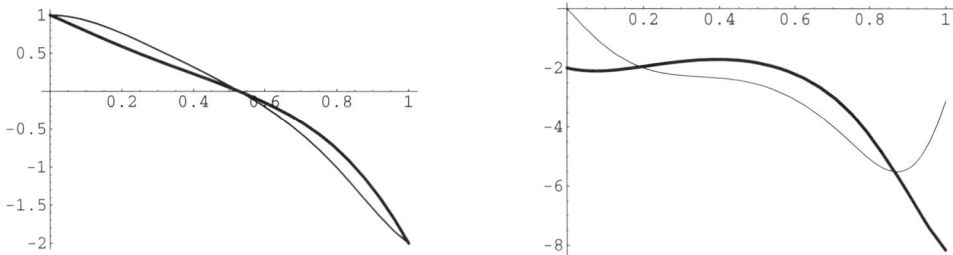

Figure 10.12 Curvature plot (left) and its derivative plot (right) of $z(t)$ in Figure 10.11.

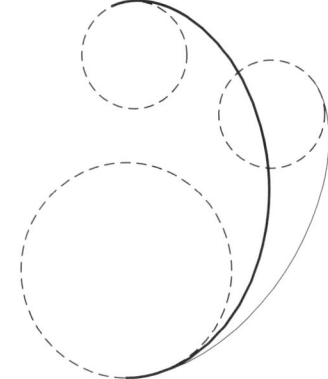

Figure 10.13 Family of PH quintic C-shaped transition $(z(t), 0 \leq t \leq 1)$.

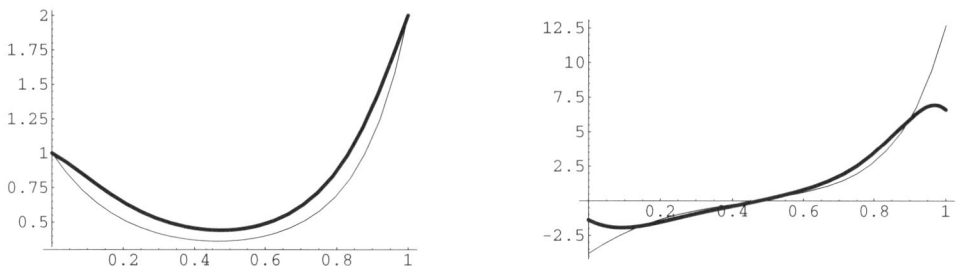

Figure 10.14 Curvature plot (left) and its derivative plot (right) of $z(t)$ in Figure 10.13.

Figures 10.11 and 10.13 show PH quintic S-shaped curves ($m = 3/4, 1$) and C-shaped curves ($m = 1, 4/3$) respectively.

The cubic case is mathematically very simple when compared to PH quintic but from some numerical results on the locus of the center of a smaller circle, we found PH quintic more flexible than cubic transition. Locus of center of a smaller circle is a circular arc for both S-type (Figure 10.15) and C-type (Figure 10.16) curves shown as a *dotted thick* arc. Its arc-length for all possible values of m has been shown in Table 10.1 to show the flexibility comparison between cubic and PH quintic transitions.

Table 10.1   Arc-length (dotted thick circular arc in Figures 10.15
and 10.16) of locus of center of smaller circle.

	Cubic transition	PH quintic transition
S-shaped	1.187 ($2/3 \leq m \leq \infty$)	1.54 ($3/4 \leq m \leq \infty$)
C-shaped	2.135 ($0.923 \leq m \leq \infty$)	3.356 ($1 \leq m \leq 3.22$)
	$(r_0 = 1, r_1 = 0.5, r = 2)$	

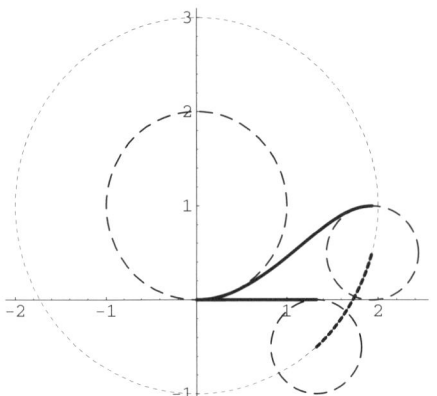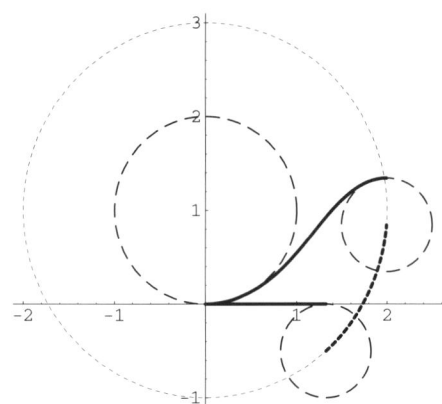

Figure 10.15   Locus of the center of smaller circle in S-case for cubic (left) and PH quintic (right).

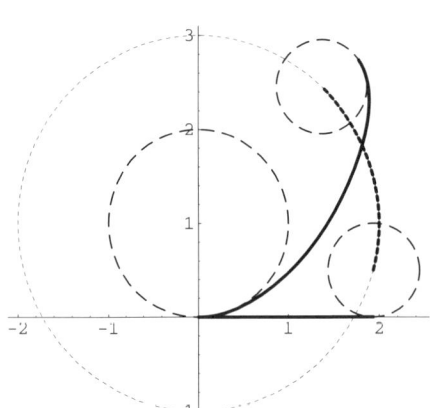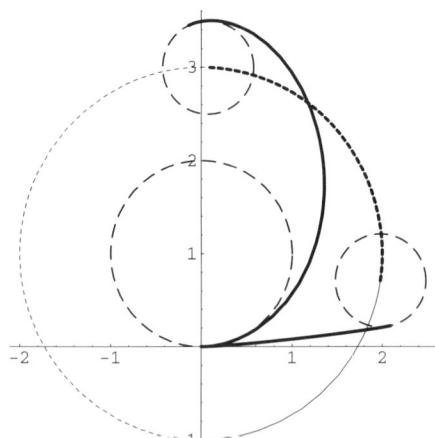

Figure 10.16   Locus of the center of smaller circle in C-case for cubic (left) and PH quintic (right).

Our constraints in cubic and PH quintic transitions are less restrictive than Walton and Meek
[18, 20] (see Table 10.2 for comparison). The tension control properties of m in cubic Bezier
transitions are demonstrated in Figures 10.17 and 10.18. The ends of S- and C-shaped curves
are shown with small circles and discs respectively.

Table 10.2 Comparison between our scheme and Walton and Meek's [18, 20] scheme.

		Our scheme	Walton and Meek scheme
Cubic transition	S-shaped	$r_0/r_1 \leq 36$ $m \geq 2/3$	$r_0/r_1 \leq 9$ $m = 2/3$
	C-shaped	$m \geq (1 + \sqrt{1 + 3\lambda})/3$	$m = 1$
PH quintic transition	S-shaped	$r_0/r_1 \leq (10/3)^3 \approx 37$ $m \geq {}^3/_4$	$r_0/r_1 \leq 8$ $m = {}^3/_4$
	C-shaped	$r \leq 15.37\, r_0$ $1 \leq m \leq 3.22$	$r \leq 3.3\, r_0$ $m = 1$

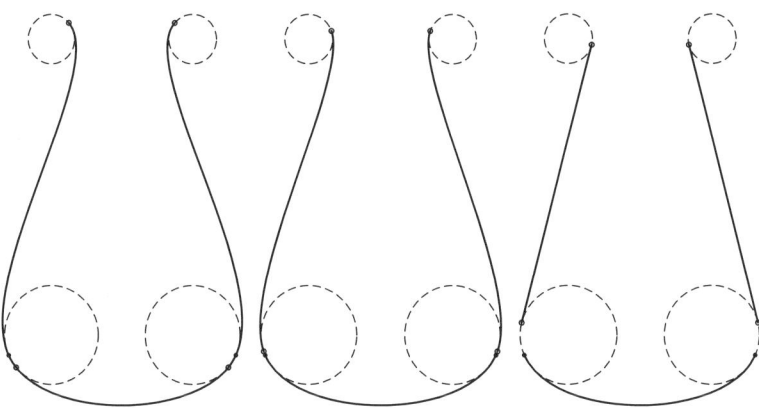

Figure 10.17 Vase profile with C-shaped: $m = 1$ (left, middle, right) and both S-shaped: $m = 2/3$ (left), 2 (middle), 50 (right).

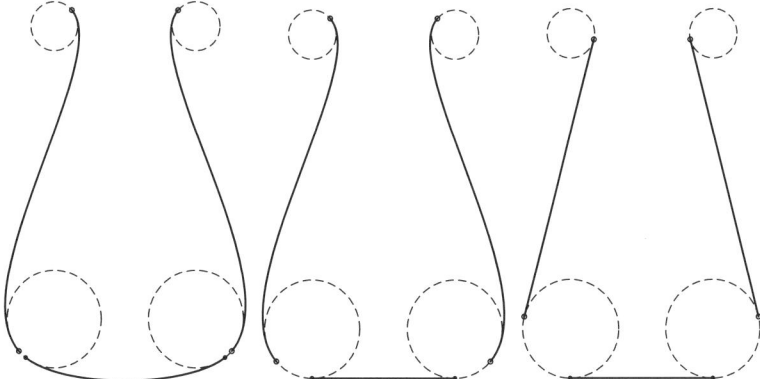

Figure 10.18 Vase profile with C-shaped: $m = 2$ (left), $m = 50$ (middle, right) and both S-shaped: $m = 2/3$ (left, middle), 50 (right).

5. Summary

The use of fair cubic Bezier and PH quintic curves for family of G^2 transitions between two circles has been demonstrated. Such blending is often desirable in CAD, CAM and CAGD applications. We not only completed the analysis of [18, 20], but also presented a very simple scheme with the proof of all shape properties offering more flexible constraints. We also guaranteed the absence of interior curvature extremum (i.e., spiral segment) in S- shaped and a single curvature extremum (at which the curvature magnitude is a minimum) in C-shaped for cubic and PH quintic transition curves.

Both cubic and PH quintic schemes presented in this chapter are important for users. Cubic transition gives simple algorithm whereas PH quintic offers more flexibility than cubic (see Table 10.1). The comparison in Table 10.2 shows that our constraints are less restrictive, more reasonable and comfortable for practical applications. Due to this high flexibility in our scheme, users can very effectively use m as a shape control parameter with wider range of radii and distance between two circles. This is quite reasonable in practical applications and users do not need to use intermediate circle which can be required in Meek and Walton's scheme.

References

[1] Baass K. G. (1984) The use of clothoid templates in highway design. Transportation Forum 1, 47–52.

[2] Farin G. (1997) *Curves and Surfaces for Computer Aided Geometric Design: A Practical Guide*. New York: Academic Press 4th edition.

[3] Gibreel G. M., Easa S. M., Hassan Y. and El-Dimeery I. A. (1999) State of the art of highway geometric design consistency. *ASCE Journal of Transporation Engineering* 125(4), 305–313.

[4] Habib Z. and Sakai M. (2002) G^2 two-point Hermite rational cubic interpolation. *International Journal of Computer Mathematics* 79(11), 1225–1231.

[5] Habib Z. and Sarfraz M. (2001) A rational cubic spline for the visualization of convex data. *Proceedings of IEEE International Conference on Information Visualization-IV'01-UK*: IEEE Computer Society Press, USA, 744–748.

[6] Hartman P. (1957) The highway spiral for combining curves of different radii. *Transactions of the American Society of Civil Engineers* 122, 389–409.

[7] Hickerson T. F. (1964) *Route Location and Design*. New York: McGraw-Hill.

[8] Hoschek J. and Lasser D. (1993) *Fundamentals of Computer Aided Geometric Design* (Translation by L. L. Schumaker). MA: A. K. Peters, Wellesley.

[9] Sakai M. (1999) Inflection points and singularities on planar rational cubic curve segments. *Computer Aided Geometric Design* 16, 149–156.

[10] Sakai M. (2001) Osculatory interpolation. *Computer Aided Geometric Design* 18, 739–750.

[11] Sakai M. and Usmani R. (1996) On fair parametric cubic splines. BIT 36, 359–377.

[12] Sarfraz M. (2003) Optimal curve fitting to digital data. *International Journal of WSCG*, Vol. 11(1).

[13] Sarfraz M. (2003) Curve fitting for large data using rational cubic splines. *International Journal of Computers and Their Applications*, Vol. 10(3).

[14] Sarfraz M. and Razzak M. F. A. (2003) A web based system to capture outlines of Arabic fonts *International Journal of Information Sciences*, Elsevier Science Inc., Vol. 150(3–4), 177–193.

[15] Sarfraz M. (2003) Some algorithms for curve design and automatic outline capturing of images. To appear in *International Journal of Image and Graphics*, World Scientific Publisher.

[16] Sarfraz, M. (2002) Visualization of positive and convex data by a rational cubic spline. *International Journal of Information Sciences*, Elsevier Science Inc., Vol. 146(1–4), 239–254.

[17] Sarfraz M. (2002) Modelling for the visualization of monotone data. *International Journal of Modelling and Simulation*, ACTA Press, Vol. 22(3), 176–185.

[18] Sarfraz M. and Razzak M. F. A. (2002) An algorithm for automatic capturing of font outlines. *International Journal of Computers & Graphics*, Elsevier Science, Vol. 26(5), 795–804.

[19] Sarfraz M. (2002) Fitting curves to planar digital data. *Proceedings of IEEE International Conference on Information Visualization IV'02-UK*: IEEE Computer Society Press, USA, 633–638.

[20] Sarfraz M. and Raza A. (2002) Visualization of data using genetic algorithm, *Soft Computing and Industry: Recent Applications*, Eds.: Roy R., Koppen M., Ovaska S., Furuhashi T., and Hoffmann F., ISBN: 1-85233-539-4, Springer, 535–544.

[21] Sarfraz M. and Raza A. (2002) Towards automatic recognition of fonts using genetic approach, *Recent Advances in Computers, Computing, and Communications*, Eds.: Mastorakis N. and Mladenov V., ISBN: 960-8052-62-9, WSEAS Press, 290–295.

[22] Walton D. J. and Meek D. S. (1999) Planar G^2 transition between two circles with a fair cubic Bezier curve. *Computer Aided Design* 31, 857–866.

[23] Walton D. J. and Meek D. S. (2001) Curvature extrema of planar parametric polynomial cubic curves. *Computational and Applied Mathematics* 134, 69–83.

[24] Walton D. J. and Meek D. S. (2002) Planar G^2 transition with a fair Pythagorean hodograph quintic curve. *Computational and Applied Mathematics* 138, 109–126.

11

Optimal Hierarchical Adaptive Mesh Construction Using FCO Sampling

Panagiotis A. Dafas

Department of Computing, School of Informatics, City University,
London EC1V OHB, United Kingdom.

Ioannis Kompatsiaris
Michael G. Strintzis

Informatics and Telematics Institute, 1st Km. Thermi – Panorama Road,
570 01 Thermi – Thessaloniki, Greece.

This chapter introduces an optimal hierarchical adaptive mesh construction algorithm using the Face-Centered Orthorhombic lattice (FCO) sampling, which is a natural extension of the quincunx lattice to the 3-dimensional case. A scheme for construction of adaptive meshes is presented. Initially, a highly detailed and densely sampled regular mesh is obtained from geometry scanning or from a non-optimal polygon mesh. The adaptive triangle mesh is constructed by using fixed position vertices along with an efficient adaptive triangulation technique. The decimation is based on FCO sampling and surface estimation filters. The result is a progressive sequence of meshes consisting of more triangles wherever sharp edges exist and fewer in uniform plane regions. Experimental results demonstrate the usage and performance of the algorithm.

1. Introduction

Today we can accurately acquire finely detailed, arbitrary topology surfaces with millions (and recently) billions of vertices. Such models place large strains on computation, storage,

Advances in Geometric Modeling. Edited by M. Sarfraz
© 2003 John Wiley & Sons, Ltd   ISBN: 0-470-85937-7

transmission and display resources. Thus, polygonal surface approximation is an essential preprocessing step in applications such as scientific visualization [1], [2], digital terrain modeling [3] and 3-D model-based video coding [4]. Compression is essential in these settings and in particular *progressive* compression, where an early, coarse approximation can subsequently be improved through the transmission of additional bits. While compression of *images* has achieved a high level of sophistication, compression of *surfaces* is a relatively new area, which is currently evolving.

Several progressive and adaptive triangle reduction techniques have been presented in the literature [5–8]. Delaunay triangulation was used in [9], where a planarity criterion was used to decide which vertices should be removed. Quadtrees [10] are adaptive data structures in which planar shapes are recursively subdivided according to certain rules. Quadtree-based methods [11] were mainly proposed for radiosity meshing, where the mesh was controlled by the illumination gradient. In [12] the Restricted Quadtree Triangulation (RQT), an adaptive hierarchical triangulation model, was first applied to terrain visualization. Its main contribution was an efficient method for screen-space error metric calculation, and efficient scene culling and vertex selection according to the error metric. In [13] two alternative vertex selection algorithms based on RQT were proposed and a more intuitive triangle strip construction method was provided. In [14] the binary tree of triangles was introduced and used for the design of an incremental mesh optimization algorithm. In [15] an algorithm was presented seeking to minimize an energy function explicitly modeling the competing requirements of conciseness of representation and fidelity to the data. In [16] the Wavelet Transform (WT) was used as an overall mathematical framework controlling the data approximation. It was assumed that the WT provides a local spectral estimate of the data and describes local variations, which define the coarseness of the surface mesh. After the simplification procedure, the resulting mesh is coded for efficient transmission or storage. In [17] the geometry to be compressed is converted into a generalized mesh form, which requires much less information, with little loss in quality. However, the coding/compression scheme is not integrated with the simplification scheme. For an excellent overview of 3-D geometry compression see [18].

Although the performance of the above algorithms has in many ways been quite satisfactory, several persistent problems remain unresolved. Particularly troublesome are the high computational costs of local triangulation, due to the requirement of extensive work on data structures and list management (as is often the case with Delaunay triangularization), and the lack of a consistent hierarchical representation of the approximated surface, which is useful in many applications [19], [20]. Compression is essential in these settings and in particular progressive compression [21], [22], where an early, coarse approximation can subsequently be improved through additional bits. In [23], progressive meshes are introduced, and complete correspondence between vertices in different levels of the hierarchy is established, something that cannot be easily achieved for triangles/faces at different levels. Improved results are obtained with the use of the compressed progressive meshes approach introduced in [24]. The construction of wavelets over arbitrary meshes has been also used for progressive compression [25].

In [26] the generalization of signal processing tools such as downsampling, upsampling and filters to irregular connectivity triangle meshes was presented. In this chapter the generalization of the well known and useful FCO subsampling interpolation scheme is presented [27], [28]. This chapter introduces the optimal hierarchical adaptive mesh construction using FCO sampling. The hierarchical representation of meshes achieved by the adaptive triangulation method provides [29], [30]:

- Mesh Compression: In order to represent accurate complex surfaces, a large number of triangles are necessary. Hence, mesh compression is essential for efficient transmission and storage.
- Mesh Simplification: The meshes created by modeling and scanning systems are seldom optimized for rendering efficiency. These meshes can be replaced by almost indistinguishable approximations, which have more triangles in regions of high detail and fewer in uniform regions.
- Progressive Transmission: Progressive mesh transmission provides, during the transmission state, low-detail mesh approximations. This is useful, for instance, in browsing through large databases and in Internet-based applications.
- Prioritized transmission of the mesh: In addition to the hierarchy of meshes, each level in the pyramid of meshes must be transmitted in a prioritized way. In this manner, the rendering performance is optimized, even if the transmission is not entirely completed for each level in the pyramid of meshes.
- Correspondence between successive scales: Straightforward correspondence between triangles of successive levels allows properties of the mesh to be calculated at lower levels and be propagated through the pyramid.
- Selective refinement: Sometimes it is desirable to adapt the level of refinement in selected regions where the interest of the user is focused.
- Computational efficiency: The computational efficiency of the algorithm is important to rapidly produce the required multiresolution representation.

This chapter is organized as follows. In the next section the construction of the initial dense regular mesh is presented. In Section 3 the decimation of the dense mesh using FCO subsampling and surface estimation filters is given. Experimental results demonstrate the efficiency of the algorithm in section 4.

2. Regular Triangle Mesh Construction

Initially, a regular mesh is constructed at the highest detail level of our mesh hierarchy. This regular mesh is obtained from unorganized points captured by three-dimensional range scanning or from a non-optimal polygonal mesh. In both cases an appropriate discrete distance d is chosen, depending on the application, and the surface is described by dense fixed position vertices inside its bounding box. In this manner, the information of the surface is described by a three dimensional binary sequence with region of definition the bounding box of the surface:

$$x[n], n = [n_1, n_2, n_3]^T, \tag{11.1}$$

where $[n_1, n_2, n_3]^T$ denotes the position of the vertices in the 3D space and $n_i \in [N_{i,\min}, \ldots, N_{i,\max}]$ is as shown in Figure 11.1. The binary sequence $x[n]$ in constructed in the following way:

$$x[n] = \begin{cases} 1 & \text{if } n \in \text{surface,} \\ 0 & \text{if } n \notin \text{ surface.} \end{cases} \tag{11.2}$$

The procedure for deriving this binary sequence from unorganized scanned points and non-optimal meshes is described elsewhere.

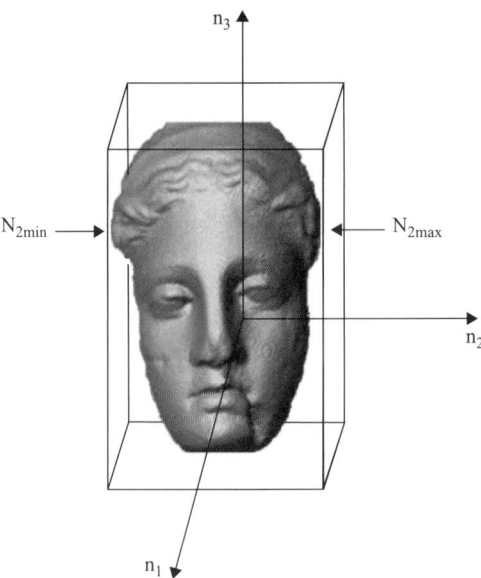

Figure 11.1   Bounding box of a 3D object.

2.1. Regular Mesh from Unorganized Points

The sequence in Equation (11.1) is obtained in a straightforward manner from the unorganized points by simply selecting the appropriate resolution distance d and rounding all points to the nearest value of an integer multiple of d. Specifically, if $\tilde{n}_i$ is a point from the set of unorganized points, then the new points n_i are:

$$n_i = round\,(\tilde{n}_i) = kd, \tag{11.3}$$

where $\|\tilde{n}_i - kd\| \le \|\tilde{n}_i - ld\|$, $\forall l$: integer $\ne k$.

2.2. Regular Mesh from Non-optimal Meshes

In this case the set of fixed position vertices will be constructed from an initial non-optimal mesh defined by a set of 3D vertices and their interconnected triangles. For each 3D point n as defined in Equation (11.1), a sequence x_{dist} is found representing the distance of the 3D point n from the surface of the model defined by the set of interconnected triangles. Each triangle defines a plane with normal vector $[a \quad b \quad c \quad d]^T$. For the estimation of $x_{dist}[n]$ the closest triangle to the point is used. Thus,

$$x_{dist}[n] = \frac{an_0 + bn_1 + cn_2 + d}{\sqrt{a^2 + b^2 + c^2}}. \tag{11.4}$$

The binary sequence (see Equation (11.1)) can be obtained as follows:

$$x[n] = \begin{cases} 1 & \text{if } x_{dist}[n] \le \dfrac{\|d\|}{2}, \\[2mm] 0 & \text{if } x_{dist}[n] > \dfrac{\|d\|}{2}. \end{cases} \tag{11.5}$$

By using the sequence $x_{dist}[n]$, a 3D point is part of x if its distance from the surface is less than $\dfrac{\|d\|}{2}$.

2.3. Surface Estimation

Since the set of fixed position 3D points describing the surface is available, an efficient triangulation algorithm is applied in order to reconstruct the surface at the highest detail. A regular triangle mesh will be constructed using only the vertices with coordinates (n_1, n_2, n_3) of the non-zero element of the $x[n]$ sequence.

The FCO is a natural extension of the quincunx lattice to the 3D case, and is in fact quincunx in all dimensions [27]. The FCO sampling matrix M is defined by [31].

$$M = \begin{bmatrix} 1 & 0 & 1 \\ -1 & -1 & 1 \\ 0 & -1 & 0 \end{bmatrix}.$$ (11.6)

Three dimensional subsampling can be depicted as follows:

$$x_{k+1}[n] = x_{k+1}[Mn]$$ (11.7)

where k denotes the stage of the subsampling procedure. In Figure 11.2 the FCO subsampling procedure is presented. At each stage k, the points marked with k leave from the lattice $LAT(M^{k-1})$. Note that a finite power of the FCO sampling matrix is an integer multiple of the identity matrix:

$$M^3 = 2^3 I.$$ (11.8)

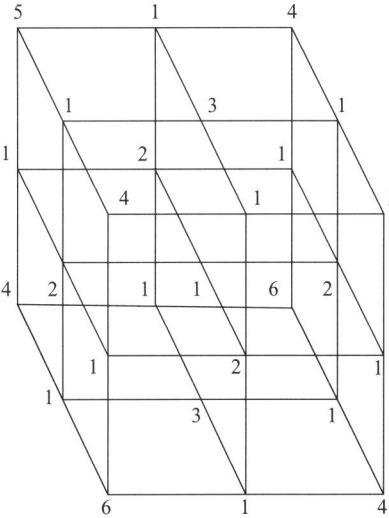

Figure 11.2   FCO subsampling. At each stage k the points marked with k leave from the lattice) *LAT* (M^{k-1}).

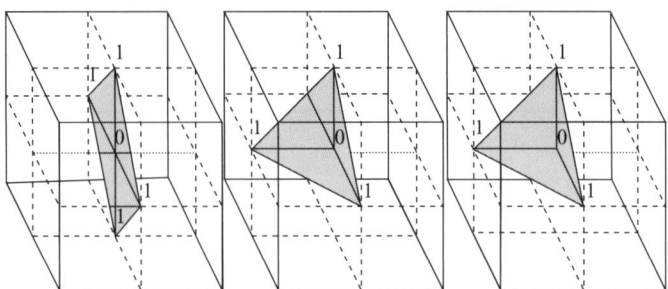

Figure 11.3   Possible triangulations using first order vertices.

The construction of the regular wireframe will be based on the the fixed position vertices and the FCO sampling matrix and is implemented in the following three steps:

- **Step 1:** In this step the triangles shown in Figure 11.3 as well those resulting from rotations of $\pi/2$ rads around the axes are constructed. The nodes '0' are retained from FCO sampling of all nodes in the bounding box and nodes '1' are first order neighbouring nodes, meaning that they are only one vertex away from node '0'. Each time $x[n] = 1$ for '0' and at least one '1' node the finite number of possible triangles as shown in Figure 11.3 are constructed. In this step both vertical and horizontal surfaces are triangulated.
- **Step 2:** In this step the triangles shown in Figure 11.4 as well those resulting from rotations of $\pi/2$ rads around the axes are constructed. Again nodes '0' are derived from FCO sampling of the initial bounding box and nodes '2' are used for triangulation. In this step oblique surfaces are triangulated.
- **Step 3:** In this step the triangles shown in Figure 11.5 as well those resulting from rotations of $\pi/2$ rads around the axes are constructed. In this step first, second and third order neighbours are used for triangulation.

The result of the triangulation procedure is a regular triangle mesh. Several examples of the triangulation algorithm are given at the experimental results section. Note that in this phase no nodes are discarded due to a FCO subsampling procedure. The FCO matrix is used only in order to define the positions for triangulation.

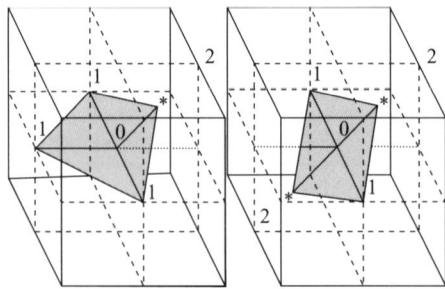

Figure 11.4   Possible triangulations using first and second order vertices.

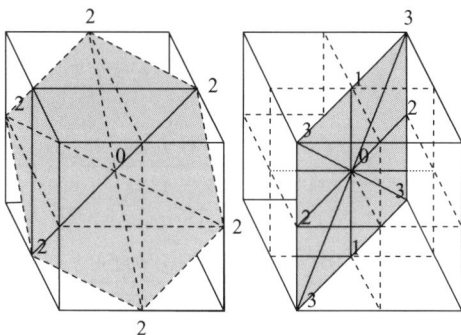

Figure 11.5 Possible triangulations using first, second and third order vertices.

3. Initial Regular Mesh Decimation

Given the initial regular mesh construction as described in the previous section, the algorithm generates successive lower density meshes consisting of more triangles wherever sharp edges exist and fewer in uniform plane regions due to the use of the surface estimation filters.

At each stage k of the subsampling FCO procedure, the 'surface estimation filters' $g_k(\boldsymbol{n})$, $g_{k+1}(\boldsymbol{n})$ and $g_{k+2}(\boldsymbol{n})$ are applied to the surface in order to identify whether planar surfaces, which can be simplified, exist. Vertices of the triangular mesh, which belong on a planar surface, are rejected according to the procedure described in the following. The surface estimation filters represent an averaging procedure around each node and their output is used in order to determine whether any planar surfaces of any orientation exist which are then simplified. The $g_{k,n_i}(\boldsymbol{n})$ is applied at stage k of the algorithm and n_i indicates the estimation of a planar surface along the n_i direction.

The information of the surface of the model is represented by the binary sequence with non-zero values being the vertices of the triangular regular mesh. In the following relations, function $f_i(n_1, n_2, n_3)$ returns the coordinate a vertex if $f_i(n_1, n_2, n_3)$ equals one for this vertex:

$$f_i(n_1, n_2, n_3) = \begin{cases} n_i & \text{if } x(n_1, n_2, n_3) = 1, \\ 0 & \text{if } x(n_1, n_2, n_3) = 0. \end{cases} \tag{11.9}$$

Each of the following relations is in fact an averaging of the neighboring to (n_1, n_2, n_3) coordinates in different directions. In all summations i and j take values $-d, +d$.

$$g_{k,n_1}(n_1, n_2, n_3) = \sum_i \frac{f_1(n_1 + i, n_2, n_3) + f_1(n_1, n_2 + i, n_3) + f_1(n_1, n_2, n_3 + i)}{x(n_1 + i, n_2, n_3) + x(n_1, n_2 + i, n_3) + x(n_1, n_2, n_3 + i)},$$

$$\tag{11.10a}$$

$$g_{k,n_2}(n_1, n_2, n_3) = \sum_i \frac{f_2(n_1 + i, n_2, n_3) + f_2(n_1, n_2 + i, n_3) + f_2(n_1, n_2, n_3 + i)}{x(n_1 + i, n_2, n_3) + x(n_1, n_2 + i, n_3) + x(n_1, n_2, n_3 + i)},$$

$$\tag{11.10b}$$

$$g_{k,n_3}(n_1, n_2, n_3) = \sum_i \frac{f_3(n_1 + i, n_2, n_3) + f_3(n_1, n_2 + i, n_3) + f_3(n_1, n_2, n_3 + i)}{x(n_1 + i, n_2, n_3) + x(n_1, n_2 + i, n_3) + x(n_1, n_2, n_3 + i)},$$

$$\tag{11.10c}$$

$$g_{k+1,n_1}(n_1, n_2, n_3) = \sum_i \sum_j \frac{f_1(n_1 + i, n_2 + j, n_3) + f_1(n_1, n_2 + j, n_3 + i)}{x(n_1 + i, n_2 + j, n_3) + x(n_1, n_2 + j, n_3 + i)},$$

(11.11a)

$$g_{k+1,n_2}(n_1, n_2, n_3) = \sum_i \sum_j \frac{f_2(n_1 + i, n_2 + j, n_3) + f_2(n_1, n_2 + j, n_3 + i)}{x(n_1 + i, n_2 + j, n_3) + x(n_1, n_2 + j, n_3 + i)},$$

(11.11b)

$$g_{k+1,n_3}(n_1, n_2, n_3) = \sum_i \sum_j \frac{f_3(n_1 + i, n_2 + j, n_3) + f_3(n_1, n_2 + j, n_3 + i)}{x(n_1 + i, n_2 + j, n_3) + x(n_1, n_2 + j, n_3 + i)},$$

(11.11c)

$$g_{k+2,n_1}(n_1, n_2, n_3) = \sum_i \sum_j \frac{f_1(n_1 + i, n_2, n_3 + j)}{x(n_1 + i, n_2, n_3 + j)},$$

(11.12a)

$$g_{k+2,n_2}(n_1, n_2, n_3) = \sum_i \sum_j \frac{f_2(n_1 + i, n_2, n_3 + j)}{x(n_1 + i, n_2, n_3 + j)},$$

(11.12b)

$$g_{k+2,n_3}(n_1, n_2, n_3) = \sum_i \sum_j \frac{f_3(n_1 + i, n_2, n_3 + j)}{x(n_1 + i, n_2, n_3 + j)}.$$

(11.12c)

For each stage k the following errors are estimated in order to be used for the determination of any planar surfaces:

$$e_{n_1} = f_1(n_1 + n_2 + n_3) - g_{k,n_1}(n_1, n_2, n_3),$$
$$e_{n_2} = f_2(n_1 + n_2 + n_3) - g_{k,n_2}(n_1, n_2, n_3),$$
$$e_{n_3} = f_3(n_1 + n_2 + n_3) - g_{k,n_3}(n_1, n_2, n_3).$$

(11.13)

These differences are an estimate of the surface roughness. More specifically:

- Point (n_1, n_2, n_3) lies on a plane if:

$$e_{n_1} = e_{n_2} = e_{n_3} = 0,$$

(11.14)

or,

$$e_{n_1} = e_{n_2}, e_{n_3} \neq 0,$$
$$e_{n_1} = e_{n_3}, e_{n_2} \neq 0,$$
$$e_{n_2} = e_{n_3}, e_{n_1} \neq 0.$$

(11.15)

- Point (n_1, n_2, n_3) lies along on edge if:

$$e_{n_1} = 0, e_{n_2} \neq 0, e_{n_3} \neq 0,$$
$$e_{n_2} = 0, e_{n_1} \neq 0, e_{n_3} \neq 0,$$
$$e_{n_3} = 0, e_{n_1} \neq 0, e_{n_2} \neq 0.$$

(11.16)

Using these differences point (n_1, n_2, n_3) is rejected if it belongs to a plane or along an edge since it carries no information. In this case, point (n_1, n_2, n_3) belong to a uniform region that can be represented with fewer and larger triangles.

An example of this operation can be presented with the help of the Figure 11.6. We assume that all nodes lie on the same plane and that there are no nodes lying outside this plane in this

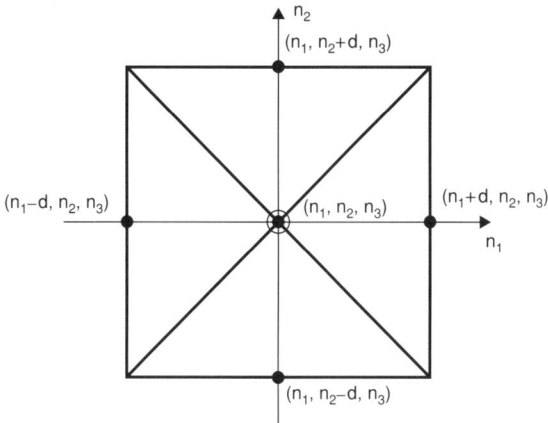

Figure 11.6   Example of application of filter $g_{k,n_1}(n_1, n_2, n_3)$.

region. In this case, filter $g_{k,n_1}(n_1, n_2, n_3)$ is applied to node (n_1, n_2, n_3):

$$g_{k,n_1}(n_1, n_2, n_3) = \frac{n_1 + d + n_1 - d + n_1 + n_1 + 0 + 0}{1 + 1 + 1 + 1} = n_1. \tag{11.16}$$

$f_1(n_1, n_2, n_3 \pm d) = 0$ since we assume that there are no nodes lying outside this plane in this region. If we perform the above operation for $g_{k,n_2}(n_1, n_2, n_3)$ and $g_{k,n_3}(n_1, n_2, n_3)$ we reach:

$$e_{n_1} = e_{n_2} = e_{n_3} = 0, \tag{11.17}$$

indication that (n_1, n_2, n_3) lies on a plane as is the case.

3.1. Hierarchical Triangulation

Every time that a vertex is rejected from the mesh the hierarchical triangulation algorithm simplifies the mesh at this specific region. In this way the final mesh will have more triangles wherever curve planes exist and fewer in uniform plane regions. The mesh simplification procedure is shown in Figure 11.7.

4. Experimental Results

The proposed hierarchical adaptive triangulation algorithm of 3D surfaces was evaluated for 3D mesh adaptive representation of the surfaces 'Sphinx' and 'Sphere'. The 'Sphere' surface is the most difficult case for the presented algorithm since it contains no planar patches making the task of adaptive triangulation very demanding. The original surfaces are shown in Figures 11.8a and 11.9a, respectively. The 'Sphinx' surface consists of 1919 vertices and 2832 triangles. The 'Sphere' surface consists of 7170 vertices and 8592 triangles.

First the regular triangle mesh was constructed for each algorithm, using the methods presented in Section 2. For the construction of the regularly distributed vertices, the technique presented in subsection 2.2 was used, since the input to the algorithm is a non-optimal mesh. After this step, the set of fixed position 3D points describing the surface was available and the

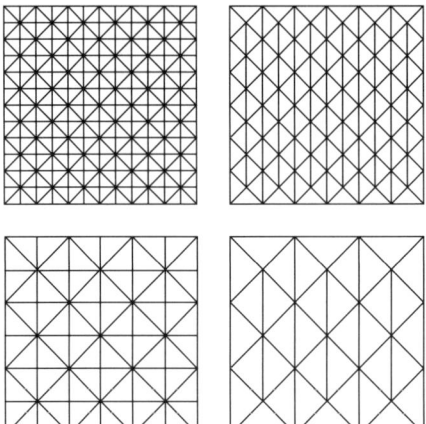

Figure 11.7   Stages of the hierarchical triangulation procedure for a plane region.

efficient triangulation algorithm presented in subsection 2.3 was applied in order to reconstruct the surface at the highest detail. Finally, the algorithm generates successive lower density meshes consisting of more triangles wherever sharp edges exist and fewer in uniform plane regions due to the use of the surface estimation filters presented in Section 3.

In Figure 11.8b a regular mesh generated for the 'Sphinx' surface is shown. In Figures 11.9b and 11.9c, a regular mesh with different value of the resolution parameters **d** is generated for the surface 'Sphere'. The mesh is still extremely good.

As a measure of the quality of the reconstructed surface, the Mean Square Error (MSE) between the original and the reconstructed surface was calculated. In order to achieve a MSE equal to 26.5×10^{-3}, for the 'Sphinx' surface, only 306 vertices and 514 triangles must be used, resulting in a decimation ration of 1:5.8. For the 'Sphere' surface, for MSE = 0.126, 3162 vertices and 4165 triangles are needed, resulting to a decimation ratio of 1:2.03. These experimental results are summarized in Table 11.1.

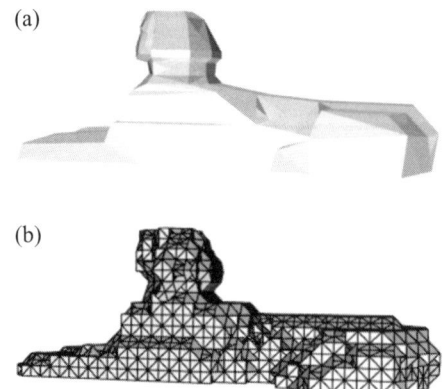

(a)

(b)

Figure 11.8   (a) Initial model 'Sphinx', (b) regular mesh of the 'Sphinx' with resolution parameter d.

Table 11.1 Experimental results of the proposed scheme for the model 'Sphinx' and 'Sphere'. (M.S.E. = Mean Square Error)

	Sphinx	Sphere
Initial Points	1919	7170
Initial Triangles	2832	8592
Transmitted Points	306	3162
Transmitted Triangles	514	4165
Compression Ration	5.8	2.03
M.S.E	$26.5 \cdot 10^{-3}$	$12.6 \cdot 10^{-2}$
Time	1.02 sec	8.94 sec
Bit Rare	14.4 Kbps	14.4 Kbps

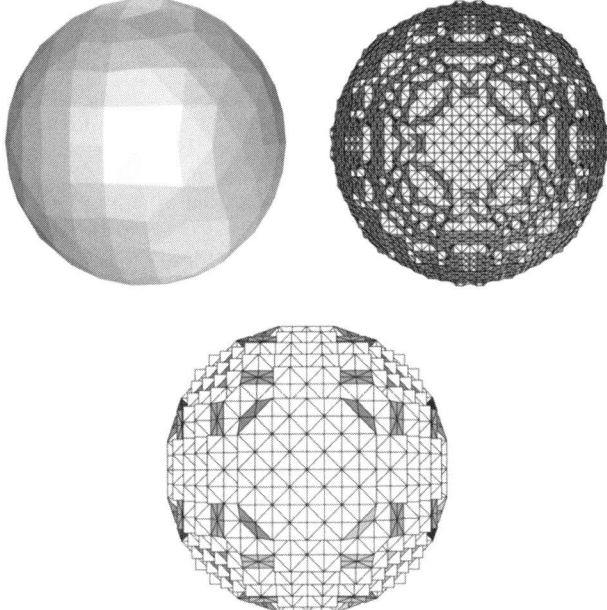

Figure 11.9 (a) Initial model 'Sphere', (b) regular mesh of the 'Sphere' with resolution parameter d, (c) regular mesh of the 'Sphere' with resolution parameter $2d$.

5. Summary

In this chapter, an optimal hierarchical adaptive mesh construction algorithm using the FCO sampling was presented. FCO is a natural extension of the quincunx lattice to the 3-dimensional case. Initially, a highly detailed and densely sampled regular mesh is obtained from geometry scanning or from a non-optimal polygon mesh. The adaptive triangle mesh is constructed by using fixed position vertices along with an efficient adaptive triangulation technique. The decimation is based on FCO sampling and surface estimation filters. The result is a progressive sequence of meshes consisting of more triangles wherever sharp edges exist and fewer in uniform plane regions. The latter property is valuable when high quality progressive coding is desired such as when browsing in large databases of 3D models, where a low-detail, high-quality preview is usually required. This procedure extends from the finer to the coarser level, until the desired detail of the mesh is reached. The resulting number of vertices is a measure of the efficiency of the decimation procedure. As can be seen from the experimental results section, the proposed algorithm achieves high decimation ratios, therefore the requirements introduced in Section 1 number 1 and 2 are met. Progressive transmission properties can be clearly seen from the presented results at different levels (requirements 3 and 4). Furthermore, precise correspondence between triangles at each level is achieved resulting in a fully hierarchical representation of the mesh (requirement 5). The triangulation mechanism results in an accurate and parsimonious representation of the surface.

References

[1] G. M. Nielson, *Modeling and Visualizing Volumetric and Surface-on-Surface Data*. New York: Springer, 1993, pp. 191–242.

[2] P. Heckbert and M. Garland, "Multiresolution modeling for fast rendering," in *Proc. Graphics Interface*, 1994, pp. 1–8.

[3] P. Lindstrom, D. Koller, W. Ribarsky, L. F. Hughes, N. Faust and G. Turner, "Real-time, continuous level of detail rendering of height fields," in *Proc. SIGGRAPH '96 Conf.*, H. Rushmeier, Ed., New Orleans, LA, Aug. 4–9, 1996, pp. 109–118.

[4] L. Haibo, P. Roivanen and R. Forcheimer, "3-D motion estimation in modelbased facial image coding," *IEEE Trans. Pattern Anal. Machine Intell.*, vol. 15, pp. 545–555, June 1993. *Commun.*, vol. 44, Jan. 1996, pp. 18–23.

[5] P. S. Heckbert and M. Garland, "Survey of polygonal surface simplification algorithms," presented at the *SIGGRAPH 97*, Course Notes 25, 1997.

[6] P. Cignoni, C. Montani and R. Scopigno, "A comparison of mesh simplification algorithms," *Comput. Graph.*, vol. 22, no. 1, 1998, pp. 37–54.

[7] L. De Floriani, P. Marzano and E. Puppo, "Multiresolution models for topographic surface description," *Vis. Comput.*, vol. 12, Aug. 1996, pp. 317–345.

[8] L. De Floriani and E. Puppo, "Hierarchical triangulation for multiresolution surface description," *ACM Trans. Graph.*, vol. 14, no. 4, 1995, pp. 363–411.

[9] W. J. Schröder, J. A. Zarge and W. E. Lorensen, "Decimation of triangular meshes," in *Proc. SIGGRAPH'92, 1992*, pp. 65–70.

[10] R. Sivan and H. Samet, "Algorithms for constructing quadtree surface maps," in *Proc. 5th Int. Symp. Spatial Data Handling*, Aug. 1992, pp. 361–370.

[11] H. Samet, "The quadtree and related hierarchical data structures," *ACM Comput. Surv.*, vol. 16, 1984, pp. 187–260.

[12] P. Lindstrom, D. Koller, W. Ribarsky, L. F. Hodges, N. Faust and G. A. Turner, "Real-time, continuous level of detail rendering of height fields," in *Proc. SIGGRAPH*, 1996, pp. 109–118.

[13] R. Pajarola, "Large scale terrain visualization using the restricted quadtree triangulation," in *Proc. Visualization '98*, vol. 515, LosAlamitos, CA, 1998, pp. 19–26.

[14] M. Duchaineau, M. Wolinsky, D. E. Sigeti, M. C. Miller, C. Aldrich and M. B. Mineev-Weinstein, "Roaming terrain: Real-time optimally adapting meshes," in *Proc. Visualization*, Los Alamitos, CA, 1997, pp. 81–88.

[15] H. Hoppe, T. DeRose, T. DuChamp, J. McDonald and and W. Stuetzle, "Mesh optimization," presented at the *SIGGRAPH '93*, Anaheim, CA, Aug. 1993.

[16] M. Gross, O. Staadt and R. Gatti, "Efficient triangular surface approximations using wavelets and quadtree data structures," *IEEE Trans. Visual. Comput. Graph.*, vol. 2, June 1996, pp. 130–143.

[17] M. F. Deering, "Geometry compression," in *Proc. SIGGRAPH '95* Conf., R. Cook, Ed., Los Angeles, CA, Aug. 6–11, 1995, pp. 13–20.

[18] G. Taubin and J. Rossignac, "3D geometry compression," in *Proc. SIGGRAPH*, 1999.

[19] William Horn Gabriel Taubin, Andre Geuziec and Francis Lazarus, "Progressive forest split compression," in *SIGGRAPH 98*. ACM SIGGRAPH, 1998, pp. 123–132, Addison Wesley.

[20] Oliver G. Staadt, Markus H. Gross and Roger Weber, "Multiresolution compression and reconstruction," in *IEEE Visualization '97*, Roni Yagel and Hans Hagen, Eds. IEEE, November 1997, pp. 337–346.

[21] A. Khodakovsky, P. Schröder and W. Sweldens, "Progressive geometry compression," in *Proc. SIGGRAPH 2000 Conf.*, 2000.

[22] C. L. Bajaj, V. Pascucci and G. Zhuang, "Progressive compression and transmission of arbitrary triangular meshes," in *Proc. Visualization '99*, Los Alamitos, CA, 1999, pp. 307–316.

[23] H. Hoppe, "Progressive meshes," in *Proc. SIGGRAPH '96* Conf., New Orleans, LA, Aug. 4–9, 1996, pp. 99–108.

[24] R. Pajarola and J. Rossignac, "Compressed progressive meshes," *IEEE Trans. Vis. Comput. Graph.*, vol. 6, Jan. 2000, pp. 79–93.

[25] D. Cohen-Or, D. Levin and O. Remez, "Progressive compression of arbitrary triangular meshes," in *Proc. Visualization '99*, Los Alamitos, CA, 1999, pp. 67–72.

[26] I. Guskov, W. Sweldens, P. Shroder, "Multiresolution Signal Processing for Meshes," in *Proc. SIGGRAPH '99*, 1999.

[27] D. Tzovaras and M. G. Strintzis, "Optimal Construction of Reduced Pyramids For Lossless And Progressive 3D Volume Data Coding," in *IMDSP '98*, Austria, Jun. 1998.

[28] Panagiotis Dafas, Ioannis Kompatsiaris and Michael G. Strintzis, "Optimal Hierarchical Adaptive Mesh Construction Using FCO Sampling", *International Conference on Information Visualisation*, London, England, 19–21 July 2000.

[29] I. Kompatsiaris, D. Tzovaras and M. G. Strintzis, "Hierarchical Representation and Coding of Surfaces using 3D Polygon Meshes", *IEEE Trans. on Image Processing*, vol 10, no. 8, August 2001.

[30] I. Kompatsiaris, D. Tzovaras and M. G. Strintzis, "Hierarchical Representation of Surfaces Using 3D Wireframes," in *8th International Conference in Central Europe on Computer Graphics, Visualization and Interactive Digital Media '2000*, Plzen, Chech Republic, February 2000.

[31] P. P. Vaidyanathan, *Multirate Systems and Filter Banks*, Prentice-Hall, 1993.

12

Virtual Sculpting-A Boundary Element Approach

Based on "Virtual Sculpting and Deformable Volume Modeling" K. C. Hui, H. C. Leung. The proceedings of the 6th International Conference on Information Visualisation 10–12 July 2002, London.

K. C. Hui and H. C. Leung

Department of Automation and Computer-Aided Engineering,
The Chinese University of Hong Kong, Shatin, Hong Kong.

Volume modeling techniques are capable of modeling objects with possible changes in topology and are widely used in virtual sculpting. A common practice are to provide shape-editing tools for adding or removing material from a volume model. However, sculpting may also involve the constant volume deformation of a model (e.g. the deformation of a clay model). This usually requires the use of physically based deformation techniques. One approach is to use the Finite Element Modeling (FEM) technique. This requires generating solid meshes from the volume data, which is a time consuming process. A modification in the volume data will require a solid mesh to be regenerated for the deformation process. In this chapter, an approach based on the Boundary Element Method (BEM) is discussed. The deformation is computed based on the iso-surface of the volume data. This eliminates the need for generating solid mesh from the volume data. By converting the deformed mesh to volume data, a deformed volume model can be further manipulated with existing volume modeling techniques.

1. Introduction

Volume modeling is known to be a powerful technique for visualizing volumetric data [1–4]. Techniques for deforming volume model extended the applications of volume modeling for applications such as virtual sculpting, and the simulation of tissue responds in medical applications. Galyan and Hughes [5] proposed using sculpting tools such as a toothpaste tube, heat gun, and sandpaper for modifying the shape of a volume model. Wang and Kaufman [6]

Advances in Geometric Modeling. Edited by M. Sarfraz
© 2003 John Wiley & Sons, Ltd   ISBN: 0-470-85937-7

extended the method for voxels with attributes such as color and texture. Other works based on the same approach have been reported [7, 8].

Early work in the modeling of elastic objects [9, 10] assumed the objects are represented as superquadrics so that a grid generated on the surface of the object can be mapped to a volume in the corresponding parametric space of the superquadrics. The grid can thus be used for the computation of deformation. This approach can be applied for objects represented as volumes in a parametric space, but may not be applied for objects in other representations. Gibson [11] simulated the deformation of a volume model by assuming adjacent voxels to be connected with springs. This produces good visual simulation of an elastic object but may not be applicable when material properties of the object are to be considered. Recent work in this area employed the finite element analysis technique for evaluating the deformation of tissue for medical applications [12, 13, 14] as well as the animation of synthetic human [15], and product design [16, 17]. A characteristic of the finite element technique is the requirement of a mesh that may have to be regenerated if there is a change in object shape. A good survey of various techniques in the deformation of objects can be found in [18]. In this chapter, the boundary element technique is adopted. This eliminates the need for regenerating the solid mesh of an object whenever the object is modified. By converting the mesh of a deformed object to a volume model, further volume operations (e.g. engraving, gluing, etc) can be performed on the deformed object. This gives a virtual sculpting system that allows physics based deformation of objects.

2. Physics Based Volume Modeling

A volume model is a representation of a closed object X defined with a function $f : R^3 \rightarrow R$. A point $\mathbf{p}$ is in the interior of the object if $f(\mathbf{p}) > k$, where k is a threshold value. A point $\mathbf{p}$ is on X if $f(\mathbf{p}) = k$, and out of X if $f(\mathbf{p}) < k$. A volume model is a three dimensional grid of points or voxels enclosing the object X. Values of the voxels are the function values $f(\mathbf{p})$ so that the voxels completely define the object X. The boundary of the object are the locations at which $f(\mathbf{p}) = k$, i.e. the iso-surface at $f(\mathbf{p}) = k$.

Adjusting the function values of the voxels can deform a volume model. This allows operations such as cutting and gluing of material. However, an elastic deformation of the object requires a physical model of the object satisfying the boundary conditions on the iso-surface of the volume model. This corresponds to the forces or displacements on the isosurface of the object. Denoting $\mathbf{u}$ as the displacement of a point $\mathbf{p} = (x, y, z)$ of the object, then the equilibrium condition of the object is given by the Navier equation:

$$\nabla^2 \mathbf{u} + \frac{1}{1 - 2v} \nabla(\nabla \cdot \mathbf{u}) + \frac{\mathbf{F}}{\mu} = 0, \qquad (12.1)$$

where $\mu = \dfrac{E}{2(1 + v)}$, and E, v are respectively the Young's modulus and Poisson's ratio of the material used, $\mathbf{F}$ is the body force at $\mathbf{p}$.

Various approaches have been developed for solving the Navier equation. The most popular one is the FEM [19]. The finite element approach evaluates the displacement and forces at the vertices of a mesh of solid elements approximating the object. Given a volume model, this requires generating the mesh of solid elements from the volume data. An alternative is to adopt the Boundary Element (BE) approach [20] in which the displacements and forces are evaluated directly on a mesh of polygons approximating the surface of the object. Since a triangular mesh of a volume model can be easily obtained using the Marching Cube method [21], the BE approach can be directly applied to the iso-surface of a volume model.

3. The Boundary Element Approach

The BE approach expresses the displacement of a surface point $\mathbf{p}$ of an object in terms of the fundament solution to Equation 12.1. Using tensor notations, and assuming there is no body forces, $\mathbf{t} = (t_x, t_y, t_z)$ is the traction (or force) at $\mathbf{p}$, the boundary integral equation can be expressed as:

$$u_{ref_j} + \int_S T_{ij} u_j dS = \int_S U_{ij} t_j dS, \qquad (12.2)$$

where S is the surface of the object, T_{ij}, U_{ij} is a fundamental solution of Equation 12.1, and $\mathbf{u}_{ref}$ is the deformation at the reference point $\mathbf{p}_{ref}$. U_{ij}, T_{ij} are the deformation and traction respectively at $\mathbf{p}$ obtained with a unit load applied to an interior point $\mathbf{p}_{ref}$ of an infinite domain. Using a point on the boundary S as a reference point, Equation 12.2 is rewritten as:

$$C_{ij} u_{ref_j} + \int_S T_{ij} u_j dS = \int_S U_{ij} t_j dS, \qquad (12.3)$$

where C_{ij} is a function of the reference point.

By approximating S with a set of node points, and applying each of the node points as reference point in Equation 12.3, a set of linear equations relating u and t can be obtained. Denoting $(u_{ix}\ u_{iy}\ u_{iz})$, $(t_{ix}\ t_{iy}\ t_{iz})$ as the deformation and traction at the ith node point, then:

$$\mathbf{AU} = \mathbf{BT}, \qquad (12.4)$$

where $\mathbf{U} = [u_{0x}\ u_{0y}\ u_{0z}\ u_{1x}\ u_{1y}\ u_{1z}..........u_{nx}\ u_{ny}\ u_{nz}]^\mathrm{T}$,
$\quad\ \mathbf{T} = [t_{0x}\ t_{0y}\ t_{0z}\ t_{1x}\ t_{1y}\ t_{1z}..........t_{nx}\ t_{ny}\ t_{nz}]^\mathrm{T}$,
$\quad\ \mathbf{A}$ is a function of C_{ij} and T_{ij},
$\quad\ \mathbf{B}$ is a function of U_{ij}.

4. Deforming the Object

Deformation is applied to a set of node points on S, while constraints are applied to selected node points on S. Denoting $\mathbf{U}_K$ as the displacements of the nodes where the deformation is applied, $\mathbf{U}_U$ as the displacement of nodes to be determined, $\mathbf{T}_K$ as the traction at the free nodes, and $\mathbf{T}_U$ is the traction to be determined, Equation 12.1 can be expressed as:

$$\begin{bmatrix} \mathbf{A}_{00} & \mathbf{A}_{01} \\ \mathbf{A}_{10} & \mathbf{A}_{11} \end{bmatrix} \begin{bmatrix} \mathbf{U}_U \\ \mathbf{U}_K \end{bmatrix} = \begin{bmatrix} \mathbf{B}_{00} & \mathbf{B}_{01} \\ \mathbf{B}_{10} & \mathbf{B}_{11} \end{bmatrix} \begin{bmatrix} \mathbf{T}_U \\ \mathbf{T}_K \end{bmatrix}.$$

Since the traction at all free nodes are zero, i.e. $\mathbf{T}_K = 0$, hence,

$$\mathbf{A}_{00}\mathbf{U}_U + \mathbf{A}_{01}\mathbf{U}_K = \mathbf{B}_{00}\mathbf{T}_U,$$
$$\mathbf{A}_{10}\mathbf{U}_U + \mathbf{A}_{11}\mathbf{U}_K = \mathbf{B}_{10}\mathbf{T}_U.$$

Eliminating $\mathbf{T}_U$ gives:

$$\mathbf{U}_U = [\mathbf{B}_{00}{}^{-1}\mathbf{A}_{00} - \mathbf{B}_{10}{}^{-1}\mathbf{A}_{10}]^{-1}[\mathbf{B}_{10}{}^{-1}\mathbf{A}_{11} - \mathbf{B}_{00}{}^{-1}\mathbf{A}_{01}]\mathbf{U}_K, \qquad (12.5)$$

Given the constraints and deformations at certain nodes of the object, the displacement of the other free nodes can be computed using Equation 12.5.

5. Converting Mesh Data to Volume Model

In order to allow volume operations on the deformed model, a surface model has to be converted into a volume model before operations such as gluing and removal of voxels can be applied. There are various approaches for converting object models to a volume model [22–24]. In this chapter, a simple approach is adopted. Each voxel $\mathbf{q}$ is classified with respect to the deformed object. The value of the voxel is 1 if $\mathbf{q}$ is *in* the object. Otherwise, the voxel is 0. This gives a rough volume model of the object because of the discrete voxel values. The iso-surface generated from these voxel values may consist of sharp jaggy corners. To obtain a smooth iso-surface, a convolution process may be applied to the volume model.

A box and a Gaussian filter are adopted for the smoothing operation. These filters are associated with a sculpting tool. By adjusting the size of the sculpting tool, the convolution process can be applied to the whole or part of the volume model.

6. Addition and Removal of Material

Altering the voxel values in the workspace can perform addition and removal of material. The same approach as stated in [5] is adopted. A sculpting tool specifies the region of the workspace where material is to be attached or removed. Consider a sculpting tool defined with a function $g(x, y, z) < 0$. In an operation for the addition of material, the value of the voxels at locations satisfying $g(x, y, z) < 0$ are changed to 1. In a material removal operation, voxels located in the intersection of the sculpting tool and the object (i.e. at locations where $g(x, y, z) < 0 \cap f(x, y, z) < 0$) are changed to 0.

7. Implementations and Results

A virtual sculpting system was implemented on a personal computer. Starting with a primitive shape such as a sphere, cylinder, etc., a volume model is generated. A user is allowed to remove voxels from the model using a spherical tool of which sizes can be adjusted. Figure 12.1a shows a volume model of a block. Figure 12.1b shows the tool position for a material removal operation. The result of the material removal operation is shown in Figure 12.1c. Similarly, voxels can be glued to the model by pointing to a location interactively (Figure 12.2). Figures 12.3 and 12.4 illustrate two engravings which are the result of applying the removal and addition tools. Figure 12.5 shows the process of deforming a plate into a flower like object. The center point at the bottom of the plate (Figure 12.5b) is fixed or constrained in position. Deformations are applied to the corners of the plate. The flower (Figure 12.5d) is obtained in three consecutive deformations.

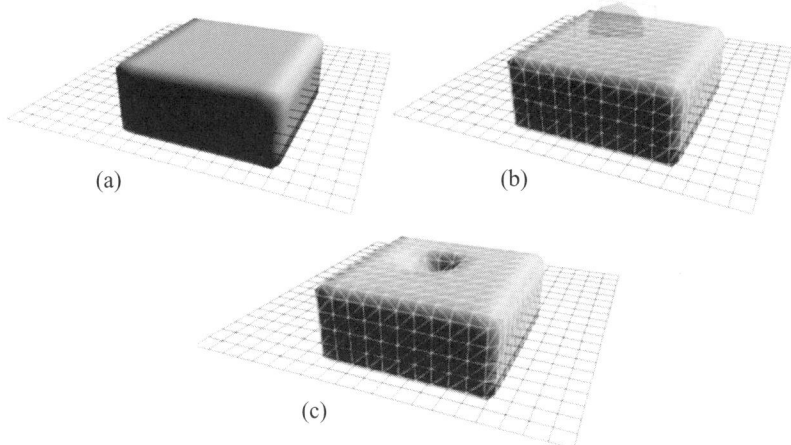

(a)   (b)

(c)

Figure 12.1 Removing material from volume model: (a) a volume model of a block, (b) positioning the removal tool, (c) result of the removal operation.

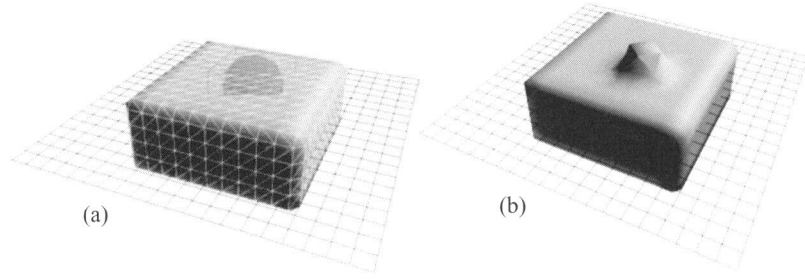

(a)   (b)

Figure 12.2 Adding material to a volume model: (a) Positioning the sculpting tool, (b) result of the addition.

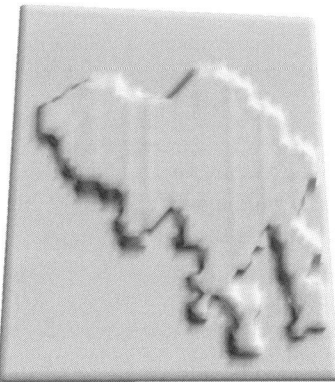

Figure 12.3 An engraving created with the removal tool.

Figure 12.4 An engraving created with the addition tool.

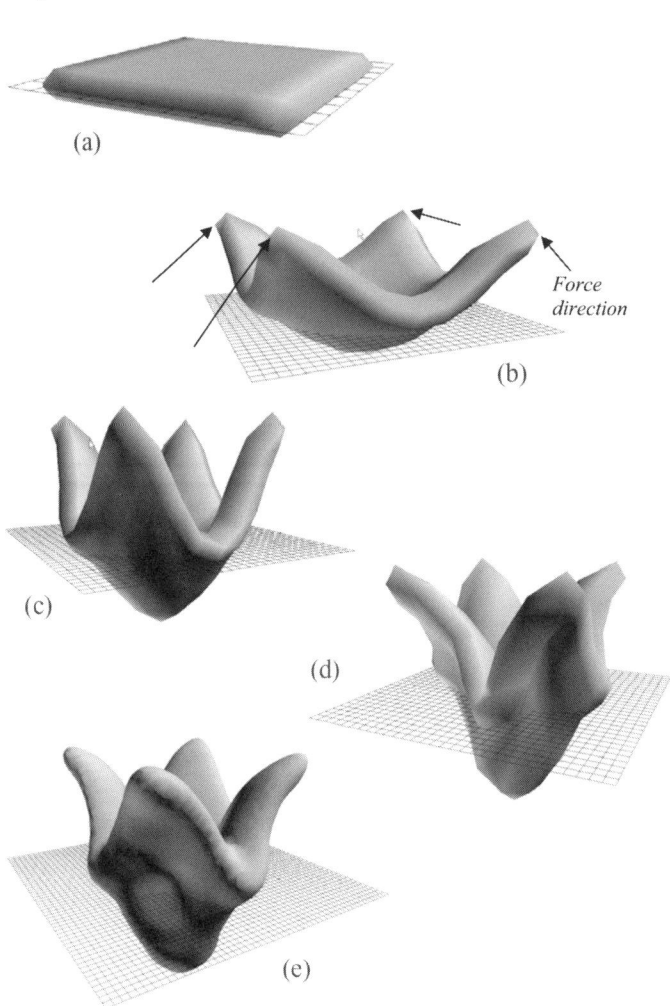

Figure 12.5   Deformation of a volume model: (a) The undeformed plate, (b) first deformation, (c) second deformation, (d) result before the smoothing operation, (e) result after the smoothing operation.

Figure 12.6a shows a jar modeled by removing material from a cylindrical block. The handle is created by adding material to the block. A deformation is applied to obtain the final shape (Figure 12.6b) of the object.

8. Summary

Existing volume sculpting techniques usually allow materials to be added to or removed from a volume model. However, a physics based deformation operation is essential for providing a clay like deformation effect in a sculpting operation. In the proposed virtual sculpting system, the boundary element technique is adopted for deforming a volume model. The boundary element technique only requires the surface mesh of an object for evaluating the deformation

(a)

(b)

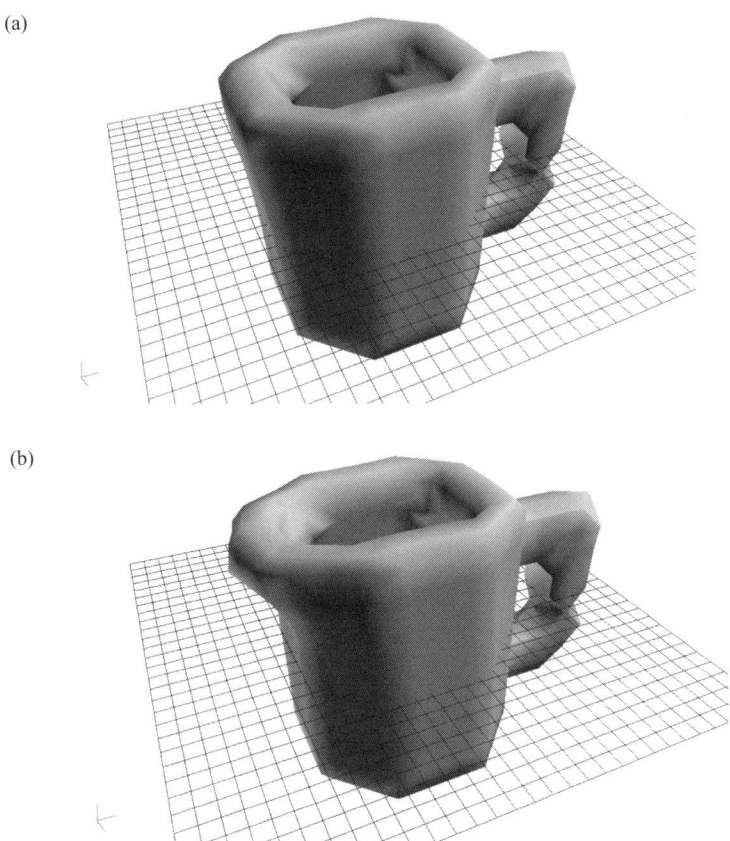

Figure 12.6   A jar model: (a) Model constructed by removing and adding material to a block, (b) the
final shape created by applying a deformation.

of an object. Since the boundary mesh of a volume model is readily available, boundary element
analysis can be directly applied to a volume model. In order to allow further volume operations
on the deformed volume model, the deformed surface mesh has to be converted into a volume
model. Although the time required for evaluating deformation increases with the mesh sizes,
experiments showed that a practical approach is to use a coarse model (model with fewer
triangles) for the deformation. The deformed object can then be smoothed to obtain the final
object at the required resolution.

Acknowledgement

This work is partially supported by a Direct Grant (ID 2050210) of the Chinese University of
Hong Kong.

References

[1] A Kaufman, *Volume Visualization*, IEEE Computer Society Press, Los Almitos, CA 1991.
[2] A Kaufman, D Cohen, R Yagel, "Volume graphics", *IEEE Computer*, Vol. 26, No. 7, July 1993, pp. 51–64.

[3] B Collins, "Data Visualization, Directions in Geometric Computing", Edited by R Martin, *Information Geometers*, 1993, pp. 31–80.

[4] G M Nielson, P Brunet, M Gross, H Hagen, S V Klimenko, "Research issues in data modeling for scientific visualization", *IEEE Computer Graphics and Applications*, March, 1994, pp. 70–73.

[5] T A Galyean, J F Hughes, "Sculpting: An interactive volumetric modeling technique", *Computer Graphics*, Vol. 25, No. 25, 1991, pp. 267–274.

[6] S Wang, A Kaufman, "Volume sculpting", *Proc. Symposium on Interactive 3D Graphics, ACM SIGGRAPH, 1995*, pp. 151–156.

[7] I Fujishiro, Y Maeda, H Sato, Y Takeshima, "Interval volume: A solid fitting technique for volumetric data display and analysis", *Proceedings of Visualization '95*, Oct-Nov. 1995, pp. 151–158.

[8] E Ferley, M P Cani, J D Gascuel, "Practical volumetric sculpting", *The Visual Computer*, Vol. 16, 2000, pp. 469–480.

[9] D Terzopoulos, K Fleischer, "Deformable models", *The Visual Computer*, Vol. 4, 1988, pp. 306–331.

[10] I Essa, S Sclaroff, A Pentland, "Physically based modelling for graphics and vision, Directions in Geometric Computing", Edited by R Martin, *Information Geometers*, 1993, pp. 161–196.

[11] S F F Gibson, "Beyond volume rendering: visualization, haptic exploration, physical modeling of voxel-based objects", *Visualization in Scientific Computing '95, Proceedings of the Eurographics Workshop* in Chia, Italy, May 35, 1995, pp. 10–24.

[12] T Kunii, "Research issues in modeling complex object shapes", *IEEE Computer Graphics and Applications*, March, 1994, pp. 80–83.

[13] M Bro-Nielsen, S Cotin, "Real-time volumetric deformable models for surgery simulation using finite elements and condensation", *Computer Graphics Forum*, Vol. 15, No. 3, (*Eurographics '96*), 1996, pp. C57–C461.

[14] M Bro-Nielsen, "Modelling elasticity in solids using Active Cubes–Application to simulated operations", *Proc. Computer Vision, Virtual Reality and Robotics in Medicine (CVRMed'95)*, 1995, pp. 535–541.

[15] J P Gourret, N M Thalmann, D Thalmann, "Simulation of object and human skin deformations in a grasping task", *Proc. SIGGRAPH'89*, 1989, pp. 21–30.

[16] H K Kang, A Kak, "Deforming virtual objects interactively in accordance with an elastic model", *Computer Aided Design*, Vol. 28, No. 4, 1995, pp. 251–262.

[17] K C Hui, N N Wong, "Hands on a virtually elastic object", *The Visual Computer*, Vol. 18, No. 3, 2002, pp. 150–163.

[18] S F Gibson, B Mirtich, "A survey of deformable models in computer graphics", Technical Report TR-97-19, Mitsubishi Electric Research Laboratories, Cambridge, MA, November 1997.

[19] K J Bathe, *Finite Element Procedures*, Prentice Hall, 1996.

[20] A A Becker, *The Boundary Element Method in Engineering – A complete course*, McGraw-Hill, 1992.

[21] W E Lorensen, H E Cline, "Marching Cubes: A high resolution 3D surface construction algorithm", *Computer Graphics*, Vol. 21, No. 4, 1987, pp. 163–169.

[22] M W Jones, "The production of volume data from triangular meshes using voxelisation", *Computer Graphics Forum*, Vol. 15, No. 5, 1996, pp. 311–318.

[23] J Huang, R. Yagel, V Filippov, Y Kurzion, "An accurate method for voxelizing polygon meshes", *Proc. 1998 Volume Visualization Symposium*, IEEE, Oct 1998, pp. 119–126.

[24] F Dachille IX, A Kaufman, "Incremental Triangle Voxelization", *Proc. Graphics Interface* 2000, May 2000, pp. 205–212.

13

Free Form Modeling Method Based on Silhouette and Boundary Lines

Jun Kamiya
Hideki Aoyama

Keio University, Department of System Design Engineering
3-14-1 Hiyoshi, Kohoku-ku, Yokohama, Japan.

In the initial process of exterior design of a product, designers often get the idea of a designing product form as silhouette lines and boundary lines. Therefore, if the 3-D model is constructed from silhouette lines and boundary lines, designers will be significantly assisted in the process of external form design. This chapter describes a system to construct a 3-D model with complex curved surfaces from silhouette and boundary lines. The silhouette and boundary lines can be easily given by a computer mouse or a tablet and are then approximated by cubic Bezier curves linked with curvature continuity at the connected points. A Bezier surface is defined from the four lines as the silhouette or the boundary by an originally developed algorithm. The 3-D model is then constructed by Bezier surfaces with curvature continuity on the connecting lines. Since the system does not obstruct a designer's thinking due to simple operations of the system, designers can elaborate and confirm the idea by considering and examining the form indicated by the system.

1. Introduction

It is not just the function which is becoming an important factor when a customer selects and buys an industrial product as the external form is also important. External form design: style design, is creative activity depending on a designer's sensibility [1], [2]. Many Computer-Aided Design (CAD) systems have been developed for modeling the form satisfying functions of a product and they have served to increase design efficiency. However, there are very few systems

Advances in Geometric Modeling. Edited by M. Sarfraz
© 2003 John Wiley & Sons, Ltd   ISBN: 0-470-85937-7

for supporting design of the external form of a product because the current CAD systems hinder a designer in getting an idea of an external form due to the complicated operations.

In the initial design process of the external form of a product, designers embody it from the image and concept of the targeted product by silhouette and boundary lines. Therefore, if a 3-D model is directly and easily constructed from such lines and the designed form can be examined and evaluated, much support is given for design activity of an external form of a product.

In this chapter, in order to support the design process of the external form of automobiles, an algorithm is proposed to easily construct a 3-D model by the silhouette lines and boundary lines given by a designer's handwriting. The effectiveness of a system implemented according to the proposed algorithm is examined.

2. Outline of Developed System

2.1. Silhouette Lines Representing Form Feature of An Automobile

The objective of this study is to develop a system to construct the 3-D model of an automobile from the silhouette lines and boundary lines. Figure 13.1 shows the silhouette lines in the side view of an automobile: the base line, shoulder line, roof line, front window line, rear window line, face line, and tail line. The lines directly and obviously represent form features of an automobile. Form expression of an automobile such as speed, stability, attractiveness, size deluxe, and so on are typically determined by the silhouette lines. All types of automobiles can be expressed by combining the silhouette lines.

2.2. Construction and Evaluation of An Automobile Model Form

The silhouette and boundary lines can be input into the developed system by writing with a mouse or a tablet. The developed system constructs the 3-D model of an automobile from the silhouette lines and boundary lines in the views from the four directions as shown in Figure 13.2. The constructed model is presented to the designer and is evaluated with the views from various directions. If the presented form does not satisfy a designer's sensibility, the silhouette and boundary lines are modified by input once more and another model is then re-formed. The processes are repeated until a designer obtains a satisfactory form. In the process, the model

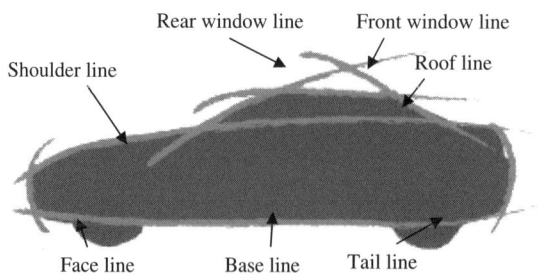

Figure 13.1   Silhouette lines in the side view of an automobile.

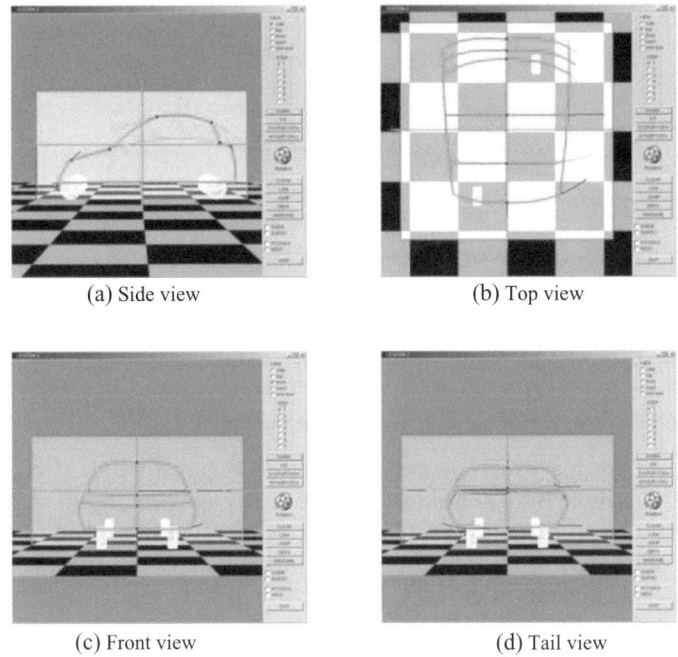

Figure 13.2   Silhouette and boundary lines to construct an automobile model.

can be evaluated with different colors, different scenes, and different types of light source. Since the modeling processes are executed by very simple operations, the developed system does not hinder designers in creating and embodying ideas.

3. Processes of Model Construction

Figure 13.3 shows the processes to construct a 3-D model from the silhouette and boundary lines. In the following, the detailed procedure of each process is described.

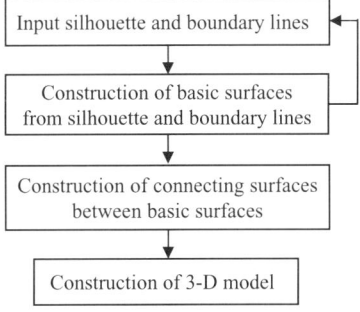

Figure 13.3   Processes to construct 3-D model.

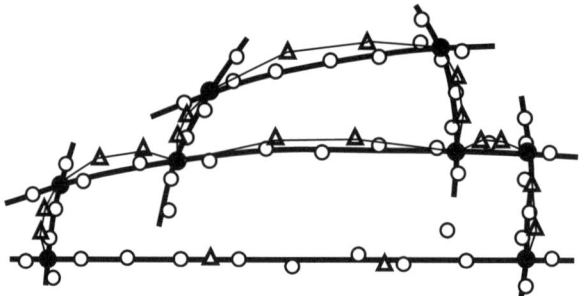

Figure 13.4 Definition of silhouette boundary lines.

3.1. Input and Definition of Silhouette and Boundary Lines

The silhouette and boundary lines shown with lines in Figure 13.4 are defined by making approximations with cubic Bézier equations from discrete point data shown with white circles in this figure input by a mouse and a tablet. The approximated Bézier curves are evaluated by the locus of the center of curvature as shown in Figure 13.5. Figure 13.5(a) shows the locus of the center of curvature of the cubic Bézier curve directly approximated from the input data. The cubic Bézier curve is modified according to the evaluation of the curvature locus. Figure 13.5(b) shows the modified Bézier curve and its locus of the center of curvature.

The intersections shown with black circles in Figure 13.4 between the approximated Bézier equations are then derived. The approximated cubic Bézier curves are fragmented with the intersections and the control points to define each fragmented cubic Bézier curves are derived. The derived control points of the fragmented Bézier curves are shown with triangles in

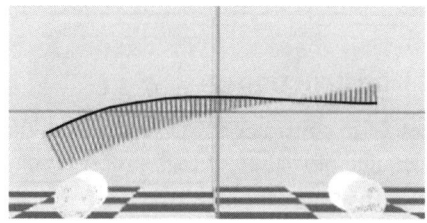

(a) Locus of curvature center of defined shape by input

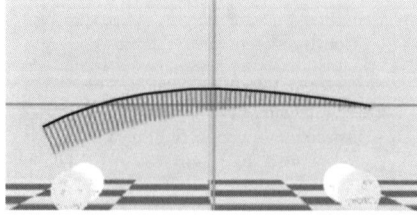

(b) Locus of curvature modified center

Figure 13.5 Evaluation of line shape by locus of curvature center.

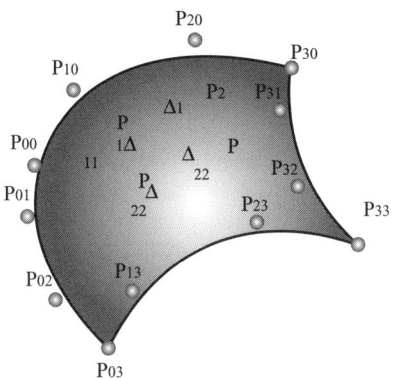

Figure 13.6   Boundary lines to define surface.

Figure 13.4. Thus, a Bézier curve defined by two intersections and two control points shown with the black circles and triangles in this figure respectively, plays a role as a boundary line to define a basic surface.

3.2. Model Construction by Basic Surfaces

The following is the procedure to define basic surfaces of a model. Figure 13.6 shows the four Bézier curves to restrict the boundary form of a surface. Each Bézier curve is defined by the four control points shown with the circles in Figure 13.6 that are derived by the procedure mentioned in Section 3.1. When a basic surface is defined by a cubic Bézier equation, another four control points P_{11}, P_{12}, P_{21}, and P_{22} shown with triangles in Figure 13.6 are needed for the twelve control points P_{00}, P_{01}, P_{02}, P_{03}, P_{10}, P_{13}, P_{20}, P_{23}, P_{30}, P_{31}, P_{32}, and P_{33}. The four control points P_{11}, P_{12}, P_{21}, and P_{22} are derived by the Equations (13.1) and (13.2).

$$\left.\begin{aligned}
P_{30c} &= P_{30} + (P_{20} - P_{30}) + (P_{31} - P_{30}) \\
P_{30d} &= P_{30} + (P_{20} - P_{30}) + (P_{32} - P_{30}) \\
P_{00c} &= P_{00} + (P_{01} - P_{00}) + (P_{20} - P_{00}) \\
P_{00d} &= P_{00} + (P_{02} - P_{00}) + (P_{20} - P_{00}) \\
P_{03a} &= P_{03} + (P_{01} - P_{03}) + (P_{13} - P_{03}) \\
P_{03b} &= P_{03} + (P_{02} - P_{03}) + (P_{13} - P_{03}) \\
P_{03c} &= P_{03} + (P_{01} - P_{03}) + (P_{23} - P_{03}) \\
P_{03d} &= P_{03} + (P_{02} - P_{03}) + (P_{23} - P_{03}) \\
P_{30a} &= P_{30} + (P_{10} - P_{30}) + (P_{31} - P_{30}) \\
P_{30b} &= P_{30} + (P_{10} - P_{30}) + (P_{32} - P_{30}) \\
P_{30c} &= P_{30} + (P_{20} - P_{30}) + (P_{31} - P_{30}) \\
P_{30d} &= P_{30} + (P_{20} - P_{30}) + (P_{32} - P_{30}) \\
P_{33a} &= P_{33} + (P_{13} - P_{33}) + (P_{31} - P_{33}) \\
P_{33b} &= P_{33} + (P_{13} - P_{33}) + (P_{32} - P_{33}) \\
P_{33c} &= P_{33} + (P_{23} - P_{33}) + (P_{31} - P_{33}) \\
P_{33d} &= P_{33} + (P_{23} - P_{33}) + (P_{32} - P_{33})
\end{aligned}\right\} \qquad (13.1)$$

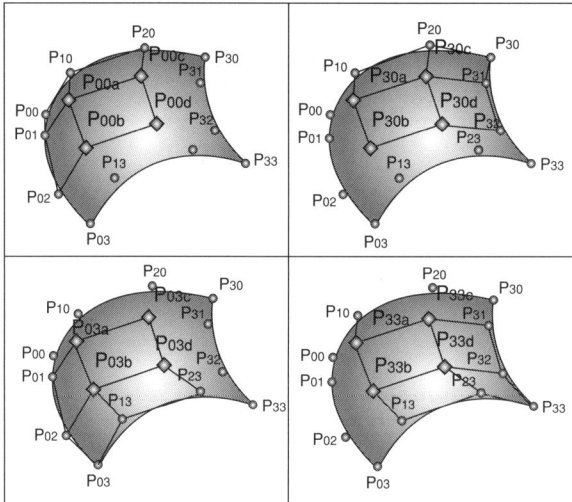

Figure 13.7   Element points.

$$
\begin{aligned}
P_{11}(u, v) &= (1.0 - v) \cdot ((1.0 - u) \cdot P_{00a} + u \cdot P_{03a}) \\
&\quad + v \cdot ((1.0 - u) \cdot P_{30a} + u \cdot P_{33a}) \\
P_{12}(u, v) &= (1.0 - v) \cdot ((1.0 - u) \cdot P_{00b} + u \cdot P_{03b}) \\
&\quad + v \cdot ((1.0 - u) \cdot P_{30b} + u \cdot P_{33b}) \\
P_{21}(u, v) &= (1.0 - v) \cdot ((1.0 - u) \cdot P_{00c} + u \cdot P_{03c}) \\
&\quad + v \cdot ((1.0 - u) \cdot P_{30c} + u \cdot P_{33c}) \\
P_{22}(u, v) &= (1.0 - v) \cdot ((1.0 - u) \cdot P_{00d} + u \cdot P_{03d}) \\
&\quad + v \cdot ((1.0 - u) \cdot P_{30d} + u \cdot P_{33d})
\end{aligned} \tag{13.2}
$$

In Equations (13.1) and (13.2), P_{ij} means the position vector of the points P_{ij}. As shown in Figure 13.7, P_{00a}, P_{00b}, P_{00c}, P_{00d}, P_{03a}, P_{03b}, P_{03c}, P_{03d}, P_{30a}, P_{30b}, P_{30c}, P_{30d}, P_{33a}, P_{33b}, P_{33c}, and P_{33d} are the position vectors of points of the originally introduced element points. And u and v are the parameters defining the Bézier surface.

A basic surface is defined with the position vectors of the twelve control points defining the boundary lines and the position vectors of the four control points P_{11}, P_{12}, P_{21}, and P_{22} derived by Equations (13.1) and (13.2) as a cubic Bézier surface by Equations (13.3) and (13.4).

When the silhouette and boundary lines shown in Figure 13.2 are given, the basic surfaces are determined according to the procedure mentioned above

$$
\begin{aligned}
S(u, v) &= [w_0(u) \quad w_1(u) \quad w_2(u) \quad w_3(u)] \\
&\quad \times \begin{bmatrix} P_{00} & P_{01} & P_{02} & P_{03} \\ P_{10} & P_{11} & P_{12} & P_{13} \\ P_{20} & P_{21} & P_{22} & P_{23} \\ P_{30} & P_{31} & P_{32} & P_{33} \end{bmatrix} \begin{bmatrix} w_0(v) \\ w_1(v) \\ w_2(v) \\ w_3(v) \end{bmatrix}
\end{aligned} \tag{13.3}
$$

$$
\begin{aligned}
w_0(t) &= (1 - t)^3 \quad w_1(t) = 3t(1 - t)^2 \\
w_2(t) &= 3t^2(1 - t) \quad w_3(t) = t^3
\end{aligned} \tag{13.4}
$$

and the constructed model is shown in Figure 13.8.

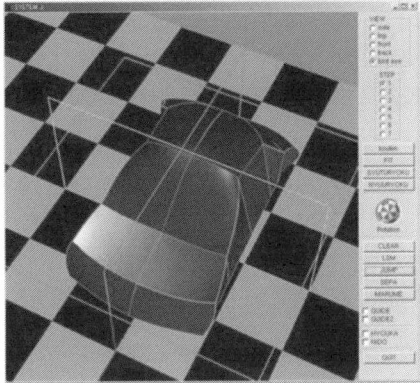

Figure 13.8   Constructed model by basic surfaces.

3.3. Construction of Smooth Surfaces Between Basic Surfaces

As shown in Figure 13.8, the basic surfaces defined from the silhouette and boundary lines are connected with only position continuity at the joined lines. In the next process, the developed system makes surfaces connecting the basic surfaces with curvature continuity.

The borders of the basic surfaces are cut as shown in Figure 13.9(a). The cut widths of the borders and the curvature of the connecting surface are determined by the required roundness at the joined lines of the basic surfaces. New surfaces are then constructed at the opening area between the basic surfaces by using the original surface equation [3], [4] as shown in Figure 13.9(b). The original surfaces can easily make smooth connection: curvature continuity, between the basic surfaces. Figure 13.10 shows an example of a constructed 3-D model.

4. Summary

This chapter summarized as follows:

(1) An easy 3-D modeling method using silhouette and boundary lines was proposed.
(2) The silhouette and boundary lines are easily given by writing using a mouse and a tablet and are defined with cubic Bézier curves by approximation.
(3) The approximated silhouette and boundary lines are evaluated by the locus of the center of curvature.
(4) Basic cubic Bézier surfaces are automatically defined from the silhouette and boundary lines.
(5) New surfaces are constructed at the border area of the basic surfaces and make a smooth connection between the basic surfaces.

The developed system can easily construct a 3-D model and can examine the form with various view directions, colors and scenes.

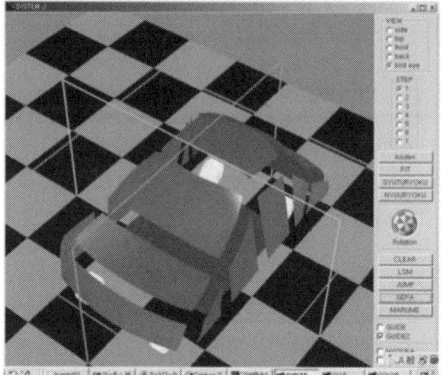

(a) Scale-down of basic surfaces

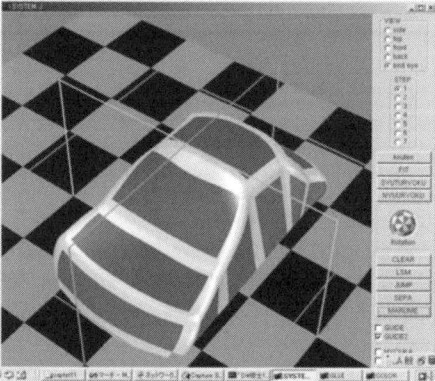

(b) Smooth connection of contracted surfaces

Figure 13.9   Construction of smooth surfaces between basic surfaces.

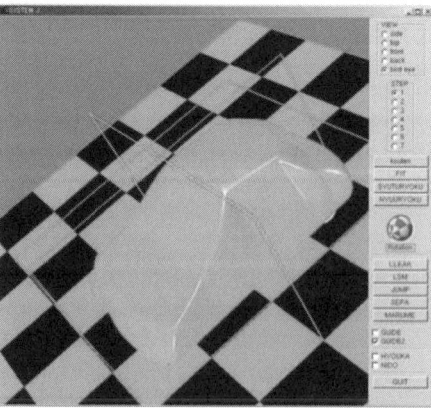

Figure 13.10   Constructed 3-D model.

References

[1] S. Yoshida, S. Miyazawa, T. Hoshino, T. Ozeki, J. Hasegawa, T. Yasuda, S. Yokoi: "Spatial Sketch System for Car Styling Design", *International Archives of Photogrammetry and Remote Sensing, XXXIII*, Part B5 (2000), 919.

[2] T. Harada: "Study of Quantitative Analysis of the Characteristics of a Curve, *Forma*, 12(1997), 55.

[3] H. Aoyama, I. Inasaki, T. Kishinami, K. Yamazaki: "A New Method for Constructing a Software Model of Sculptured Surfaces with C^2 Continuity from a Physical Model", *Advancement of Intelligent Production*, Japan Society for Precision Engineering, 1994, 7.

[4] M. Ota, H. Aoyama: "Aesthetic Design Based on KANSEI LANGUAGE", *Initiatives of Precision Engineering at the Beginning of a Millennium (10th International Conference on Precision Engineering (ICPE))*, Japan Society for Precision Engineering, 2001, 917.

14

Intuitive and Precise Solid Modeling in A Virtual Reality Environment

Yongmin Zhong
Wolfgang Müller-Wittig

Centre for Advanced Media Technology, Department of Computer Engineering,
Nanyang Technological University, Nanyang Avenue, Singapore.

Weiyin Ma

Department of Manufacturing Engineering and Engineering Management,
City University of Hong Kong, Hong Kong, China.

This chapter presents an efficient approach for solid modeling in a Virtual Reality (VR) environment. A hierarchically structured constraint-based data model is developed to support solid modeling in the VR environment. Solid modeling in the VR environment is precisely performed in an intuitive manner through constraint-based manipulations. Constraint-based manipulations are accompanied with automatic constraint recognition and precise constraint satisfaction to establish the hierarchically structured constraintbased data model and are realized by allowable motions for precise 3D interactions in the VR environment. The allowable motions are represented as a mathematical matrix for conveniently deriving allowable motions from constraints. A procedure-based degree-of-freedom incorporation approach for 3D constraint solving is presented for deriving the allowable motions. A rule-based constraint recognition engine is developed for both constraint-based manipulations and implicitly incorporating constraints into the VR environment.

1. Introduction

Virtual Reality (VR) technology is regarded as a natural extension to 3D computer graphics with advanced input and output devices. It brings a completely new environment to the CAD

Advances in Geometric Modeling. Edited by M. Sarfraz
© 2003 John Wiley & Sons, Ltd   ISBN: 0-470-85937-7

community. VR offers many advantages compared with traditional CAD. It provides intuitive 3D interaction, direct manipulation and 3D immersion. The use of VR enhances the user's understanding of a virtual object. It helps to speed up the design process and permits a designer to smoothly develop their concepts, thus fully developing their creativity. However, most of the existing VR systems [8, 12, 17, 18] only offer very limited tools for solid modeling and lack sophisticated modeling and modification tools for creating complex solid models in a VR environment. Among others, the finite resolution of virtual objects without topological information is not suitable for representing solid models for design purposes. The limited accuracy and reliability of 3D input and output devices also prevents users from precise design activities. This chapter presents an efficient approach for intuitive and precise solid modeling in a VR environment. A hierarchically structured constraint-based data model is developed to support solid modeling in the VR environment. The data model integrates a high-level constraint-based model for precise object definition, a mid-level CSG/Brep hybrid solid model for hierarchical geometry abstractions and object creation, and a low-level polygon model for real-time visualization and interaction in the VR environment. Constraints are embedded in the solid model and are organized at different levels to reflect the modeling process from features to parts. Solid modeling in the VR environment is precisely performed in an intuitive manner through constraint-based manipulations. Constraint-based manipulations are accompanied with automatic constraint recognition and precise constraint satisfaction to establish the hierarchically structured constraint-based data model and are realized by allowable motions for precise 3D interactions in the VR environment. The allowable motions are represented as a mathematical matrix for conveniently deriving allowable motions from constraints. A procedure-based Degree-Of-Freedom (DOF) incorporation approach for 3D constraint solving is presented for deriving the allowable motions. A rule-based constraint recognition engine is developed for both constraint-based manipulations and implicitly incorporating constraints into the VR environment. A prototype system has been implemented for precise solid modeling in an intuitive manner through constraint-based manipulations in the VR environment.

The use of VR for CAD is not totally new. In the area of 3D modeling, one of the earliest systems was 3DM that allows users to interactively create simple geometric objects such as cylinders and spheres in the VR environment [2]. 3DM includes several grid and snap functions. It however lacks many other aids and constraints that are necessary to accomplish precise modeling work. JDCAD also tackled many issues for interactive 3D object modeling [13]. Users could directly interact in 3D space using a 6-DOF input device. However, only simple solids can be created in JDCAD. Fa *et al.* introduced some results on 3D objects placement [4]. Fernando *et al.* further extended these results into a shared virtual environment [5] and presented a software architecture of a constraint-based virtual environment [6]. The most important contribution of their method is the concept in constraining 3D direct manipulations through the allowable motions of the object being manipulated for precise locations and operations. However, only simple geometries and constraints are treated and simple solid models can be created in the VR environment. Complex models are still created from CAD systems and then are imported into the VR environment. Dani and Gadh [3] presented a COVIRDS system for concept design in a VR environment. This system performs object modeling based on design features and the geometric modeling kernel ACIS has been used for their development. The precise interactions mainly rely on voice commands other than 3D direct manipulations. Stork and Maidhof also reported some work on interactive and precise solid modeling using a 3D input device [16]. Precise solid modeling is realized through 3D grids, grid snapping and discretized dimensions. Constraint-based interactions for feature-based modifications depend on some pre-defined rules. Although precise solid modeling can be ensured, the constraint-based interactions are too rigid for extensive use. Kiyokawa *et al.* also reported a two-handed modeling environment VLEGO for efficient manipulations in the VR environment [11]. The result was extended to a collaborative VR environment for object creation [12]. However, only simple objects can be created in this system and complex models are created by placing some simple objects together. Nishino reported some work on gesture-based 3D shape creation [15]. A 3D modeler is developed for creating complex shapes by combining the defined hand actions while excluding any discussions on precise interactions. Gao reported a method on constraint-based solid modeling in a VR environment [8]. In this method, the manipulations to a primitive depend on some Shape Control Points (SCP) on the primitive other than the primitive itself and a 3D mouse must be set to the SCP for manipulating the SCP. Furthermore, the SCP cannot sufficiently reflect the natural behaviors of the geometric elements of a primitive. Therefore, the interactions in the virtual environment are unintuitive and inconvenient.

We also reported some preliminary results on precise solid modeling in the VR environment [14, 19, 20]. Some results (in [14]) reported on creating assemblies with embedded constraints between mating features through direct manipulations while [19] reported some results on creating parts by features through direct manipulations. However, the details on the model representation for supporting solid modeling in the VR environment, the constraint solving approach for deriving constraint-based manipulations and the process of solid modeling through constraint-based manipulations were not discussed in [14, 19]. A hierarchically structured constraint-based data model for supporting solid modeling in the VR environment was reported in [20] while the details on how to derive constraint-based manipulations from the constraints in this model and how to establish this model through constraint-based manipulations were not discussed.

This chapter is based on our previous work and presents the details on solid modeling in the VR environment. Constraint-based manipulations are elaborated for precise solid modeling in the VR environment. A procedure-based DOF incorporation approach for solving 3D constraints is presented for deriving the allowable motions. A rule-based constraint recognition engine is developed for automatic constraint recognition. Furthermore, the details on solid modeling through constraint-based manipulations and the establishment of the model representation during the modeling process are also discussed.

2. Hierarchically Structured Constraint-Based Data Model

A fundamental problem for solid modeling in a virtual environment is model representation. In the graphics and VR community, active research on model decimation, multi-resolution, level-of-detail management and zone culling are currently being carried out [1, 9, 10]. Comparatively little research has been conducted for accommodating precise CAD models in a VR environment [7]. If CAD formats were directly used in a VR environment, the online processing time for visualizing a typical CAD model would make it impossible to interact in real-time. The polygon model used in most VR systems provides the illusion of being immersed, but it may not be able to precisely define the object geometry. The use of a high-resolution model in a VR environment can increase model precision. The system may however not be able to respond in real-time either. On the other hand, it is difficult to perform modeling because of the lack of topological relationships and constraint information in the polygon model. Therefore,

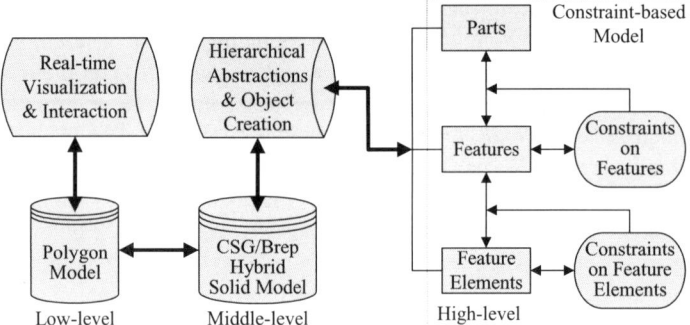

Figure 14.1   The hierarchically structured constraint-based data model.

it is necessary to develop a suitable model representation to support solid modeling in a VR environment. The model representation not only needs to support real-time visualization and interaction in a VR environment, but it also needs to support modeling activities as well as reflect the modeling process.

A hierarchically structured constraint-based data model is presented to support solid modeling in the VR environment (Figure 14.1). This data model includes five levels of information, i.e. parts, features, feature elements, geometric and topological relationships, and polygons. The data model integrates a high-level constraint-based model for precise object definition, a mid-level (Conlogical Solid Geometry/Boundary representation) CSG/Brep hybrid solid model for hierarchical geometry abstractions and object creation, and a low-level polygon model for real-time visualization and interaction in the VR environment. The information in the high-level model used for modeling can be divided into two types, i.e. the object information on the different levels and the constraint information on the different levels. The object can be a part, a feature or a feature element. The constraints on each object level that summarize the associativities between the individual objects on the same level not only provide precise object definition, but also provide a convenient way to realize precise 3D interactions. The mid-level solid model is the geometric and topological description of an object and is represented as a CSG/Brep hybrid structure. It not only provides the geometric and topological information of an object to support the hierarchical geometry abstractions and object creation, but also provides a convenient way for interactive feature-based solid modeling. The low-level polygon model provides the polygon data that corresponds to the mid-level Brep solid model for real-time visualization and interaction in the VR environment. It describes each face in the Brep model as a common array of vertices together with a connection list that defines each facet as a set of indices into the array of vertices.

Constraints are embedded in the solid model and are organized at different levels to reflect the modeling process from features to parts (Figure 14.2).

Level 1 is the feature-based part model representation. A part consists of features and the constraints between these features. The constraints on this level represent the spatial position relationships between the different features and they are called the external feature constraints. An external feature constraint has one direction and this direction is dependent on those of the external element constraints included in this external feature constraint.

Level 2 and Level 3 are the feature element based part model representations. The constraints on Levels 2 and 3 are those between feature elements. Since an external feature constraint

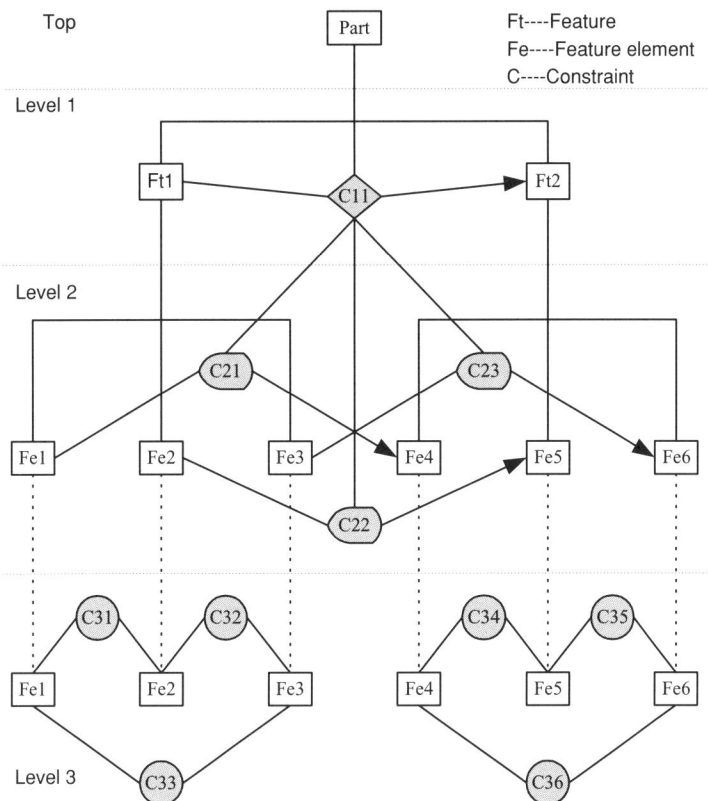

Figure 14.2   The hierarchical structure in the constraint-based model.

between features is difficult to represent, we subdivide a feature into a set of feature elements and the constraints between these feature elements. Correspondingly, an external feature constraint on Level 1 is broken up into a set of constraints between the feature elements that individually belong to different features.

The constraints on Level 2 represent the spatial position relationships between the feature elements that individually belong to different features and they are called the external element constraints. An external feature constraint on Level 1 is subdivided into a set of external element constraints on this level. An external element constraint has one direction and this direction points to the feature element that has been constrained. Typical external element constraints include against, alignment and distance, etc.

Level 3 is the feature model representation. A feature consists of feature elements and the constraints between these feature elements. The constraints on this level represent the spatial position relationships between the feature elements that belong to a feature and they are called the internal element constraints. The internal element constraints define the shape of a feature and are non-directional. They can be further divided into internal element geometric constraints and internal element topological constraints according to their properties. The internal elements geometric constraints represent the spatial position relationships between the feature elements that belong to a feature and are described as a face-based representation, such as parallel faces, perpendicular faces, distance faces and angular faces, etc. The internal element topological constraints represent the topological relationships between the feature elements that belong to a feature and are described as an edge-based representation, such as co-edge and cocircle.

The details on how to represent the objects and constraints at the different levels can be found in [20].

3. Constraint-Based Manipulations

The framework of constraint-based manipulations is shown in Figure 14.3. For every object in the VR environment, such as a feature element, a feature and a part, an event list is regarded as the attribute of this object and is attached to this object. An action list is connected to every event in the event list of the object. This action list shows the actions that will be done as soon as the event occurs. The constraint-based manipulations are realized by these basic interactive events and the actions being performed when these events occur. These basic interactive events are attached to every object. Examples for the basic interactive events are the grasping event,

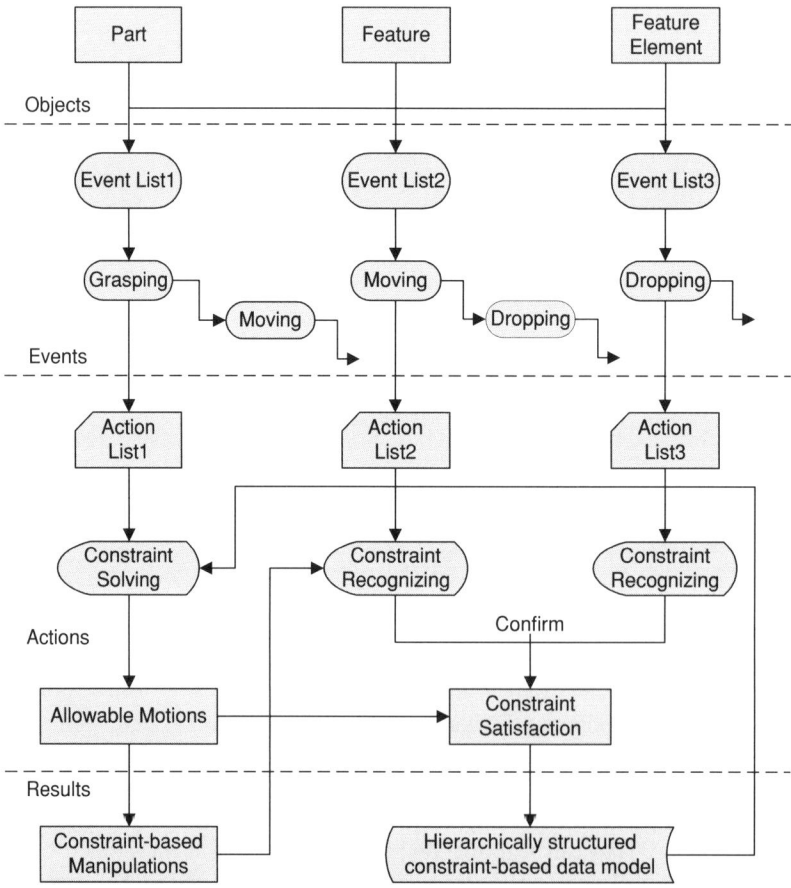

Figure 14.3   The framework of constraint-based manipulations.

the moving event and the dropping event. An action for acquiring the current allowable motions of an object is attached to the grasping event. An action for recognizing the constraints between objects is attached to the moving event and the dropping event. As soon as the user grasps an object, the grasping event occurs and the current allowable motions of this object are derived from the hierarchically structured constraint-based data model through constraint solving. The constraint-based manipulations are acquired by constraining the motions of 3D hands to the allowable motions. This is done by transferring 3D motion data from the 3D input devices into the allowable motions of the object. The constraint-based manipulations not only ensure that the precise positions of an object can be obtained, but also guarantee that the existing constraints will not be violated during future operations.

Once a constraint is recognized during the constraint recognition, it will be highlighted and will await the user's confirmation. Once it is confirmed, the recognized constraint will be precisely satisfied under the current allowable motions of the object and will be inserted into the constraint-based data model. The satisfied constraint will further restrict the subsequent motions of the object.

3.1. Representation of Allowable Motions

The constraints between objects are implicitly created by the constraintbased manipulations with automatic constraint recognition and precise constraint satisfaction. The newly created constraint reduces the DOFs of the object being manipulated and implicitly provides a confinement to the future operations applied to the object. The remaining DOFs define the allowable motions of the object. The allowable motions explicitly describe the next possible operations and ensure that future operations will not violate the existing constraints. The allowable motions are represented as a mathematical matrix for conveniently deriving the allowable motions of an object from the constraints applied to this object.

For every object in free space, its configuration space has six DOFs: 3 translational DOFs and 3 rotational DOFs. To simplify the computation and to clarify the presentation of the allowable motions, we divide the configuration space along three linear independent directions: X-axis, Y-axis and Z-axis. Therefore, some basic DOFs, i.e. 3 translational DOFs and 3 rotational DOFs can be obtained. Furthermore, the 3 basic translational or rotational DOFs are linear-independent among each other. Any remaining DOFs used to define the allowable motions can be described by these basic DOFs, therefore the allowable motions can be represented by these basic DOFs as the following matrix:

$$\begin{bmatrix} T_x & R_x & T_{x\,\min} & T_{x\,\max} & R_{x\,\min} & R_{x\,\max} \\ T_y & R_y & T_{y\,\min} & T_{y\,\max} & R_{y\,\min} & R_{y\,\max} \\ T_z & R_z & T_{z\,\min} & T_{z\,\max} & R_{z\,\min} & R_{z\,\max} \end{bmatrix} \tag{14.1}$$

where the first column elements T_x, T_y and T_z are the linear translations along X-axis, Y-axis and Z-axis respectively and the second column elements R_x, R_y and R_z are the rotations about the corresponding axis respectively. The values of these elements in the matrix are either 0 or 1. Integer 1 indicates that the motion is allowable in the direction along the corresponding axis. Integer 0 indicates that the motion is not allowable in the corresponding axial direction. The third and fourth column elements are the allowable ranges of the 3 translations, which are defined by the minimum and maximum values of the 3 translations. For example, $T_{x\,\min}$

and $T_{x\,\text{max}}$ are the minimum and maximum values of the translation along X-axis. The fifth and sixth column elements are the allowable ranges of the 3 rotations, which are defined by the minimum and maximum values of the 3 rotations. For example, min $R_{x\,\text{min}}$ and $R_{x\,\text{max}}$ are the minimum and maximum values of the rotation about the X-axis. If the translation or rotation along some axis is not allowable, the corresponding minimum and maximum values are zero.

3.2. Constraint Solving for Deriving Allowable Motions

Since most constraints are geometric constraints and they are shown as the limitation of relative geometric displacements between objects, i.e. the limitation of DOFs, the constraints applied to an object can correspond to the DOFs of the object. In fact, the correspondence from constraints to DOFs can be extended to the correspondence from a set of constraints to the incorporation of DOFs. Therefore, the representaion of constraint relationships can be obtained by analyzing and reasoning the DOFs of an object. Furthermore, constraint solving can also be regarded as a process of analyzing and reasoning the DOFs of an object. Based on this, we present a procedure-based DOF incorporation method for 3D constraint solving (Figure 14.4). This method has an intuitive manner for constraint solving since it combines DOF analysis with 3D direct manipulations in the VR environment.

As shown in Figure 14.4, the current allowable motions of an object are derived from the current remaining DOFs of this object. The action of grasping an object is interpreted by the constraint solver as requesting the current remaining DOFs of the object. The current constraints applied to the object can be obtained from the hierarchically structured constraint-based data model. Initially, the object is unconstrained and has six remaining DOFs. If there is only one constraint applied to the object, the current remaining DOFs can be directly obtained by DOF analysis. If there are multi-constraints (more than one) applied to the object, the current remaining DOFs of the object can be obtained by DOF incorporation. The DOF incorporation for solving multi-constraints is based on the DOF analysis for solving individual constraints. Under the limitation of the current remaining DOFs determined by the current constraints, the object aims to satisfy a new constraint recognized by the current constraint-based manipulations

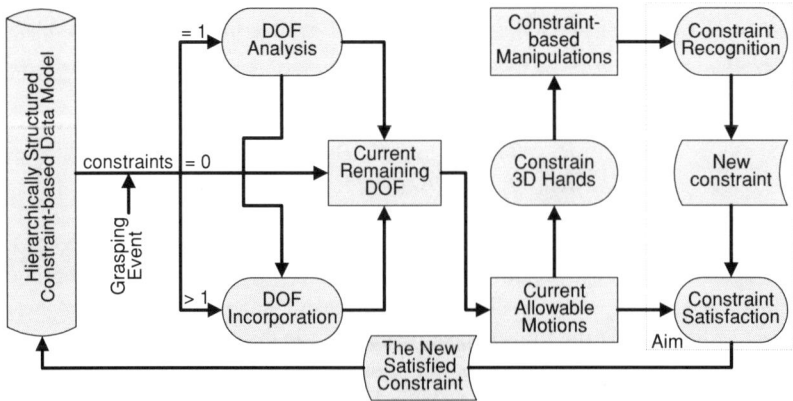

Figure 14.4   The procedure-based DOF incorporation method for constraint solving.

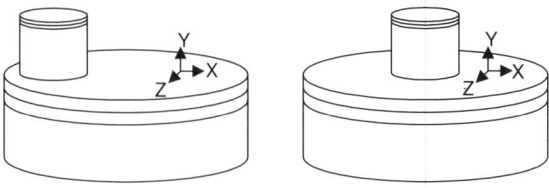

Figure 14.5   The 'against' and 'line-alignment' constraints between two cylinders.

applied to the object. The new constraint is precisely satisfied under the current allowable motions of the object and is further inserted into the hierarchically structured constraint-based data model to update the current constraints applied to the object. The update of the current constraints results in the update of the current remaining DOFs of the object and further results in the update of the current allowable motions of the object. Finally, the constraint-based manipulations applied to the object are updated correspondingly.

3.2.1 DOF Analysis

Since DOFs are divided into the 3 basic translational DOFs and the 3 basic rotational DOFs, it is easy to connect a constraint with remaining DOFs by analyzing the remaining basic translational and rotational DOFs that correspond to this constraint. On the other hand, the allowable motion matrix introduced in Section 3.1 is described by the 3 basic translational DOFs and the 3 basic rotational DOFs. Therefore, the allowable motion matrix that corresponds to a constraint can be directly obtained by analyzing the remaining basic translational and rotational DOFs that correspond to this constraint.

For example, if a small cylinder is placed on a big cylinder and they are also needed to be axis-aligned (Figure 14.5), the constraints between two cylinders are the 'against' and 'line-alignment' constraints. By DOF analysis, the allowable motion matrices that correspond to the two individual constraints are (14.2) and (14.3) respectively.

$$
\begin{bmatrix}
1 & 0 & -10.0 & 10.0 & 0 & 0 \\
0 & 1 & 0 & 0 & 0 & 2\pi \\
1 & 0 & -10.0 & 10.0 & 0 & 0
\end{bmatrix}
\tag{14.2}
$$

$$
\begin{bmatrix}
0 & 0 & 0 & 0 & 0 & 0 \\
1 & 1 & -10.0 & 10.0 & 0 & 2\pi \\
0 & 0 & 0 & 0 & 0 & 0
\end{bmatrix}
\tag{14.3}
$$

Similarly, the allowable motions matrices that correspond to other individual constraints can also be obtained by DOF analysis.

3.2.2 DOF Incorporation

The DOF incorporation is used to represent the remaining DOFs that correspond to multi-constraints. It refers to the intersection within DOF of the allowable motions that correspond to individual constraints respectively. The DOF incorporation can be divided into the incorporation of translational DOFs and the incorporation of rotational DOFs since translational and rotational DOFs are a closed set respectively.

Furthermore, any translational DOFs can be represented by the 3 basic translational DOFs due to the linear independence of the 3 basic translational DOFs. Therefore, the incorporation of translational DOFs can be further divided into the individual incorporations of the 3 basic translational DOFs. Similarly, the incorporation of rotational DOFs can be further divided into the individual incorporations of the 3 basic rotational DOFs. Accordingly, the DOF incorporation can be further regarded as the individual incorporations of the 3 basic translational DOFs and the individual incorporations of the 3 basic rotational DOFs.

On the other hand, an allowable motion matrix is described by the 3 basic translational DOFs and the 3 basic rotational DOFs. Therefore, the DOF incorporation can be finally represented as the incorporation of the allowable motion matrices that correspond to individual constraints respectively. The incorporation of the allowable motion matrices that correspond to individual constraints respectively can be realized by the 'AND' Boolean operation of the allowable motion matrices that correspond to individual constraints respectively. For example the 'AND' Boolean operations of the corresponding elements with the same position at the first and the second columns and the intersections of the allowable ranges of the translations or rotations along the same axis in the allowable motion matrices correspond to individual constraints respectively.

In this way, the remaining DOFs of an object that correspond to multi-constraints can be obtained and the allowable motion matrix that corresponds to multi-constraints can also be acquired. For example, for the 'against' and 'line-alignment' constraints shown in Figure 14.5, the allowable motion matrices that correspond to the two individual constraints are (14.2) and (14.3) respectively. By DOF incorporation, the final allowable motion matrix that corresponds to the two constraints is (14.4).

$$
\begin{bmatrix}
0 & 0 & 0 & 0 & 0 & 0 \\
0 & 1 & 0 & 0 & 0 & 2\pi \\
0 & 0 & 0 & 0 & 0 & 0
\end{bmatrix}
\tag{14.4}
$$

3.3. Rule-Based Constraint Recognition

Constraints are implicitly incorporated into the VR environment by automatic constraint recognition. Constraint recognition refers to the verification of the current positions and orientations between two objects to see if they satisfy a particular type of constraint within a given tolerance.

The framework of constraint recognition is shown in Figure 14.6. While performing direct manipulations, as soon as a moving event or a dropping event occurs, an automatic constraint recognition process is triggered to detect all the possible constraints between the related objects. The system recognizes the constraints between objects from the current position and orientation of the manipulated object according to a rule base. The rule base defines some of the rules applied to the constraint recognition for recognizing some specified constraints (Table 14.1). The constraints include against, alignment, parallelism, perpendicularity and distance, etc. If the current positions and orientations between two objects satisfy the conditions of some constraint within the given tolerance, the correspondent constraint is recognized. Once a constraint is recognized within the given tolerance, it is highlighted and awaits the user's

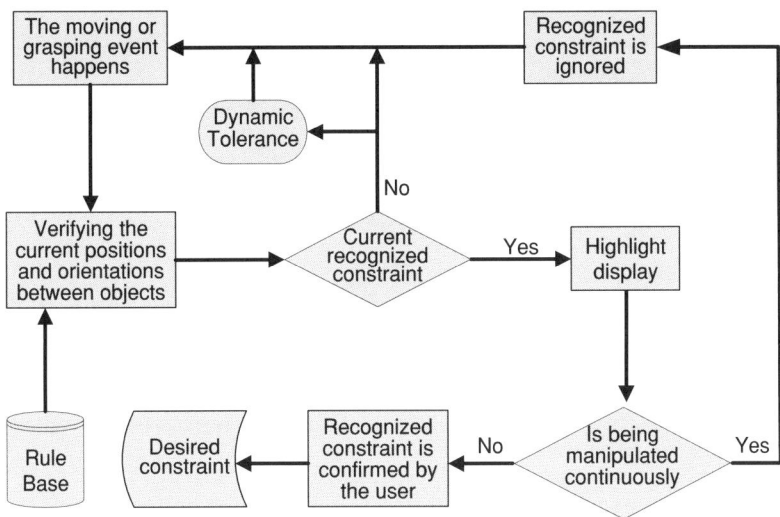

Figure 14.6   The framework of rule-based constraint recognition.

confirmation. If the object is further manipulated continuously within the given time, the current recognized constraint is ignored and the constraint recognition is restarted. Otherwise, the current recognized constraint is confirmed and the desired constraint is obtained. Furthermore, a dynamic tolerance is adopted into the constraint recognition to improve the efficiency of constraint recognition. If the desired constraint is not recognized within the given tolerance, the tolerance is enlarged according to the given step until the desired constraint is recognized.

3.4. Some Special Constraint-Based Manipulations

To reduce the searching time for detecting various types of constraints from various objects and to enhance the modeling efficiency, some special constraint-based manipulations are also implemented as modeling operations for solid modeling in the VR environment. These operations include <placement>, <alignment>, <distance> and <insertion>. For each of the operations, the constraint recognition process is triggered to detect a particular pair of elements that satisfies some special constraint within the given tolerance.

The <placement> operation is responsible for locating an object relative to another object and is used as the initial locating operation of an object. It refers to an action of placing one object onto another object or placing two objects together. The constraint involved in this operation is an 'against' constraint. If the recognized 'against' constraint is precisely satisfied, the <placement> operation is stopped.

The <alignment> operation is responsible for locating an object relative to another object and is used as the precise locating operation of an object. The constraint involved in this operation is an 'alignment' constraint. The <alignment> operation can be classified into two types according to the elements involved in it, i.e., <face-alignment> and <line-alignment>.

The <distance> operation is also responsible for locating an object relative to another object and is used as the precise locating operation of an object. The constraint involved in

Table 14.1   Some typical rules for constraint recognition.

Rules for detecting two against planar faces:
- Parallel: the product of the two unit normal vectors approaches to 0.0;
- Direction: the dot product of the two unit normal vectors approaches to 1.0;
- Close: the distance between a point on one facet to the projected point on the other facet is smaller than a given tolerance;
- Overlapping: the project of one facet on the other facet is not zero.

Rules for detecting two aligning planar faces:
- Parallel: the product of the two unit normal vectors approaches to 0.0;
- Close: the distance between a point on one facet to the projected point on the other facet is smaller than a given tolerance.

Rules for detecting two distance planar faces:
- Parallel: the product of the two unit normal vectors approaches to 0.0;
- Distance: calculate the distance between a point on one facet to the projected point on the other facet.

Rules for detecting two parallel lines:
- The product of the two unit vectors of the line segments approaches to 0.0.

Rules for detecting two perpendicular lines:
- The dot product of the two unit vectors of the line segments approaches to 0.0.

Rules for detecting two co-linear lines/axis:
- Parallel: the product of the two unit vectors of the line segments approaches to 0.0;
- Close: the distance between a point on one line to the projected point onto the other line is smaller than a given tolerance.

Rules for detecting two distance lines/axis:
- Parallel: the dot product of the two unit vectors of the line segments approaches to 0.0;
- Distance: calculate the distance between a point on one line to the projected point onto the other line.

Rules for detecting the face-linear distance:
- Parallel: the product between the unit normal vectors of the face and the unit vectors of the line segment approaches to 0.0;
- Distance: calculate the distance between a point on one line to the projected point onto the face.

Rules for detecting two parallel faces:
- The product between the unit normal vectors of the two faces approaches to 0.0.

Rules for detecting two perpendicular faces:
- The dot product between the unit normal vectors of the two faces approaches to 0.0.

Rules for detecting two co-edge faces:
- Co-edge: all deviations between selected sample points on the two edges approach to 0.0.

Rules for detecting two co-circle faces:
- Co-circle: two circles with the same orientation and dimension.

---

this operation is a 'distance' constraint. The <distance> operation can be classified into three types according to the elements involved in it, i.e., <face-face distance>, <line-line distance> and <face-line distance>. The value of the distance is displayed nearby the object being manipulated for the user to acquire the precise distance. A toolbox with cursor and displaying number is also provided for the user to acquire the precise distance according to a given step.

The operations mentioned before are responsible for the precise location of an object before modeling and are called the locating operations. However, the <insertion> operation is used for performing a specific modeling task and is responsible for determining the final position

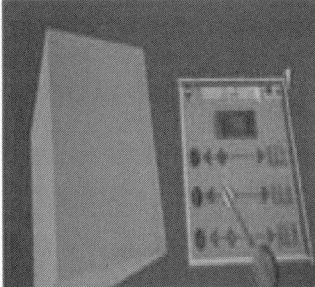

Figure 14.7   Resizing a feature primitive by constraint-based manipulations and a toolbox.

of an object. The basic motion of this operation is a translation. The constraint involved in the insertion operation is a 'face-alignment' constraint.

4. Solid Modeling through Constraint-Based Manipulations

4.1. Creation and Modification of Feature Primitives

Feature primitives, such as blocks, spheres, cylinders and cones, are regarded as the basic building blocks for solid modeling. A user can create a primitive through directly selecting the icon of this primitive in a 3D menu by the laser beam emitted from the hands. When the icon of a feature primitive is selected, the feature solid with standard sizes stands in 3D space. At the same time, the corresponding feature elements and the internal element constraints between these feature elements are established and stored in the hierarchically structured constraint-based data model. A user can directly resize the feature solid through constraint-based manipulations applied to the feature elements of the feature (see the left image in Figure 14.7).

The constraint-based manipulations are derived from the internal element geometric constraints applied to the feature element being manipulated. The actual dimensions of the feature are dynamically displayed nearby the manipulated elements for a user's reference. A user can also resize the feature solid by changing the parameters of the feature through a toolbox (see the right image in Figure 14.7). During the resizing process, the feature shape is dynamically updated through solving the internal element topological constraints applied to the feature element being manipulated.

4.2. Locating Feature Primitives

Feature primitives are located by some kinds of feature locating ways. These locating ways are provided by the combinations of the locating operations introduced in Section 3.4. The location of a feature primitive relative to other features in a part can be completely determined by each of the locating ways. Some typical locating ways are:

- «one <placement> operation + two <face-face distance> operations»
- «one <placement> operation + two <face-alignment> operations»

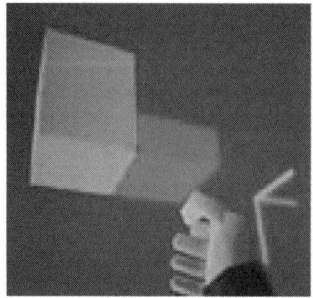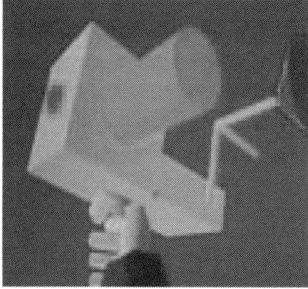

Figure 14.8   Two examples for locating feature primitives.

- «one <placement> operation + two <face-line distance> operations»
- «one <placement> operation + one <face-alignment> operation + one <face-face distance> operation»
- «one <placement> operation + one <face-line distance> operation + one <line-line distance> operation»
- «one one <placement> operation + one <line-alignment> operation»

Two examples for locating feature primitives are shown in Figure 14.8. The left image shows that the small block is located by the locating way «one <placement> operation + two <face-alignment> operations». The right image shows that the cylinder is located by the locating way «one <placement> operation + one <line-alignment> operation».

4.3. Part Creation

The process of creating a part is that of establishing the external feature constraints between the features that constitute this part. In fact, the process of creating a part is that of establishing the external element constraints since an external feature constraint is represented as a set of external element constraints.

A part is created by Boolean operations after a feature primitive is located. The feature generated by the union operation directly inherits the constraints involved in the locating operations and these constraints are also inserted into the hierarchically structured constraint-based data model. For the Boolean subtraction operation, the <insertion> operation is used to further determine the final position of the feature primitive before the Boolean subtraction. Since the feature primitive is located by the selected locating way, the <insertion> operation can be triggered after the constraint involved in the initial locating operation <placement> is released. It means that the constraint involved in the <insertion> operation replaces the constraint involved in the <placement> operation to further locate the feature primitive and the corresponding constraints are directly inherited by the feature generated from the Boolean subtraction. After Boolean operations, the system automatically checks if there are other newly satisfied constraints between the generated feature and the reference features except for the directly inherited constraints. The newly satisfied constraints are also inserted into the hierarchically structured constraint-based data model. Figure 14.9 gives two parts that have been intuitively created by precise constraint-based 3D direct manipulations in the VR environment.

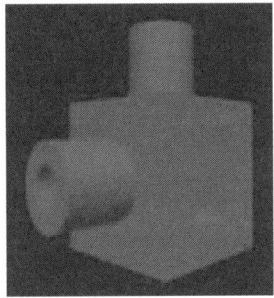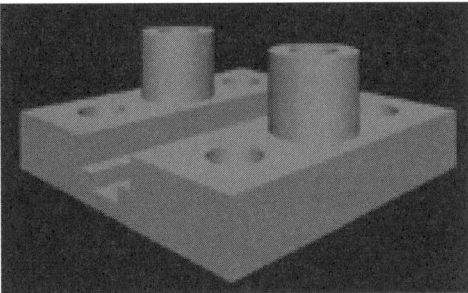

Figure 14.9   Two parts created by constraint-based manipulations in the VR environment.

4.4. *Visual Cues for Constraint-Based Manipulations*

Visual cues are given to a user to obtain the desired constraint-based manipulations during the solid modeling process. This is done by visualizing the allowable motions of an object. The allowable motions can be clearly visualized to convene a message to a user for obtaining the desired constraint-based manipulations. As shown in Figure 14.10, a coordinate frame with a set of allowable motion flags and other visual cues are used to display the allowable motion information. The origin of the coordinate frame is located nearby the object being manipulated and the three axes respectively represent the X-axis, Y-axis and Z-axis. The normal arrow at the end of each axis indicates an allowable translation along the axis and the inverse arrow at the end of each axis indicates an allowable rotation along the axis.

5. Implementation

A prototype system is implemented on the Division VR software based on a SGI Onyx2 graphics workstation. In this system, a user can intuitively create solid models in the VR environment by precise constraint-based 3D direct manipulations. The system framework is shown in Figure 14.11. The body actor communicated with other actors handles all aspects of user interaction. It receives and processes the information from the input actor, monitors and processes the events and actions that happened in the VR environment and output the processed results to the visual actor and the audio actor. A 3D mouse controlled by the input actor is mainly used as the input device to carry out 3D manipulations. A six-degree-of-freedom head tracker

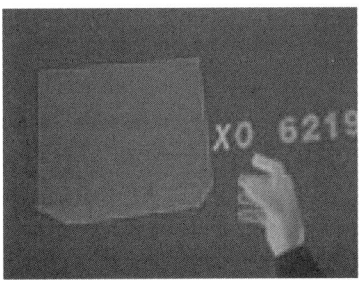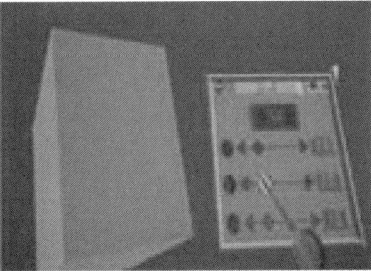

Figure 14.10   Visual cues for constraint-based manipulations.

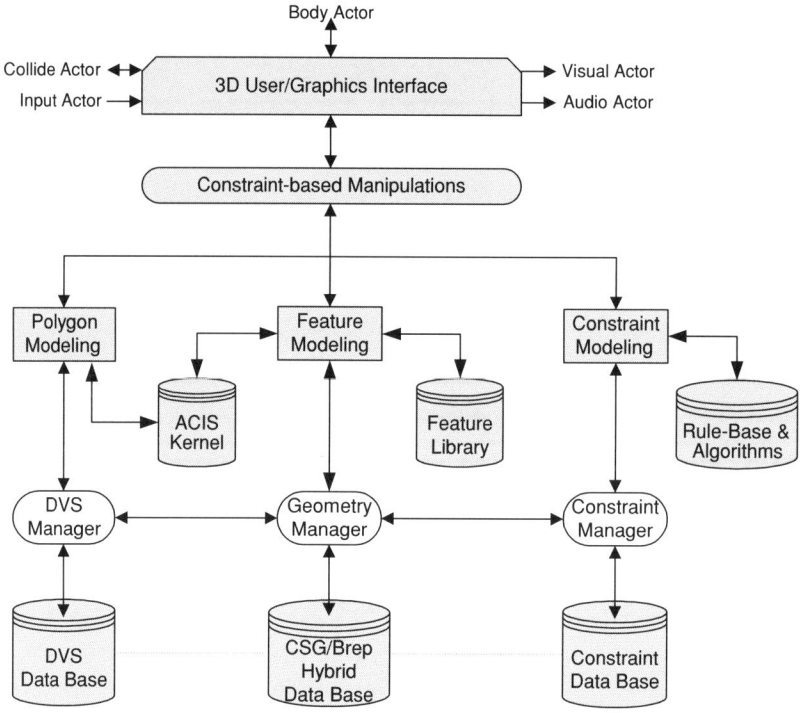

Figure 14.11   System framework.

and CrystalEyes shutter glasses controlled by the visual actor are used for stereo display. Two sound blasters controlled by the audio actor are used for audio. The collide actor resides in the system to detect the possible collisions between the objects in the VR environment. A rule base and related algorithms for constraint processing are developed to support constraint modeling. A feature library for providing some basic primitives is developed to support feature modeling. A geometric kernel ACIS is employed to support CAD-related operations. The ACIS kernel is also employed to support polygon modeling through its triangulation function.

For efficiently integrating CAD with VR, a triple database concept is adopted for solid modeling in the VR environment. The variation in the DVS database is propagated into the CSG/B-rep hybrid database to change the geometries of objects and further results in the variation in the constraint database to change or maintain the constraints between objects. On the other hand, the variation in the constraint database is propagated into the CSG/B-rep hybrid database to change the geometries of objects and further results in the variation in the DVS database.

6. Summary

This chapter presents an efficient approach for intuitive and precise solid modeling in a VR environment. A hierarchically structured constraintbased data model is developed to support solid modeling in the VR environment. The data model integrates a high-level constraint-based model for precise object definition, a mid-level CSG/Brep hybrid solid model for hierarchical geometry abstractions and object creation, and a low-level polygon model for real-time visualization and interaction in the VR environment. Constraints are embedded in the solid model and are organized at different levels to reflect the modeling process from features to parts. Solid modeling in the VR environment is performed in an intuitive manner through precise constraint-based manipulations. Constraint-based manipulations are accompanied with automatic constraint recognition and precise constraint satisfaction to establish the hierarchically structured constraintbased data model and are realized by allowable motions for precise 3D interactions in the VR environment. The allowable motions are represented as a mathematical matrix for conveniently deriving allowable motions from constraints. A procedure-based DOF incorporation approach for 3D constraint solving is presented for deriving the allowable motions. A rulebased constraint recognition engine is developed for both constraint-based manipulations and implicitly incorporating constraints into the VR environment. A prototype system has been implemented for precise solid modeling in an intuitive manner through constraint-based manipulations in the VR environment.

References

[1] Andujar, C., Saona-Vazquez, C. and Navazo, I. (2000) LOD visibility culling and occluder synthesis. *Computer-Aided Design*, Vol. 32, No. 13, 773–783.

[2] Butterworth, J., Davidson, A., Hench, S. and Olano, T.M. (1992) 3DM: a three dimensional modeler using a head-mounted display. *ACM Computer Graphics: Proc. 1992 Symp. on Interactive 3D Graphics*, Vol. 25, No. 2, 197–208.

[3] Dani, T. and Gadh, R. (1997) COVIRDS: Shape modeling in a virtual reality environment. *ASME 1997 Computers in Engineering Conference*, Sacramento, California.

[4] Fa, M., Fernando, T. and Dew, P.M. (1993) Interactive constraint-based solid modeling using allowable motion. *Proc. of 2nd ACM Symposium on Solid Modeling and Applications*, Montreal, Canada, 243–252.

[5] Fernando, T., Dew, P.M. and Fa, M. (1995) A shared virtual workspace for constraint-based solid modeling. *Virtual Environment'95*: Selected papers of the Eurographics workshops in Barcelona, Spain, Springer Wien, New York, 185–198.

[6] Fernando, T., Muttay, N., Tan, K. and Wimalaratne, P. (1999) Software architecutre for a constraint-based virtual environment. *Proceedings of the ACM Symposium on Virtual reality software and technology*, London, UK, 147–154.

[7] Figueiredo, M. and Teixeira, J. (1994). Solid modeling as a framework in virtual environments. *Proc. of the IFIP WG 5.10 Workshops on Virtual Environments and Their Applications and Virtual Prototyping*, Rix, J., Haas, S. and Teixeira, J. (eds.), 99–112.

[8] Shuming, G. Wan, H. and Peng, Q. (2000) An approach to solid modeling in a semi-immersive virtual environment. *Computer & Graphics*, No 24, 191–202.

[9] Gobbetti, E. and Bouvier, E. (2000) Time-critical multiresolution rendering of large complex models. *Computer-Aided Design*, Vol. 32, No. 13, 785–803.

[10] Kahler, K., Rossl, C., Schneider, R., Vorsatz, J. and Seidel, H.-P. (2001) Efficient processing of large 3D meshes. *Proc. of International Conference on Shape Modeling and Applications*, Genova, Italy, 228–237.

[11] Kiyokawa, K., Takemura, H. and Katayama, Y. (1998) VLEGO: A simple two-handed 3D modeler in a virtual environment. *Eletronics and Communications in Japan*, Part 3, Vol. 8, No. 11, 1517–1526.

[12] Kiyokawa, K., Takemura, H. and Yokoya, N. (2000) SeamlessDesign for 3D object creation. *IEEE Multimedia*, No.1, Vol. 7, 22–33.

[13] Liang, J. and Green, M. (1994) JDCAD: a highly interactive 3D modeling system. *Computer & Graphics*, Vol. 18, No. 4, 499–506.

[14] Ma, W., Tso, S.-K. and Zhong, Y. (1998) Constraint-based modeling in a virtual environment. *Proc. of CIRP Design Seminar on New Tools and Workflows for Product Development*, Berlin, 221–232.

[15] Nishino, H., Fushimi, M. and Utssumiya, K. (1999) A virtual environment for modeling 3D obejcts through spatial interaction. *1999 IEEE International Conference on Systems, Man, and Cybernetics*, Oita University, Japan, 81–86.

[16] Stork, A. and Maidhof, M. (1997) Efficient and precise solid modeling using a 3D input device. *Proc. of Fourth Symposium on Soild Modeling and Applications*, Altanta, 181–194.

[17] Tan, J., Liu, Z. and Zhang, S. (2001) Intelligent assembly modeling based on semantics knowledge in virtual environment. *The Sixth International Conference on Computer Supported Cooperative Work in Design*, London, Ontario, Canada, 568–571.

[18] Whyte, J., Bouchlaghem, N., Thorpe, A. and McCaffer, R. (2000) From CAD to virtual reality: modeling approaches, data exchange and interactive 3D building design tools. *Automation in Construction*, No.10, 43–55.

[19] Zhong, Y., Yang, H. and Ma, W. (1999) A constraint-based approach for intuitive and precise solid modeling in a virtual reality environment. *The Sixth International Conference on Computer Aided Design & Computer Graphics,* Shanghai, China, 1164–1171.

[20] Zhong, Y., Mueller-Wittig, W. and Ma, W. (2002) A model representation for solid modeling in a virtual reality environment. *IEEE International Conference on Shape Modeling and Applications*, Banff, Alberta, Canada, 183–190.

15

Efficient Simplification of Triangular Meshes

Muhammad Hussain
Yoshihiro Okada
Koichi Niijima
Graduate School of Information Science and Electrical Engineering,
Kyushu University, 6-1, Kasuga-koen, Kasuga, Fukuoka 816, Japan.

We have proposed a new edge collapse simplification algorithm that can efficiently produce high quality approximations of closed manifold surface models. To reduce the number of triangular faces in a polygonal model, a sequence of edge collapses is performed and to choose the appropriate sequence of edge collapses, we have introduced a new error metric based on a quantity proportional to the volume of a tetrahedron. Our proposed algorithm is simple, fast and memory efficient and can efficiently reduce very large polygonal surface models. Moreover, simplified models created using our method preserve the essential features of a model and compare favorably with many well-known published simplification techniques in terms of maximum geometric error and mean geometric error and bear high visual reliability even after significant simplification.

1. Introduction

A polygonal surface model M consists of a fixed set of vertices $V = \{v_0, v_1, v_2, \ldots, v_r\}$ and a fixed set of faces $F = \{f_0, f_1, f_2, \ldots, f_n\}$. Without loss of generality, we can assume that the faces constituting the surface model are triangular. The objective of polygonal simplification is to produce a simplification M' of an original model M such that M' contains fewer polygons than M and is as similar as possible to M. Advanced technological systems such as laser range scanners, computer vision systems, medical imaging devices, and CAD systems have given rise to vast databases of polygonal surface models which are often very complex and highly detailed. Models consisting of millions of polygons are commonplace. Different areas like

Advances in Geometric Modeling. Edited by M. Sarfraz
© 2003 John Wiley & Sons, Ltd   ISBN: 0-470-85937-7

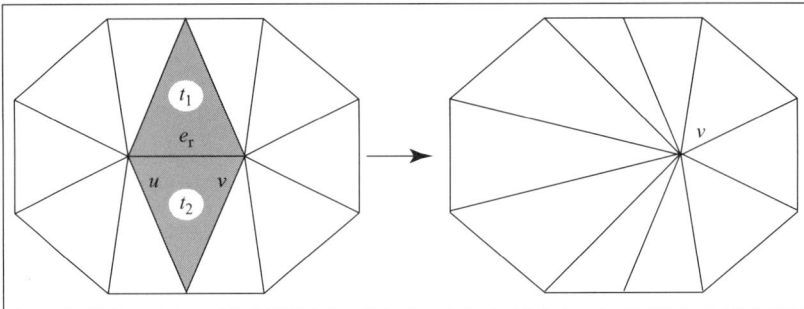

Figure 15.1   In an edge collapse operation, edge e_r will be substituted with vertex v and triangles t_1 and t_2 will be eliminated.

distributed virtual environments and finite element methods involve such models and require them to be simplified to achieve acceptable running times. In recent years, the polygonal simplification problem attracted the attention of researchers and many algorithms have been developed. One of the major goals of research in this area involves computing a multiresolution model, a surface representation which can accommodate a wide range of viewing contexts, of an object.

Among various approaches to polygonal simplification, edge collapse based techniques have emerged as the most popular one. Edge collapse operation is attractive because it allows us to position the new vertex in a manner that helps preserve the location and shape of the original surface, and does not need the invocation of a triangular algorithm. One of the earliest algorithms based on this approach was proposed by Hoppe *et al.* [7] and it provides a foundation for most existing simplification algorithms of this category. An algorithm belonging to this class usually involves two ingredients: *edge collapse operation*, a topological operation to modify the topology of a surface model, and an *error metric* which reflects the error measure between the simplification and the original model. To be specific an edge collapse operation substitutes two end vertices of an edge with a single vertex, thereby eliminating the collapsed edge and its incident triangles, see Figure 15.1.

The general edge collapse algorithm has two decisions to make: where to place the substitute vertex resulting from an edge collapse operation, and choosing the order of edges to collapse using appropriate error measure. Existing edge collapse algorithms differ only in making these two decisions. As far as the positioning of the substitute vertex is concerned, two approaches are in common use: *subset placement* or *half-edge collapse*, and *optimal placement*. Subset placement causes one of the endpoints to be selected as the target position and is the simplest strategy one can adopt. In optimal placement, the position of the substitute vertex is allowed to float freely in space in order to minimize some error metric. In our algorithm, we have opted for subset placement because it results in a more concise representation of polygonal surface model which is useful for progressive compression as well as for view-dependent dynamic level of detail management. To determine the sequence of edge collapses, we have proposed an error metric defined using the area measure swept out by an edge when an edge collapse operation is applied, weighted by the angle between the old and new positions of an adjacent triangular face.

The subsequent arrangement of this chapter is as follows. In Section 2, we give an overview of related edge collapse algorithms. Notation and terminology adopted throughout is presented in Section 3. Section 4 gives an overview of our simplification method. Error metric used in our method has been detailed in Section 5. Sections 6 deals with quality check respectively. Results of our simplification algorithm have been reported in Section 7 and Section 8 concludes the chapter.

2. Related Work

In this section, we give an overview of some of the related edge collapse algorithms. During last few years many researchers have had an interest in developing algorithms of this type due to the simplicity and effectiveness of this approach. The basic differentiating factor among different algorithms of this kind is how to define an error metric reflecting the error introduced when an edge collapse occurs.

Some authors have defined error metrics based on an elaborate measure of error to decide the order in which the edges will be collapsed . For example, the progressive mesh of Hoppe [7] uses an error metric that is defined as the average distance from the proposed new triangles in the mesh to a set of sample points on the original model. The distance to the sample points is also used to define a quadratic energy functional which is minimized to select a new vertex position. This algorithm produces high quality results, but several distance-to-surface measurements make it quite slow. Simplification of a very large model might take several hours. Gueziec [5, 6] defines a tolerance volume as a convex combination of spheres located at each vertex of the simplification. He selects edges based on shortest edge length and then chooses a new vertex position such that the original surface is guaranteed to lie within that volume. This algorithm also produces good quality results, and appears to be slow, but it is faster than Hoppe's [7].

Another common and generally less expensive approach is to define error metric using local surface properties. Ronfard and Rossignac [13] assign to each vertex the set of planes associated with its incident triangles. As a result of one edge collapse operation, two vertices are merged into one and the new vertex inherits the planes of the merged vertices. The maximum distance from the new vertex to its supporting planes is used as an error metric to measure the edge collapse cost. Garland and Heckbert [4] used this work as the starting point of their own simplification algorithm. Instead of maintaining a list of planes, they measure the squared distance from the collection of planes associated with triangles incident on a vertex and store them as a symmetric 4×4 matrix, one matrix per vertex. This error metric called quadric error metric is used both to select the new vertex position and the edge to collapse. While their approach is fast and gives high quality approximations, it is not memory efficient. For each vertex it stores ten floats and for a polygonal model consisting of some million polygons a very large amount of memory is occupied. The memoryless algorithm recently developed by Lindstrom and Turk [8, 9] uses linear constraints, based primarily on the conservation of volume, to decide the edge collapse sequence and the position of the new vertex. The most interesting aspect of this algorithm is that it makes decisions based purely on the current approximation alone; all other mentioned algorithms keep track of some kind of geometric history. It produces good quality simplifications and is fairly efficient, particularly in memory consumption. But it is rather slow when compared to that by Garland and Heckbert [4].

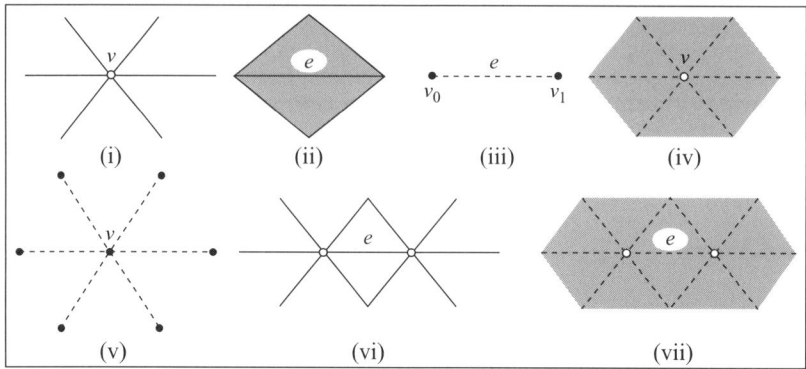

Figure 15.2   (i) $\lceil v \rceil$, edges incident on vertex v (ii) $\lceil e \rceil$, triangles incident on edge e. (iii) $\lfloor e \rfloor$, vertices
associated with e (iv) $\lceil\!\lceil v \rceil\!\rceil$, triangles incident on vertex v (v) $\lfloor\!\lfloor v \rfloor\!\rfloor$, vertices associated
with vertex v (vi) $\lfloor\!\lceil e \rceil\!\rfloor$, edges associated with edge e (vii) $\lceil\!\lfloor\!\lceil e \rceil\!\rfloor\!\rceil$, triangles associated with
edge e.

3. Notation and Terminology

Before presenting the details of our simplification algorithm, a brief description of terminology
and notation is in order. In computer graphics, a 3D model is usually represented by a triangular
mesh. A triangular mesh is specified by a pair (P, K), where P is a set of n point positions
$P = \{v_i \in R^3 \mid 1 \le i \le n\}$ and K is an abstract simplicial complex, which contains all the
topological information. In simple words, P and K describe the geometry and topology of a
polygonal mesh. The complex K is the set of subsets of $\{1, 2, 3, \ldots, n\}$. These subsets are
known as simplices and three types of them are in common use: 0-simplex, 1-simplex and
2-simplex which are usually known as a vertex, an edge and triangular face respectively. We
denote 0-simplex or vertex by v with its geometric counterpart as a 3D vector v. An edge e
or 1-simplex is a subset $\{v_0^e, v_1^e\}$. We represent an oriented edge as an order pair (v_0^e, v_1^e) and
denote it by $\vec{e}$. A triangle t or 2-simplex is a set of oriented edges i.e. $t = \{\vec{e}_0, \vec{e}_1, \vec{e}_2\}$.

Following the definitions of simplex operators $\lceil\ \rceil$ and $\lfloor\ \rfloor$ as adopted in [8], $\lceil v \rceil$, $\lceil\!\lceil v \rceil\!\rceil$, $\lfloor\!\lfloor v \rfloor\!\rfloor$,
$\lfloor e \rfloor$ stand for edges incident on v, triangles incident on v, neighboring vertices of v, vertices
of e respectively and, $\lfloor\!\lceil e \rceil\!\rfloor$ and $\lceil\!\lfloor\!\lceil e \rceil\!\rfloor\!\rceil$ represent edges and triangles respectively incident upon
the end vertices of e as shown in Figure 15.2.

A mesh is known to be a closed manifold (open manifold) if for a vertex v, $\lceil\!\lceil v \rceil\!\rceil$ is topologically equivalent to a disk (half disk respectively). A mesh is referred to as non-manifold if
it does not satisfy this criterion.

4. Overview of Our Algorithm

Our simplification method, like most related algorithms, is a simple greedy procedure. It
iteratively selects an edge with minimum cost, collapses it to one of the end vertices, and then
re-evaluates the cost of the edges affected by this edge collapse operation. More precisely, it
involves the following steps.

- It computes the cost of collapse for each edge in the surface model using our proposed error metric and determines the sequence of edge collapses in increasing order of the magnitude of the cost of edge collapse.
- It chooses an edge $e = \{v_0, v_1\}$ with the minimum cost of simplification and substitutes it either with v_0 or v_1. During this operation triangles $\lceil e \rceil$ become singular and are discarded. The remaining edges $\lfloor e \rfloor - \{e\}$ and triangles $\lceil\lceil e \rceil\rceil - \lceil e \rceil$ incident upon v_0 and v_1 are updated such that all occurrences of v_1 (or v_0) are replaced with v_1 (or v_1).
- The cost of collapse for the edges $\lfloor e \rfloor - \{e\}$ is re-evaluated and the sequence of edge collapses is updated

Most of the simplification algorithms based on iterative edge collapse involve this basic structure. We prefer to use subset-placement or half-edge collapse as a topological operator in our algorithm because we believe that a topological operator must be as simple as possible. It is a common observation that the choice of a particular topological operator has no significant effect on the results. What matters the most is the criterion which decides where to apply the next simplification operator. We have not only adopted subset replacement because of its simplicity, but also because of its other advantages. It does not involve any undefined degree of freedom which would have to be determined by local optimization. It does not invent new geometry by allowing some heuristic to decide about the position of the substitute vertex. The vertices of the simplified mesh always form a proper subset of the original vertices. This makes the progressive transmission of meshes more effective and is crucial for an integrated level of detail.

The most important task of a simplification algorithm is to assign a cost of collapse to each edge in the mesh to form a sequence of edge collapses. In fact, this cost of collapse reflects an error that will be introduced into the simplification as a result of an edge collapse. The way of determining this cost is the basic differentiating factor among algorithms of this family. Many authors have proposed various error metrics. Some of them measure global error and are very sophisticated [5, 6, 7]. These result in high quality simplifications, but have very high computational cost. Some others are based on local error and give rise to very fast algorithms [10, 11, 12, 14], but they result in approximations which deviate substantially from the original ones. Recently some error metrics have been developed which stand in between these two extremes [4, 8, 9]. The primary motivation behind our work is also to develop an error metric which leads to an algorithm which has a good trade off between accuracy, time complexity and memory consumption. In Section 5, we have proposed a new error metric based on the area of an error triangle defined by the old and the expected new position, resulting from an edge collapse, of an edge and the edge to be collapsed and the angle between the positions of an associated triangle before and after edge collapse, see Figure 15.3.

The basic achievements of our algorithm are as follows:

- *Memory consumption*: It is one of the important factors which affects the efficiency of an algorithm. Our algorithm does not consume extra memory other than that is necessary to store the basic geometrical and topological information of a mesh. So it is capable of reducing very big models quite easily. Memory consumption of our algorithms is less than half of that consumed by Garland's algorithm [4].
- *Computational time*: Our algorithm is little bit slower than that by Garland, but it is faster than Peter's Algorithm [8, 9], which is also a memory less algorithm.

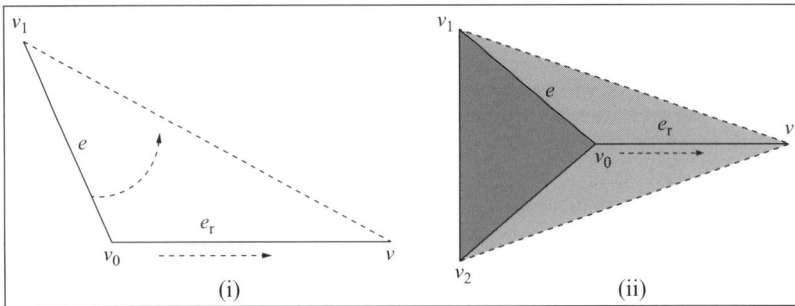

Figure 15.3   (i) Error triangle, (ii) two positions of an adjacent triangular face before and after an edge collapse.

- *Accuracy*: Simplifications resulted from our algorithm are comparable with those by standard methods such as quadric error metric method and memoryless algorithm. It preserves the essential features of a model even after significant reduction.

5. Error Metric

We have defined an error metric based on the observation that when an edge collapse operation takes place, some edges from $\lceil e \rceil$ undergo displacement and some triangles from $\lceil\lceil e \rceil\rceil$ undergo angular displacement, see Figure 15.4(i). The area of the triangle defined by the old and displaced positions of an adjacent edge and the collapsed edge e_r and the angle between the old and the rotated adjacent triangles reflect the error which will be introduced in the model as a result of an edge collapse.

To be precise, when edge e_r collapses, v_0 will move towards v and will coincide with v, see Figure 15.3(i). At the same time, edge $e_{01} = \{v_0, v_1\}$ will sweep an area equal to the area

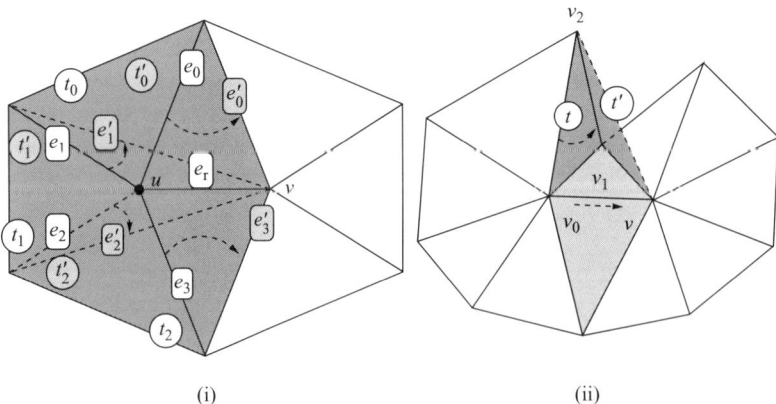

Figure 15.4   (i) When edge e_r is collapsed, triangles t_0, t_1, t_2 are moved to t_0', t_1', t_2' and edges e_0, e_1, e_2, e_3 are displaced to e_0', e_1', e_2', e_3' respectively. (ii) Triangle t will fold over when edge e will collapse.

of triangle $t'' = (v_0, v_1, v_2)$. This area reflects the introduced error. Thus error associated with edge e_{01} is as follows:

$$Q_{e_{01}}(v) = \frac{1}{2}|\mathbf{a} \times \mathbf{b}|,$$

where $\mathbf{a} = v_0 - v_1$, and $\mathbf{b} = v - v_1$.

In some cases, this may not measure simplification error accurately. For example, when all triangles in $[\![v]\!]$ are coplanar or nearly coplanar, then edge collapse operation will only reduce redundancy in the mesh and will not introduce any error. In this situation $Q(v)$ must assume zero value but practically it does not happen.

So as a remedy to this drawback, we weight $Q(v)$ with the angle θ between the triangles $t = (v_0, v_1, v_2)$ and $t' = (v_1, v_2, v)$ see Figure 15.3(ii).

This choice of weights has another added advantage. Note Figure 15.4(ii), foldings may appear when an edge to be collapsed is surrounded by a very concave polygon. When edge $e_r = \{v_0, v\}$, collapses and v_0 coincides with v, triangle $t = (v_0, v_1, v_2)$ will fold over, thus creating folds in the mesh. In this situation, the angle between the triangles $t = (v_0, v_1, v_2)$, and $t' = (v_1, v_2, v)$ will bear greater value and will cause greater value to be added to the cost of edge collapse thereby preventing this edge collapse.

The cost of collapse of an edge will be the weighted sum of errors associated with each edge in $\lceil v \rceil - \{e_r\}$, i.e.

$$Cost(e_r) = \sum_{e \in \lceil v \rceil - \{e_r\}} \theta_e Q_e(v).$$

To preserve the geometry of a model, it is necessary that the *flat vertices* i.e. the vertices for which $[\![v]\!]$ are nearly coplanar, must be removed first; then the *edge vertices* v_i, i.e. the vertices along a feature edge of an object and for which triangles $[\![v_i]\!]$ can roughly be divided into two groups according to their orientation, can be removed and their collapse must be along the feature edge. The removal of a *corner vertex*, i.e. the vertex at a sharp corner of the surface, will certainly affect the geometry of the object model and so it should be the last to be removed. Our proposed metric causes us to keep track of this hierarchy and preserve the visual appearance of the model, as is obvious from Figure 15.5.

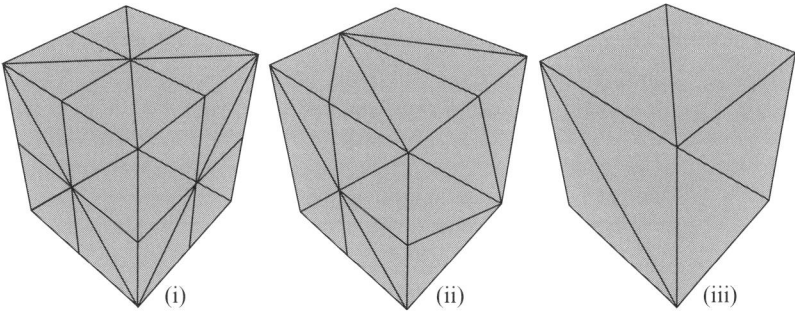

(i)   (ii)   (iii)

Figure 15.5   Face vertices will be collapsed first, then edge vertices and lastly corner vertices.

6. Metro

To assess the quality of our method, more rigorous error measures are needed. One of the most well-known metrics for making geometric comparison between two surfaces is *Hausdorff Metric*. Its asymmetric form is defined as follows:

$$\bar{d}_\infty(X_1, X_2) = \max_{x \in X_1} \min_{y \in X_2} d(x, y),$$

where X_1 and X_2 are the sets of points on the two surfaces to be compared and $d(x, y)$ is Euclidean metric. Its symmetric form is as follows:

$$d(X_1, X_2) = \max\{\bar{d}_\infty(X_1, X_2), \bar{d}_\infty(X_2, X_1)\}.$$

Simplification envelopes [3] is based on bounding this error measure. Another well-known metric for the geometric comparison of surfaces is mean geometric error. Its discrete version is given as follows:

$$d_1(X_1, X_2) = \frac{1}{|X_1| + |X_2|} \left(\sum_{x \in X_1} d(x, X_2) + \sum_{x \in X_2} d(x, X_1) \right),$$

where $|X|$ stands for the number of elements in X and $d(x, X) = \min d(x, y)$.

Metro [2] is a geometric comparison tool which has been developed to evaluate the quality of simplified models. It uses the above mentioned measures of error to compare surfaces. To eliminate bias, we use the metro tool to determine the quality of simplified models created by our method.

7. Results and Discussion

We have tried our implementation of SFME on several large triangular models and have achieved encouraging results. Our method can simplify very large models consisting of millions of triangular faces in a fairly short amount of time and the simplified models bear good visual resemblance with the originals. To evaluate our method, we make a comparison with QSlim [4] and memoryless simplification [8, 9] because among the existing standard simplification algorithms, the former is the fastest one and the latter generates better quality simplifications and can reduce very large models efficiently. We choose the Stanford bunny and hand as test models because of their complex structure.

Table 15.1 lists the computation time taken by QSlim and SFME to simplify hand and horse models shown in Figure 15.8. Notice that our algorithm is almost two times slower than QSlim. We run both the algorithms on 800MHz Intel Pentium III machine with 384 MB of

Table 15.1   Time taken in seconds to reduce to one face.

Model	Model Size (faces)	SFME	QSlim
Horse	96,966	9.8	4.5
Hand	654,666	71.33	39.9

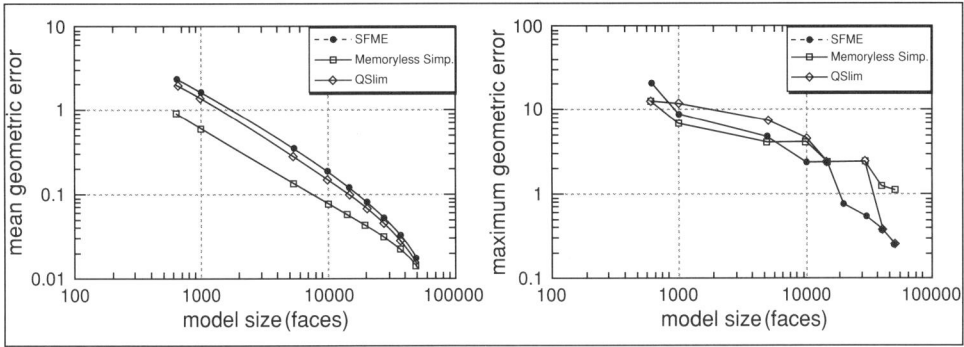

Figure 15.6   Maximum and mean geometric errors for horse model.

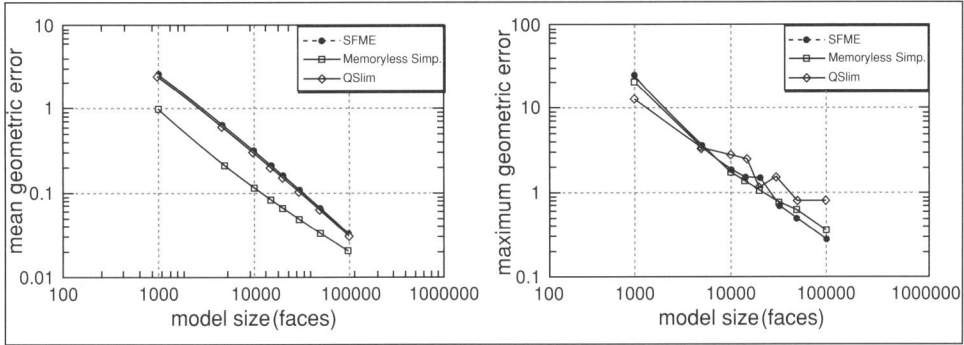

Figure 15.7   Maximum and mean geometric errors for hand model.

main memory. From the results reported in [8] (see Table 15.1), it is obvious that memoryless simplification is about 5 times slower than QSlim; so we can safely conclude that our method is about 2.5 times faster than memoryless simplification.

Graphs shown in Figures 15.6 and 15.7 illustrate the mean geometric and maximum geometric errors between the original and the simplified models created by SFME, QSlim and memoryless simplification. We plotted 1000 times the ratio of the error and the bounding box of the original model along logarithmic y-axis and the number of faces along logarithmic x-axis. It is apparent that our algorithm is almost as accurate as QSlim in terms of mean geometric error and compares well with both QSlim and memoryless simplification in terms of maximum geometric error. Keeping in view these and the results reported in [8], we can claim that our method compares well with standard simplification algorithms in terms of mean and maximum geometric error.

Simplified models shown in Figure 15.8 demonstrate how fairly our method preserves the essential features of a model. All the major detail of the original model remains even after significant simplification. Observe the model depicted in Figure 15.8 (top right), it is the simplified model of hand consisting of 4266 triangular faces, 0.65% of the original model which consists of 654,666 faces. It can be seen that the major features of the original model still remain in spite of being highly simplified.

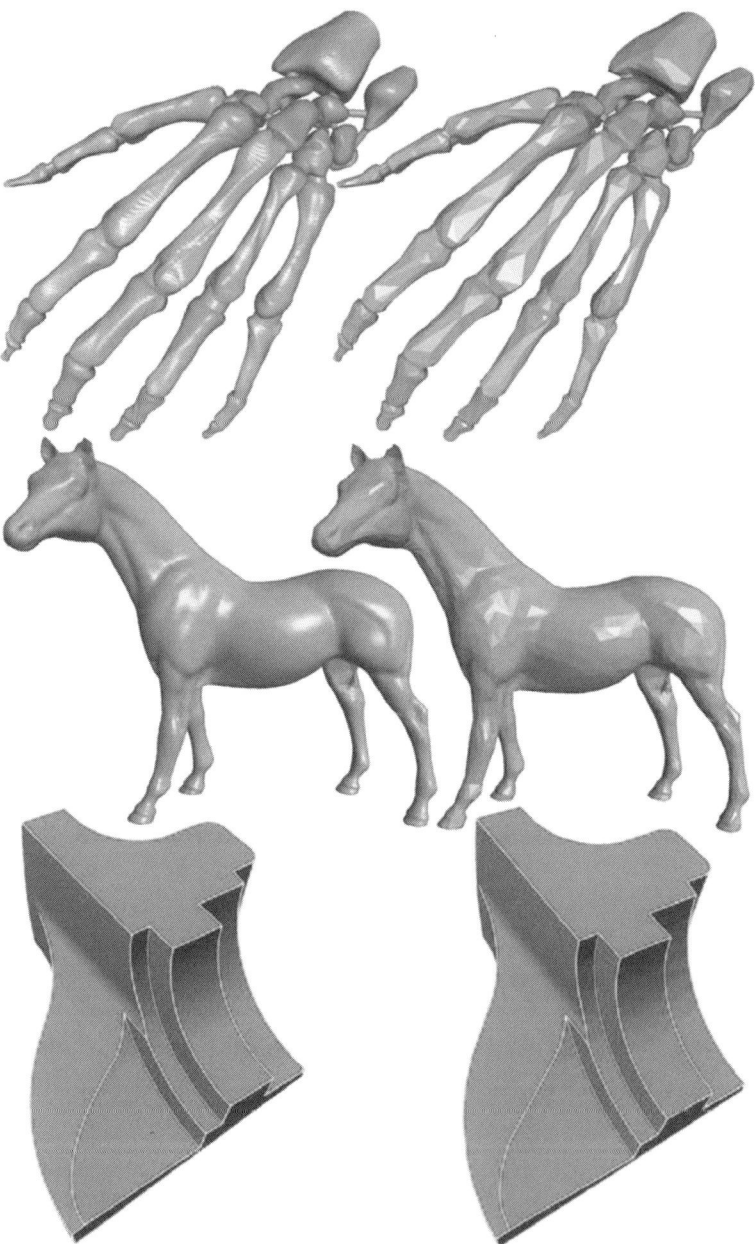

Figure 15.8   Original hand (top left), horse (middle left) and fandisk (bottom left) models and the corresponding reduced models on the right side of each original model.

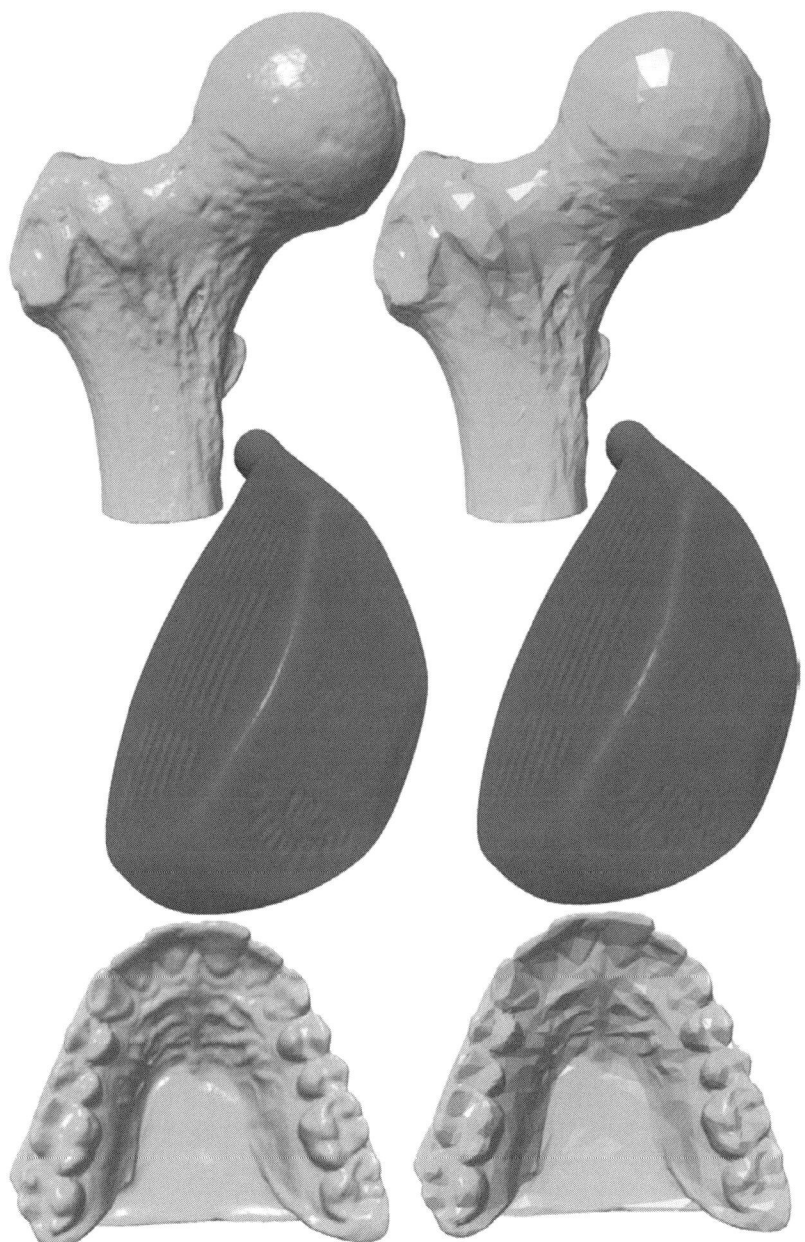

Figure 15.9 Original balljoint (top left), club (middle left), and teeth models and corresponding simplified models on the right side of the original model.

The model shown in Figure 15.8 (middle left) is a horse model consisting of 96,966 triangular faces. Figure 15.8 (middle right) shows its simplified version comprising only 3996 faces, just 4.1% of the original size. Despite drastic simplification, ears, nostrils, hoofs and the contour on the rear leg are finely preserved. Also observe balljoint, club and teeth models, consisting of 274,120 faces, 419,554 faces and 233,204 faces respectively, shown in Figure 15.9, corresponding simplified models, containing 5,196 faces, 5,396 faces and 4,396 faces respectively, although being drastically reduced yet show the essential detail of the original ones.

The Fandisk model, originally consisting of 12,946 faces and simplified containing 496 faces, has been shown in Figure 15.8. Even after 96.1% reduction, features lines are apparent.

The polygonal simplification is NP hard problem, so no one of the existing methods can be regarded as the best one. Each of these has some added advantages over the others. Our algorithm is faster than memoryless simplification and can simplify very large models consisting of millions of triangular faces. It is useful for applications that require good visual fidelity, but not tight error bounds.

8. Summary

We have presented a polygonal simplification method based on a new measure of approximation error which we derive from the area swept by an edge when an edge collapse takes place. Our method has very good trade off between memory consumption, computation time and accuracy. It can simplify huge models consisting of millions of triangular faces in relatively short time. The quality of simplifications is good. It preserves the essential features of an object even after significant reduction. We intend to extend it to include surface attributes.

Acknowledgement

The authors would like to thank Dr. Sarfraz for his valuable comments to improve the quality of this document.

References

[1] A. Ciampalini, P. Cignoni, C. Montani and R. Scopigno. Multiresolution Decimation based on Global error. *The Visual Computer*, 13:228–246, 1997.

[2] P. Cignoni, C. Rocchini and R. Scopigno. Metro: Measuring error on simplified surfaces. *Computer Graphics Forum*, 17(2):167 174, June 1998.

[3] J. Cohen, A. Varshney, D. Manocha, G. Turk, H. Weber, P. Agarwal, F. Brooks and W. Wright. Simplification Envelopes. In *Proc. SIGGRAPH'96*, pages 119–128.

[4] M. Garland and P. S. Heckbert. Surface Simplification using Quadric error metric. In *Proc. SIGGRAPH'97*, pages 209–216, August 1997.

[5] A. Guéziec. Surface Simplification inside a Tolerance volume. Technical report, York Town Heights, NY 10598, May 1997. IBM Research Report RC 20440(90191).

[6] A. Guéziec. Locally Toleranced Surface Simplification. *IEEE Transactions on Visualization and Computer Graphics*, 5(2):168–189, April–June 1999.

[7] H. Hoppe. Progressive Meshes. In *Proc. SIGGRAPH'96*, pages 99–108, August 1996.

[8] P. Lindstrom and G. Turk. Fast and Memory efficient Polygonal Simplification. In *Proc. IEEE Visualization'98*, pages 279–286, 544 Oct. 1998.

[9] P. Lindstrom and G. Turk. Evaluation of Memoryless Simplification. *IEEE Transactions on Visualization and Computer Graphics*, 5(2):98–115 April–June 1999.

[10] Maria-Elena Algori and F. Schmitt. Mesh Simplification. *Computer Graphics Forum*, 15(3), August 1996, *Proc. Eurographics'96*.

[11] S. Melax. A. Simple Fast and Efficient Polygon Reduction Algorithm. *Game Developer*, pages 44–49, November 1998.

[12] M. Reddy. SCROOGE: Perceptually-Driven Polygon Reduction. *Computer Graphics Forums*, 15(4):191–203, 1996.

[13] R. Ronfard and J. Rossignac. Full Range Approximation of Triangular Polyhedra. *Computer Graphics Forum*, 15(3), 1996. *Proc. Eurographics'96*.

[14] J. C. Xia and A. Varshney. Dynamic View-Dependent Simplification for Polygonal Models. In *Proc. Visualizaion'96*, pages 327–334. Oct. 1996.

[15] R. Qu, and M. Sarfraz. A New Approach to the Improvement of Surface Triangulations using Local Algorithms. *Parallel & Scientific Computations* (Special Issue on Computer Aided Geometric Design), 5(1–2), Dynamic Publishers, USA, 221–238, 1997.

[16] M. Sarfraz. Designing of 3D Rectangular Objects, *Lecture Notes in Computer Science 1024: Image Analysis Applications and Computer Graphics*, Eds.: R. T. Chin, H. H. S. Ip, A. C. Naiman, and T-C. Pong, 1995, Springer-Verlag, 411–418.

[17] M. Sarfraz. Curves and surfaces for CAD using C2 rational cubic splines. *Engineering with Computers*, Springer-Verlag, Vol.11(2), 94–102, 1995.

[18] M. Sarfraz. Designing of Curves and Surfaces using Rational Cubics. *Computers and Graphics*, Elsevier Science, Vol. 17(5), 529–538, 1993.

16

Multiresolution and Diffusion Methods Applied to Surface Reconstruction Based on T-Surfaces Framework

Gilson A. Giraldi
Rodrigo L. S. Silva
Walter H. Jiménez

Department of Computer Science, National Laboratory for Scientific Computing,
Av. Getulio Vargas, 333, 25651-070, Petrópolis, RJ, Brazil.

Edilberto Strauss

Federal University of Rio de Janeiro, Department of Electronics Engineering –
DEL/EE, Rio de Janeiro, RJ, Brazil.

Antonio A. F. Oliveira

Federal University of Rio de Janeiro, Computer Graphics Laboratory,
Mail Box 68511, CEP 21945-970, Rio de Janeiro, RJ, Brazil.

In this chapter we present a new approach, which integrates the T-Surfaces framework and a multiresolution method in a unified methodology for segmentation and surface reconstruction. For noise images, we can improve the result by anisotropic diffusion. Despite this improvement, some manual intervention may be required to complete the reconstruction. Thus, we take advantage of the topological capabilities of T-Surfaces to enable the user to modify the topology of a surface. Besides, we discuss the utility of diffusion-reaction schemes for vector fields in our approach. Finally, we present some results for both synthetic and actual medical image volumes.

Advances in Geometric Modeling. Edited by M. Sarfraz
© 2003 John Wiley & Sons, Ltd   ISBN: 0-470-85937-7

1. Introduction

Parametric Deformable Models, which include the popular *snake models* [6] and deformable surfaces [13], are well known techniques for boundary extraction and tracking in both 2D and 3D images.

In this chapter we focus on parametric surface models. These models basically consist of an elastic surface, which can dynamically conform to the object shapes, in response to both internal and external forces (image forces and constraint forces).

Recently, McInerney and Terzopoulos [8,9] have proposed the T-Surfaces/T-Snakes model to add topological capabilities (splits and merges) to a parametric model. The basic idea is to embed a discrete deformable model within the framework of a simplicial domain decomposition (triangulation) of the image domain. Also, T-Surfaces depend on some threshold to define a normal force, which is used to drive the model towards the targets [8].

Based on these elements (threshold and simplicial decomposition framework) we proposed in [2] a segmentation approach for 2D images based on multiresolution methods and the T-Snakes model.

In this work we firstly extend that approach for 3D through the T-Surfaces. Thus, we also assume a scale restriction for the targets. In a first stage, we use this restriction to define the coarsest image resolution that guarantees not to split the objects. From the corresponding grid, we make a simple CF-triangulation of the image domain. The low-resolution image field is thresholded to get a binary function, which we call an *Object Characteristic Function*. Then, a simple continuation method is used to extract a set of closed polygonal surfaces, which contain the anatomical structures.

The grid resolution is application dependent. However, an important point of our method is its multiresolution/multigrid nature: having resolved (segmented) the image in a coarser (grid) resolution, we can detect regions where the grid has to be refined and then recursively apply the method only over these regions.

The polygonal surfaces so extracted are in general rough approximations of the surfaces of interest. We improve these approximations by using the T-Surfaces model [8] whose framework is the basic one for this chapter. For noisy images, we can increase the efficiency by pre-processing the original image through anisotropic diffusion [10].

If the segmentation remains incomplete at the finest resolution, we propose an interactive procedure based on the T-Surfaces framework to cut a surface and complete the segmentation.

In the following text we first present the multiresolution method used. In Section 4 we describe the T-Surfaces model. The segmentation and geometry extraction framework is presented in Section 5. In Section 6 we analyze scalars and vector diffusion methods in the context of our work. Section 7 discusses experimental results. Conclusions and future works are presented in Section 8.

2. Multiresolution

In this chapter we are interested in applications where the intensity (grey level) patterns of an object O (or of the background) can be characterized by a threshold T, or some statistics (mean μ and variance σ) of the image field I [2, 7].

That means $p \in O$ if:

$$I(p) > T \ or \ |I(p) - \mu| \le \sigma. \tag{16.1}$$

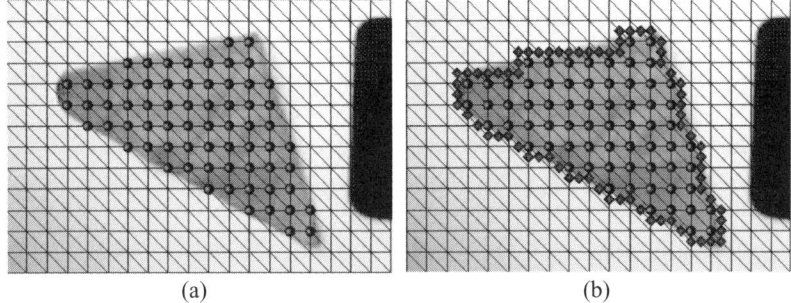

Figure 16.1 (a) Original image and characteristic Function. (b) Boundary approximation.

First, we assume a local scale property: Given a point $p \in O$ let r_p be the radius of a hyperball B_p which contains p and lies entirely in the object region. We assume that, for all $p \in O$, we have $r_p > 1$.

The local scale property guarantees that the image resolution can be reduced without losing the object of interest. In this way, we incorporate the basic philosophy of some nonparametric multiresolution techniques used in image segmentation [5]: as the resolution is decreasing, small background artifacts become less significant relative to the object(s) of interest.

Observe a simple example pictured in Figure 16.1. In this case, the object is easily segmented by thresholding ($T < 150$). In the Figure 16.1(a) we have a CF triangulation whose grid resolution is 10×10.

So, we can define a simple function, called a *Object Characteristic Function*, as follows:

$$\chi \colon \Re^2 \to \{0, 1\}, \tag{16.2}$$

where $\chi(p) = 1$ if $I(p) < T$ and $\chi(p) = 0$ otherwise, where p is a node of the triangulation.

A step further can be done as shown in Figure 16.1(b). The picture describes a curve that belongs to the triangles in which the characteristic function (marked nodes) changes its value. Observe that this curve approximates the boundary we seek.

This scheme is adaptive in the sense that resolution can be increased inside the extracted curve. To increase the resolution we just refine the coarser grid and sample the image over the corresponding grid nodes.

The generation of that curve (Figure 16.1(b)) is a process, which fits well in the subject of Piecewise Linear Manifolds (*PL Manifolds*) and is discussed next.

3. PL Methods

Firstly, let's see some useful definitions: We call an edge τ of a triangle σ *completely labeled* in respect to χ if this function changes its value in τ. A triangle σ in $\Re^3$ is called *transverse,* with respect to χ, if it contains a completely labeled edge.

A fundamental concept in this theory is the *Piecewise Linear Manifold* (PL Manifold). For this discussion, given a bidimensional manifold M implicitly defined, it is enough to say that a PL Manifold $\hat{M}$ is a polygonal (*cell*) representation of that manifold with the following properties: (1) The intersection $\sigma_1 \cap \sigma_2$ of two cells $\sigma_1, \sigma_2 \in \hat{M}$ is empty, or a common edge/node of both cells; (2) An edge τ is common to at most two cells of $\hat{M}$; (3) $\hat{M}$ is locally finite, that is, any compact subset of M meets only finitely many cells of $\hat{M}$.

The following definition will be useful next. Let's suppose two simplices σ_1, σ_2, which have a common face, and the vertices, $v_1 \in \sigma_1$ and $v_2 \in \sigma_2$, both opposite the common face. The process of obtain v_2 from v_1 is called *pivoting*.

Each connected component of the bidimensional PL Manifold can be generated by the following algorithm [1]:

PL Generation Algorithm:
```
σ₀ = FindTransverseTriangle();
∑ = {σ₀};
V(σ₀) = GetVertices(σ₀);
while V(σ) ≠ 0 for some σ ε ∑ do
get σ ε ∑ such that V (σ) ≠ 0 do
v = getVertice(V(σ));
σ' = getPivoted(σ , v);
if IsTransverse(σ') = 0 do
  dropVertice(v, V(σ));
else
  if σ' ε ∑ do
    dropVertice(v, V(σ));
    dropVertice(v', V(σ'));
else
  ∑ = ∑ + σ';
  V(σ') = GetSetOfVertices(σ');
  dropVertice(v, V(σ));
  dropVertice(v', V(σ'));
end while
```

In this pseudocode, the procedure 'getPivoted' performs the pivoting of vertex v into v', as defined above, and returns the new simplex σ' generated by the common face and v'.

The obtained PL manifolds are in general not smooth, and do not fit well to the desired boundary, as we can verify in Figure 16.1(b). Next, we present the deformable model which we use to improve the result.

4. T-Surfaces

The T-Surface approach is composed basically by three components [8]: a triangulation of the domain of interest, in our case a closed subset $D \subset R^3$, a particle model of the deformable surface and a *characteristic function* defined similarly in Equation (16.2) but distinguishing the interior (*Int* (S)) from the exterior (*Ext* (S)) of a surface $S : \chi(p) = 1$ if $p \in Int\ (S)$ and $\chi(p) = 0$ otherwise, where p is a node of the triangulation.

In this framework, the reparameterization of a surface is done by [8, 9]: (1) Taking the intersection points of the surface with the triangulation grid; (2) Carrying out topological changes by applying the PL Generation Algorithm to get the transverse triangles; (3) For each completely labeled edge chose an intersection point. These points will be used to define the surface cells.

4.1. Discrete Model

A T-Surface, can be seen as a discrete form of the classical parametric deformable surfaces [13].

It is defined as a closed elastic mesh, consisting of a set of nodes which are the vertices of a PL Manifold defined by the corresponding characteristic function (defined above).

When a cell is quadrilateral, we can go a step further and subdivide it in two triangles to get a triangular mesh. In this case, each triangle of the mesh is called a *triangular element*, each node is called a *node element* and each pair of nodes v_i, v_{i+1} is called a *model element*.

The node elements are linked by a springs with a null natural length. Hence, given the deformations $r_{ij} = \|v_i - v_j\|$, we define a tensile force given by:

$$\vec{\alpha}_i = c \sum_j r_{ij}, \tag{16.3}$$

where c is a scale factor. The model also has a normal force which can be a weight like in [8, 9]:

$$\vec{F}_i = k \cdot sign_i \frac{M\vec{n}_i}{\|M\vec{n}_i\|}, \quad M\vec{n}_i = \sum \frac{\vec{n}_i}{l}, \tag{16.4}$$

where $\vec{n}_i$ is the normal vector at node i, $sign_i = +1$ if $I(v_i) > T$ and $sign_i = -1$ otherwise (T is defined in expression (16.1) and $I(v_i)$ is the image intensity in v_i), and k is a scale factor. This force is used to push the model towards image edges until it is opposed by external image forces.

The external (image) force is given by:

$$\vec{f}_i = -\gamma_i \nabla \|\nabla I\|^2, \tag{16.5}$$

The evolution of the surface is governed by the following dynamical system:

$$v_i^{(t+\triangle t)} = v_i^t + h_i \left(\vec{\alpha}_i^t + \vec{F}_i^t + \vec{f}_i^t \right), \tag{16.6}$$

where h_i is an evolution step. During the T-Surfaces evolution some grid nodes become interior to a surface. Such nodes are called *burnt nodes* and its identification is fundamental to update the characteristic function [8, 9]. To deal with self-intersections of the surface the T-Surface model incorporates an entropy condition: *once a node is burnt it stays burnt*. A termination condition is obtained based on the number of deformation steps that a triangle has remained a transverse one.

The threshold T used in the normal force (16.4) plays an important role in the T-Surfaces model. If it was not properly chosen, the T-Surface can stop in a region far from the target(s) [2]. The choice of T is more critical when two objects to be segmented are too close, as shown in Figure 16.2.

For the T-Surface (T-Snake) to separate the objects pictured, it has to burn the grid nodes marked. To accomplish this, force parameters in Equations (16.3)–(16.5) should be chosen properly to advance the T-Snake over these nodes which is a nontrivial task.

The basic point to propose the following framework is that if we have a local scale property, we can use the threshold T to initialize the T-Surface closer to the structures of interest. To

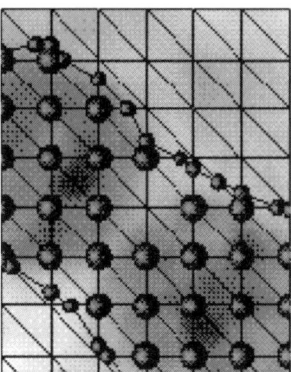

Figure 16.2   T-Snake and grid nodes marked.

accomplish this, we just take the object characteristic function (expression (16.2)) as the start characteristic function of T-Surfaces model.

Moreover, once a T-Surface is also a PL manifold, the polygonal surfaces extracted by the algorithm of Section 3 can be used to initialize the T-Surfaces model.

5. Segmentation Framework

The segmentation method proposed by this chapter is based on the following steps: (1) Extract region based statistics; (2) Coarser image resolution and triangulation; (3) Define the *Object Characteristic Function*; (4) PL Manifold extraction by the algorithm of Section 3; (5) Apply T-Surfaces model.

We assume a *local scale property* for the structures of interest. So, according to Section 2, we do not need to concern ourselves with merge of regions. Also, the following properties are supposed by the objects boundaries: (a) Closedness, (b) Orientedness, (c) Connectedness.

These topological constraints are satisfied by the PL Generation Algorithm results and, as a consequence, are consistent with the T-Surfaces reparameterization of Section 4. Due to these constraints, when a connected component is generated, we should not apply the PL Generation Algorithm inside it. In this way, we can maintain a spherical topology without the need of topological preservation methods [8, 9].

Among the surfaces extracted, there may be open surfaces which start and end in the image frontiers, small surfaces corresponding to artifacts or noise in the background. The former is discarded by a simple automatic inspection. To discard the latter, we need a set of pre-defined features (volume, surface area, etc), and corresponding lower bounds [11]. For instance, from the local scale property and the triangulation used we can set the volume lower bound as $8r_l^3$.

Besides, some polygonal surfaces may contain more than one object of interest (see Figure 16.3). Now, we can use upper bounds for the features. These upper bounds are application dependent (anatomical elements can be used). The surfaces whose interior have volumes larger than the upper bound will be processed in a finer resolution. It is important to stress that the upper bound(s) is not an essential point for the method. It's role is only to avoid expending time computation in regions were the boundaries enclose only one object.

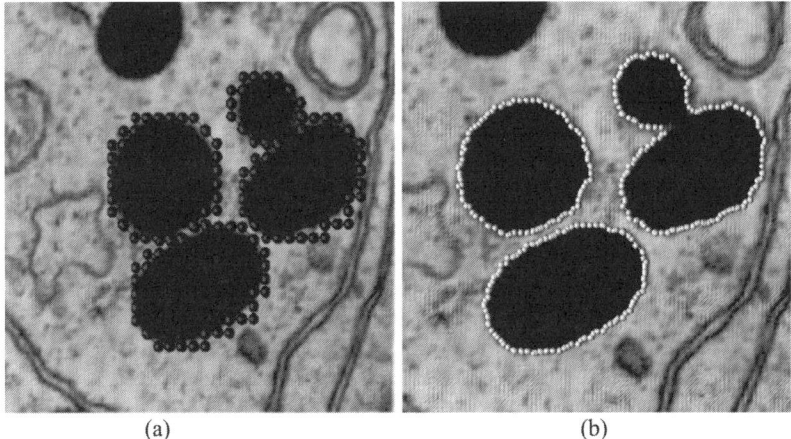

(a)   (b)

Figure 16.3   (a) OPL manifolds for resolution 3×3. (b) Result with the highest (image) resolution.

For instance, in images like in Figure 16.3, the *outer* scale corresponding to the separation between the objects may be finer than the local scale property of the objects of interest. Hence, the coarsest resolution could not *separate* the objects. This happens for the bottom-left cells on Figure 16.3(a). To correct that result we increase the resolution in those regions to account for more details (Figure 16.3(b)).

The result of step (4) can be improved by anisotropic diffusion (see the next section) [10]. Such approaches enables us to blur small discontinuities (improving the surface extraction) as well as to enhance edges (improving the T-Surface result). Mathematical morphological operators could also be used [12].

However, for images like those in Figure 16.3, some manual intervention may be required, even with these improvements. To free hand split a T-Snake/T-Surface represented in two components, it is only a matter of taking the following steps: (a) Define a cutting plane; (b) Sct to zcro thc grid nodes belonging to the triangles that the plane cuts and that are interior to the T-Surface; (c) Apply steps (4) and (5) above.

The grid nodes set to zero become and burnt nodes. Thus, the entropy condition will prevent intersections of the two T-Surfaces generated during the evolution. Hence, we can efficiently guarantee that these surfaces will not merge again.

It is important to highlight that T-Surfaces models can deal naturally with the self-intersections that may happen during the evolution of the surfaces obtained by step (4). This is an important advantage of T-Surfaces.

Also, we must emphasize that our approach is multigrid/multiresolution when segmenting the image but not when applying the T-Surfaces. This deformable model is used at the end of the boundary extraction stage.

We do not need to use multiscale relaxation methods [3] in that final step because we take the full resolution of the image for evolving T-Surfaces. Besides, when the grid resolution of T-Surfaces is increased we just reparameterize the model through the finer grid and evolve the corresponding T-Surfaces.

6. Diffusion Methods

In image processing, diffusion schemes for scalar and vector fields have been successfully applied [4, 14]. Gaussian blurring is the most known one [4].

Other approaches are the anisotropic diffusion and the Gradient Vector Flow [10, 14]. Below, we summarize these methods and conjecture their unification.

In the next section, we will apply anisotropic diffusion to improve the results obtained through the methodology described in Section 5.

Anisotropic diffusion is defined by the following general equation:

$$\frac{\partial I(x, y, t)}{\partial t} = div(c(x, y, t) \nabla I), \tag{16.7}$$

where I is the gray level image intensity [10].

In this method, the blurring over parts with high gradient can be made much smaller than in the rest of the image.

To show this property, we follow Perona-Malik's work [10]. Firstly, we suppose that the edge points are oriented in the x direction. Thus, Equation (16.7) becomes:

$$\frac{\partial I(x, y, t)}{\partial t} = \frac{\partial(c(x, y, t) \cdot I_x(x, y, t))}{\partial x}. \tag{16.8}$$

If c is a function of the image gradient, $c(x, y, t) = g(I_x(x, y, t))$, we can define $\phi(I_x) \equiv g(I_x) \cdot I_x$ and then rewrite Equation (16.8) as:

$$\frac{\partial I(x, y, t)}{\partial t} = \frac{\partial I}{\partial x}(\phi(I_x)) = \phi'(I_x) \cdot I_{xx}. \tag{16.9}$$

We are interested in the time variation of the slope $\frac{\partial I_x}{\partial t}$. If $c(x, y, t) > 0$ we can change the order of differentiation and with a simple algebra to demonstrate that:

$$\frac{\partial I_x}{\partial t} = \frac{\partial I_t}{\partial x} = \phi'' \cdot I_{xx}^2 + \phi'(I_x) \cdot I_{xxx}. \tag{16.10}$$

At the edge points we have $I_{xx} = 0$ and $I_{xxx} << 0$ as these points are local maxima of the image gradient intensity [4, 10]. Thus, there is a neighborhood of the edge point in which the derivative $\frac{\partial I_x}{\partial t}$ has a sign opposie to $\phi'(I_x)$. If $\phi'(I_x) > 0$ the slope of the edge point decrease in time. Otherwise it increases, that means, the border becomes sharper.

So, the diffusion scheme given by Equation (16.7) allows us to blur small discontinuities and to intensify the stronger ones.

In this work, we have used ϕ as follows:

$$\phi = \frac{\nabla I}{1 + [\|\nabla I\|/K]^2}, \tag{16.11}$$

where ϕ is given by expression (16.11) with $K = 300$. The number of interactions, in the numerical scheme used to solve this equation [10], was 4. Figure 16.4(c, d) shows the cross section corresponding to the slice 40.

We observe that with anisotropic diffusion the result is closer to the boundary. Also, the final result is more precise when pre-processing with anisotropic diffusion (Figure 16.4(f)) where the constant K can be determined by a histogram of the gradient magnitude.

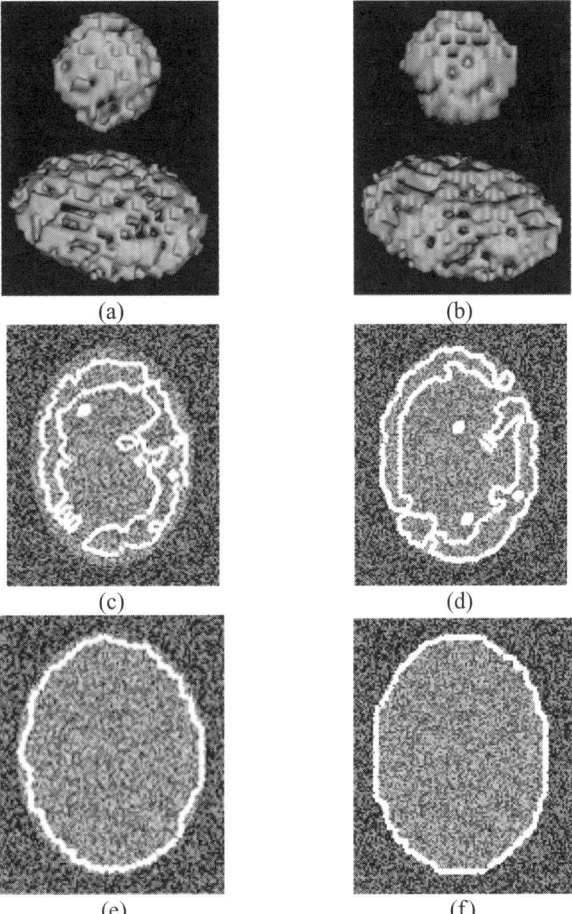

Figure 16.4   (a)–(b) Result for steps (1)–(4) with Gaussian and Anisotropic Diffusion, respectively.
(c)–(d) Cross section of (a),(b) for slice 40, respectively. (e)–(f) Final solution for (c),(d)
respectively.

In the above scheme, I is a scalar field. For vector fields, a useful diffusion scheme is
the Gradient Vector Flow (GVF). It was introduced in [14] and can be defined through the
following diffusion-reaction equation:

$$\left.\begin{aligned}
\frac{\partial u}{\partial t} &= \nabla \cdot (g \, \nabla u) + h(u - \nabla f), \\
u(x, 0) &= \nabla f.
\end{aligned}\right\} \qquad (16.12)$$

where f is a function of the image gradient (for example, $f = \|\nabla I\|^2$) and $g(x), h(x)$ are
nonnegative functions defined on the image domain.

The field obtained by solving the above equation is a smooth version of the original one,
which tends to be extended far away from the object boundaries. When used as an external
force for deformable models it makes the methods less sensitive to initialization [14].

As the result of steps (1)–(5), of Section 5, is in general close to the target we could apply this method to push the model towards the boundary when the grid is turned-off.

However, for noisy images some kind of diffusion (smoothing) must be used before applying GVF. Gaussian diffusion has been used [14] but precision may be lost due to the nonselective blurring [4, 10].

The anisotropic diffusion scheme presented above is an alternative smoothing method. Such observation points forward the possibility of integrating anisotropic diffusion and the GVF in an unified framework.

A straightforward way of doing this is allowing g and h to be dependent upon the vector field u. The key idea would be to combine the selective smoothing of anisotropic diffusion with the diffusion of the initial field obtained by GVF. Besides, we expect to get a more stable numerical scheme for noisy images.

7. Experimental Results

The first point to be demonstrated is the utility of multiscale methods in our work. We take a synthetic $150 \times 150 \times 150$ image volume composed by a sphere with radius 30 and an ellipsoid with axes 45, 60 and 30 inside an uniform noise specified by the image intensity range 0–150.

Figure 16.4 shows the result for steps (1)–(4) applied to this volume after (a) gaussian diffusion; and (b) the anisotropic diffusion defined by the equation:

$$\frac{\partial I}{\partial t} = div(\phi), \tag{16.13}$$

This is in accordance with the discussion in Section 6: Equation (16.13) enables us to blur small discontinuities (gradient magnitude bellow K) as well as enhancing edges (gradient magnitude above K) [10].

Another point becomes clear in this example: the topological abilities of T-Surfaces enable us to correct the defects observed in the surface extracted through the steps (1)–(4). Hence, after few interactions, the method gives one connected component which is a better approximation of the target.

The T-Surface parameters are: $c = 0.65$, $k = 1.32$ and $r = 0.01$. The grid resolution is $5 \times 5 \times 5$, the freezing point is set to $T \in (120, 130)$ and threshold $T \subset (120, 134)$ in Equation (16.4). The number of deformation steps for T-Surfaces was 17. The extracted surfaces approximates the real ones with an error below 4.38 which we consider fine in this case.

Figure 16.5(a) shows an example where the steps (1)–(5) where not able to complete the segmentation. This can be resolved by user interaction through the method described in the last section (steps (a)–(d)). Figure 16.5(b) shows the final result. The parameters are the same as the last example.

Finally, we segment an artery from an $155 \times 170 \times 165$ image volume obtained from the Visible Human project. The T-Surfaces parameters are: $c = 0.75$, $k = 1.12$ and $\gamma = 0.3$, grid resolution is $4 \times 4 \times 4$ and freezing point is set to 10. The (correct) result is pictured in Figure 16.6(b) which was obtained through anisotropic diffusion.

An important point to highlight is that by initializing the T-Surfaces with steps (1)–(4) we achieve speed ups even for finer grid resolutions.

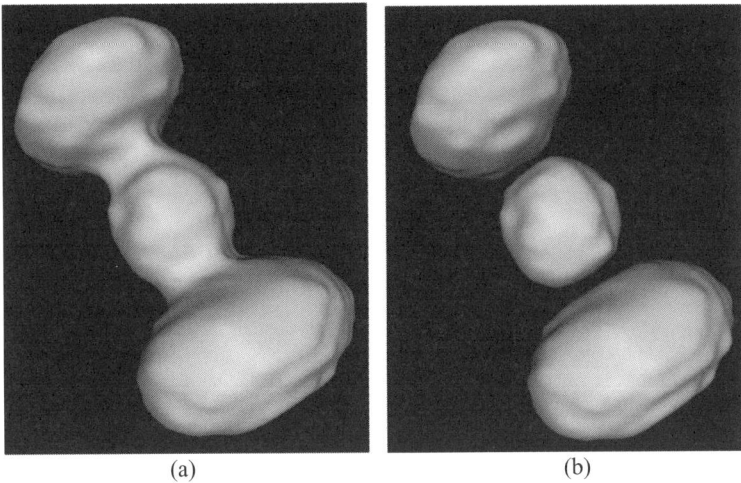

(a)   (b)

Figure 16.5   (a) Partial result. (b) Final solution after manual cut.

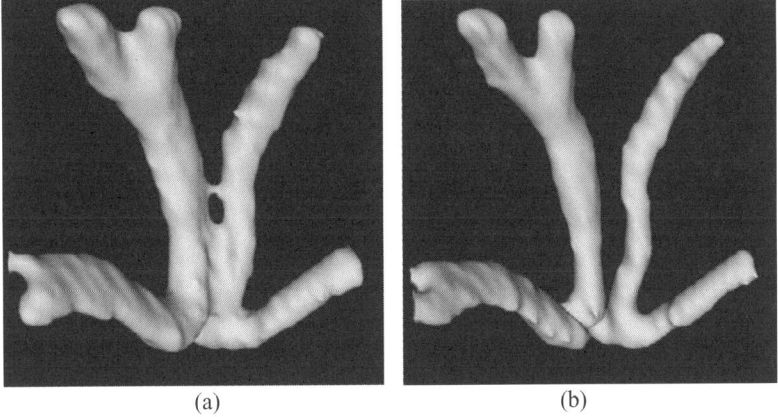

(a)   (b)

Figure 16.6   (a) Result without and (b) after anisotropic diffusion.

8. Summary

The implemented PL Generation algorithm shares the basic elements used in isosurface generation methods.

However, these methods in general do not incorporate the scale and topological restrictions (connectedness and closedness) which formalize our *prior knowledge* of the structures of interest.

The topological abilities of T-Surfaces enable an efficiently initialization through steps (1)–(4) as well as an interactive procedure to change the topology of a surface. Anisotropic diffusion can improve the step (4) as well as the T-Surfaces result.

Future directions for this work will be to generalize the user interaction method by substituting the plane by a scalpel and allowing the user to drag the scalpel. Besides, we aim to apply GVF variants to improve the results of our method.

References

[1] Allgower, E. L. and Georg, K. 1990. *Numerical Continuation Methods: An Introduction*. Springer-Verlag Berlin Heidelberg.

[2] Giraldi, G. A.; Strauss, E.; Oliveira, A. F. (2000). A Boundary Extraction Approach Based on Multi-resolution Methods and the T-Snakes Framework. In: *International Symposium on Computer Graphics, Image Processing and Vision (SIBGRAPI'2000)*.

[3] Heitz, F.; Perez, P.; Bouthemy, P. (1994). Multiscale Minimization of Global Energy Functions in Some Visual Recovery Problems. CVGIP: *Image Understanding*, Vol. 59, No. 1, pp. 125–134, January.

[4] Jain, A. (1989). *Fundamentals of Digital Image Processing*. Prentice-Hall.

[5] Jolion, J. M. and Montanvert, A. (1992). The Adaptive Pyramid: A framework for 2D Image Analysis. CVGIP: *Image Understanding*, 55(3), 339–348.

[6] Kass, M.; Witkin, A.; Terzopoulos, D. (1987). Snakes: Active contour models, *Proc. First Int. Conf. Comput. Vision*, pp. 259–268, London, 1987.

[7] Lorensen, W. and Cline, H. (1987). Marching Cubes: A High Resolution 3D Surface construction Algorithm, *Computer Graphics*, Vol 21, No. 4, July 1987.

[8] McInerney, T. and Terzopoulos, D. (1999). Topology Adaptive Deformable Surfaces for Medical Image Volume Segmentation. *IEEE Transactions on Medical Imaging*, 18(10), 840–850.

[9] McInerney, T. (1997). Topologically Adaptable Deformable Models for Medical Image Analysis. *Ph.D. thesis*, Department of Computer Science, University of Toronto.

[10] Perona, P. and Malik, J. (1990). Scale-Space and Edge Detection Using Anisotropic Diffusion. *IEEE Trans. on Patter Analysis and Mach. Intell.*, 12(7), 629–639.

[11] Samtaney, R.; Silver, D.; Zabusky, N.; Cao, J. (1994). Visualizing Features and Tracking their Evolution, *IEEE Computer* 27, No. 7, pp. 20–27, July 1994.

[12] Sarti, A.; Ortiz, C.; Lockett, S.; Malladi, R. (1998). A Unified Geometric Model for 3D Confocal Image Analysis in Cytology. *Proc. International Symposium on Computer Graphics, Image Processing, and Vision (SIBGRAPI'98)*, 69–76.

[13] Singh, A.; Goldgof, D.; Terzopoulos, D. (1998). *Deformable Models in Medical Image Analysis*. IEEE Computer Society Press.

[14] Xu, C., and Prince, J. (1998). Snakes, Shapes, and Gradient Vector Flow. *IEEE Trans. Image Proc.*, March, pp. 359–369.

17

A Multiresolution Framework for NUBS

Muhammad Sarfraz
Mohammed Ali Siddiqui

Department of Information and Computer Science, King Fahd University of Petroleum and Minerals, Dhahran 31261, Saudi Arabia.

The piecewise polynomial B-spline representation is a flexible tool in Computer Aided Geometric Design (CAGD) for representing and designing the geometric objects. In the field of Computer Graphics (CG), Computer Aided Design (CAD), or Computer Aided Engineering (CAE), a very useful property for a given spline model is to have locally supported basis functions. This allows localized modification of the shape. Unfortunately this property can also become a serious disadvantage when the user wishes to edit the global shape of a complex object. A multiresolution representation, for Non-uniform B-splines (NUBS), is proposed as a solution to alleviate this problem. The proposed model has features that it uses control point decimation strategy for decomposing NUBS curves and it is efficient in both time and space utilization. A comparative study of the proposed work is also made with an alternate approach in the literature, which is based upon knot decimation.

1. Introduction

In the field of geometric modeling, the construction of efficient, intuitive, and interactive editors for geometric objects is a fundamental objective, but it is still a difficult challenge. In many freeform geometric modeling systems the users are allowed to work in the framework of a specific data model, e.g. Bezier or non-uniform rational B-splines [5]. This imposes constraints on the set of geometric manipulation operations that can be performed, the man-machine interface and the type of objects which can be modeled.

There are various curve manipulation techniques, which have been proposed in current literature. For example, the Euclidean distances between the point of modification and the

Advances in Geometric Modeling. Edited by M. Sarfraz
© 2003 John Wiley & Sons, Ltd   ISBN: 0-470-85937-7

control points of a B-spline curve were used as weights to affect the control points in [3]. The difficulty with this approach appears when the two separate portions of the curve are close. To alleviate the difficulty in editing freeform shapes while matching engineering specifications, constraint based approaches were proposed in [1, 18]. Direct and interactive manipulation tools of freeform curves and surfaces are investigated in [4].

In the field of computer graphics or Computer Aided Design (CAD), a very useful property for a given spline model is to have locally supported basis functions in order to allow localized modifications of the shape. Unfortunately this property can also become a serious disadvantage when the user wishes to edit the global shape of a complex object. Piecewise polynomial B-spline representation is common in many contemporary geometric modeling systems. While this is a powerful tool with many desirable properties, the same properties impose some undesirable constraints on the user. For example, the most attractive property, *locality*, restricts the user to perform global operations on the object being modeled. To perform a global operation, it has to be transformed into a series of local operations affecting only a small portion of the curve, which makes the process time wasting and precision hazardous [10]. The ability to simultaneously perform both local and global operations at will would add significant functionality to any modeling system.

Multiresolution representation is a possible solution, which addresses this problem, because it allows the user to edit objects at different resolution levels. Both local as well as global operations can be performed on curves by representing them using multiresolution decomposition. Several approaches [7, 15–17] have been proposed for multiresolution representation of splines, mostly based on wavelets. All these approaches involve expensive pre-calculations and, in the case of open curves and surfaces, often require specific treatment of boundary control points. Moreover, these approaches depend on the given spline model they manipulate; the whole scheme has to be redefined when it comes to manipulating other spline models, only the philosophy of the calculus can potentially be reused [10].

Among the type of B-splines, NUBS [2, 6, 9, 11] have been receiving considerable attention in the areas of computer graphics and geometric modeling. NUBS are industry standard tools for the representation and design of geometry. The term NUBS is given to it because they are defined on a knot vector where the interior knots spans are not equal. NUBS are useful because:

- By manipulating the control points and knot vector, NUBS provide the facility to design a variety of shapes.
- They offer a common mathematical form for representing and designing freeform curves and surfaces.
- Evaluation is reasonably fast and computationally stable.
- NUBS have clear geometric tool kit (knot insertion/deletion, degree elevation etc.), which can be used to design, analyze, process and interrogate objects.

In this work, a multiresolution representation has been proposed for NUBS based on control point decimation. Section 2 describes the general theory of B-splines and NUBS. In Section 3, we review the multiresolution representation method [5] for B-splines. Section 4 is about the proposed method for multiresolution representation for NUBS. In Section 5, the proposed method is demonstrated by means of some pictures and graphics. This section also contains a comparison analysis of the proposed method with the method reviewed in Section 3. Finally, we conclude in Section 6.

2. Theory of NUBS

The mathematical or natural spline is a piecewise polynomial of degree p with continuity of derivatives of order $p-1$ at common joints between segments. A spline curve is specified by a given set of coordinates positions, called *control points*, indicating the shape of a curve. Spline curve is defined, modified and manipulated with operations on the control points. Control points are then fitted with piecewise continuous parametric polynomial functions in one of two ways. The first type of splines is called *approximatory splines* in which the polynomials are fitted to the general control point path without necessarily passing through any control point and the resulting curve is said to approximate the set of control points. The second type of splines is called *interpolatory splines* in which the curve passes through each control point and the resulting curve is said to interpolate the set of control points.

The general expression for the calculation of coordinate positions along a B-spline curve in a blending function formulation is of the form:

$$S(t) = \sum_{i=0}^{n} s_i B_{i,p}(t), t_{\min} \leq t < t_{\max}, 2 \leq p \leq n+1,$$

where s_i, $i = 0, 1, \ldots, n$, is an input set of $n+1$ control points and the B-spline blending functions $B_{i,p}$ are polynomials of degree p. The Cox Deboor [11] recursive formula for the B-spline basis can be defined as:

$$B_{i,1}(t) = \begin{cases} 1, & if \quad t_i \leq t \leq t_{i+1}, \\ 0, & otherwise, \end{cases}$$

and

$$B_{i,p}(t) = \frac{(t - t_i) \, B_{i,p-1}(t)}{t_{i+p-1} - t_i} + \frac{(t_{i+p} - t) \, B_{i+1,p-1}(t)}{t_{i+p} - t_{i+1}}.$$

NUBS are Non-Uniform B-Splines and is the term given to curves that are defined on a knot vector where the interior knot spans are not equal. As an example, we may have interior knots with spans of zero. Some common curves require this type of non-uniform knot spacing. The use of this option allows better shape control and the ability to model a larger class of shapes. The shape of NUBS not only depends on the control points but also on the *knot vector* associated with the set of control points.

3. Multiresolution of NUBS Using Knot Decimation

In this section, for the sake of completeness and comparative study, a review of an already existing approach for the multiresolution representation of B-splines is briefly presented. This deals with the multiresolution control for NUBS, which uses the knot decimation and least squares approximation. This method, for the multiresolution representation of NUBS, has been presented in [5] in detail. The multiresolution decomposition of the freeform NUBS curve is computed using least-squares approximation based on existing data reduction techniques. The least-squares decomposition allows the support of NUBS curves, but it also imposes some processing penalties in both time and space compared to techniques for multiresolution uniform B-spline curves.

Let $C_k(t)$ be a B-spline curve of order n and l_k control points, defined over the knot vector τ_k, where $k \in Z^+$. Let V_k be the space induced by τ_k, and let $\tau_{k-1} \subset \tau_k$. The new space induced by τ_{k-1}, denoted by V_{k-1} is clearly a strict subspace of V_k. Now, suppose $C_{k-1}(t) (\in V_{k-1})$ be the least-squares approximation of $C_k(t)$ in the space V_{k-1}, and their difference be the *detail* $D_{k-1}(t) \in V_k$, given by:

$$D_{k-1}(t) = C_k(t) - C_{k-1}(t),$$

This process of decomposing a curve into two parts, one low resolution approximation and one high resolution detail can be applied recursively. $C_k(t)$ could then be expressed as:

$$C_k = C_0(t) + \sum_{i=0}^{k-1} D_i(t),$$

where $C_0(t) \in V_0$ and $D_i(t) \in V_{i+1}$.

In order to construct a multiresolution decomposition of a NUBS curve as in the earlier equation, the knot sequence τ_i, inducing the subspaces V_i must first be defined. τ_k is the knot vector of the original curve, the subsequent knot vectors τ_i, $0 \leq i < k$, can be constructed such that $\tau_i \subset \tau_{i+1}$ and $2|\tau_i| \approx |\tau_{i+1}|$, where $||$ denotes the size of the knot vector. The end conditions of the original curve must be preserved, hence the knots $\tau_j \in \tau_i, 0 \leq j < n$ and $l_i \leq j < l_i + n$, $\forall\, 0 \leq i < k$ are unmodified, where l_i denotes the number of control points defining $C_i(t)$ over τ_i. In general, $l_i = |\tau_i| + n$. This knot decimation process defines the function space hierarchy and is independent of the specific curve being decomposed.

For a B-spline curve with knot vector τ_k of size 2^k, k subspaces will be constructed; each induced by approximately half the knots of the previous level. The lowest resolution approximation $C_0(t)$ will be a single polynomial curve. That is, the knot vector τ_0 has no interior knots ($\tau_0 = 2n$). Least-squares techniques are employed to find the curve $C_i(t) \in V_i$, defined over τ_i, best approximating $C_k(t)$.

Knots are selected so as to minimize the local effect on the curve due to removals from level i to level $i + 1$. Hence, consecutive knots should not be removed in one step. Removing every n^{th} knot, where n is the order of the curve will cause the least change from one level to the next, yet affecting the entire curve. As the degree of a NUBS curve is increased, the curve becomes smoother and smoother due to the low pass property of the basis functions of the representation. Therefore, as n increases, by selecting every n^{th} knot for removal, the knots are removed at larger intervals yet the curve becomes smoother. In practice, it is found that removing every alternate knot still retains a sufficient number of resolution levels to enable an effective multiresolution control. Moreover, the computational overhead required for the algebraic summation is kept at interactive speeds.

4. Multiresolution of NUBS Using Point Decimation

By using the ability to control a B-spline curve by changing the position and order of the control points, we can come up with a multiresolution representation for NUBS. In this work we use the control point decimation for the purpose of multiresolution representation of NUBS.

Let $C_k(t)$ be a NUBS curve, defined over the set of polygon vertices or control points P_k (consisting of corresponding weight values for each point in addition to X and Y co-ordinate values) containing l_k points, using the knot vector T_k, where k is a positive integer, greater than zero. There are various methods proposed for the calculation of non-uniform knots, a popular method is to calculate the knot vector proportional to the chord lengths between the defining polygon vertices. We use the same knot calculation method. The NUBS curve $C_k(t)$ is calculated from the control points P_k as described in detail in Section 2.

Let V_k be the space of all the curves that can be defined using control points P_k. Now, we find a subset P_{k-1} of P_k ($P_{k-1} \subset P_k$), clearly the space V_{k-1} induced by P_{k-1} is a subset of V_k. Let $C_{k-1}(t) \in V_{k-1}$ be a curve defined over the control points P_{k-1}, and we found out that it is the approximation to the higher resolution curve $C_k(t)$. To find P_{k-1} from P_k, we use the process of decimation.

Let a unary operator $\mathbf{d}_j$ is defined for decimation, where j denotes the interval that is used to decimate the control points. If j is 2 then every 2nd (alternate) control point is decimated, if j is 3 then select every 3rd control point (i.e., control points numbered 3, 6, 9, ...) for removal. Similarly, if j is i then decimate every i^{th} control point. Mathematically control point decimation is given by:

$$P_{k-1} = \mathbf{d}_j(P_k).$$

To minimize the local effect on the resulting curve $C_{k-1}(t)$, consecutive control points from P_k should not be removed to obtain P_{k-1}. It is observed that removing every alternate point causes the acceptable amount of local effect and still retains a sufficient number of resolution levels to enable an effective multiresolution control. The lost control points can be captured as Q_{k-1}.

Let another unary operator $\mathbf{c}_j$ is defined to capture the decimated control points. Here also j denotes the interval used to decimate the points. Mathematically Q_{k-1} can be computed as:

$$Q_{k-1} = \mathbf{c}_j(P_k).$$

The process of decomposition can be applied recursively until P_0, which contains only n control points, where n is the order of the B-spline curve. The algorithm in Figure 17.1 summarizes the multiresolution decomposition process and the flow chart in Figure 17.2 shows it pictorially.

The reconstruction of P_i from P_{i-1} and Q_{i-1} is carried out by merging the sets P_{i-1} and Q_{i-1}. A binary operator $\mathbf{r}_j$ is defined for the process of reconstructing P_i from P_{i-1} and Q_{i-1}.

INPUT:
    $Ck(t)$, a NUBS Curve.
OUTPUT:
    $P0$, Qi, $0 \leq i < k$, the multiresolution decomposition of $C_k(t)$.
AlGORITHM:
-   $P_k \Leftarrow$ Control Points of $C_k(t)$;
-   for $i = k\text{-}1$ to 0 step -1 do
    begin
        $P_i = \mathbf{d}_j(P_{i+1})$;
        $Q_i = \mathbf{c}_j(P_{i+1})$;
    end;

Figure 17.1 The algorithm.

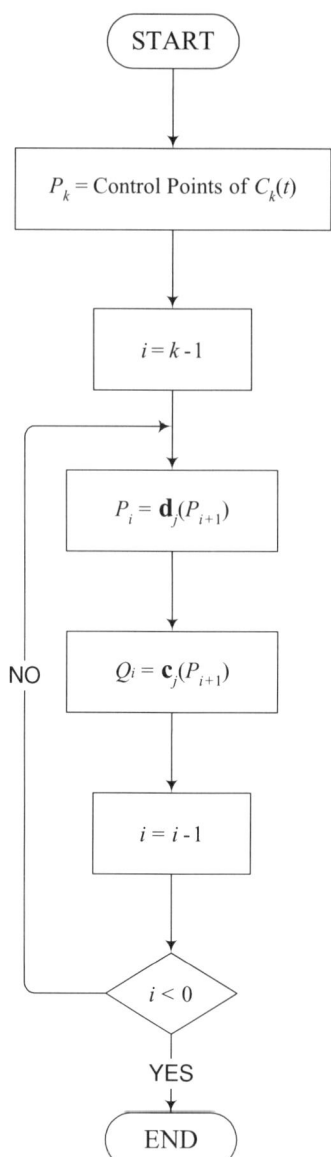

Figure 17.2   Flow chart of the multiresolution decomposition process.

The reconstruction is mathematically represented as:

$$P_i = \mathbf{r}_j(P_{i-1}, Q_{i-1}).$$

While reconstructing, the criteria used for the decomposition should be followed. For example, if every j^{th} point is decimated during decomposition, then the reconstruction of P_i is obtained by rearranging P_{i-1} and Q_{i-1} while placing $(j-1)$ points from P_{i-1} and one point from Q_{i-1} in the same order and so on.

By means of recursively applying the reconstruction operator the original set of control points can be represented in terms of its multiresolution components as:

$$P_k = \mathbf{r}_j(P_0, Q_0, Q_1, Q_2, \ldots, Q_{k-1}). \tag{17.1}$$

The above recursion comes through the following procedure:

$$
\begin{aligned}
P_k &= \mathbf{r}_j(\mathbf{r}_j(P_0, Q_0), Q_1, Q_2, \ldots, Q_{k-1}), \\
&= \mathbf{r}_j(P_1, Q_1, Q_2, \ldots, Q_{k-1}), \\
&= \mathbf{r}_j(\mathbf{r}_j(P_1, Q_1), Q_2, \ldots, Q_{k-1}), \\
&\quad\vdots \\
&= \mathbf{r}_j(P_{k-1}, Q_{k-1}).
\end{aligned}
$$

5. Demonstration

In this section, the proposed multiresolution representation is demonstrated by applying it to NUBS curves. Figure 17.3 shows a Star Shaped NUBS curve, whose details are given in Table 17.1. Figure 17.4 shows all its decomposition levels using the knot decimation multiresolution method of Section 3. In this figure, five lower resolution levels are obtained. Figure 17.4(a) is the original curve. Figures 17.4(b) to 17.4(f) show its lower level curves. In each figure the original curve is shown by thin line and the decomposed curve is shown by thick line. The average execution time is recorded as 1.73 seconds. Figure 17.5 shows

Figure 17.3   A Star Shaped NUBS curve.

Table 17.1   Attributes of Figure 17.4.

Property	Value
Type of Curve	NUBS
Degree	2
No. of Control Points	100

the multiresolution decomposition levels of the same curve by the proposed scheme (control point decimation method) in this paper. With this method, five lower resolution levels are also possible. The average execution time is recorded as 0.07 seconds.

Another example, for the implementation of the proposed method, is also considered. Figure 17.6 shows another NUBS curve drawn with 259 control points. After applying the multiresolution decomposition, the decomposed curves are obtained as shown in Figure 17.6. Figures 17.7(a) through 17.7(f) contain 130, 66, 34, 18, 10, and 6 control points respectively.

Figure 17.8 shows a NURBS curve of degree 3 consisting of 319 control points with default weight values. In total, six multiresolution levels are obtained for this curve, as shown in Figure 17.9. The original curve is shown in thin lines and the curves in thick lines are the decomposed versions at each level of multiresolution. The original curve is said to be at *level* 6, the curve in Figure 17.9(a) consists of 159 control points and is at *level* 5. Similarly the curves in Figure 17.9(b) through Figure 17.9(f) contain 80, 40, 20, 10, and 5 control points and are at decomposition levels 4, 3, 2, 1, and 0 respectively.

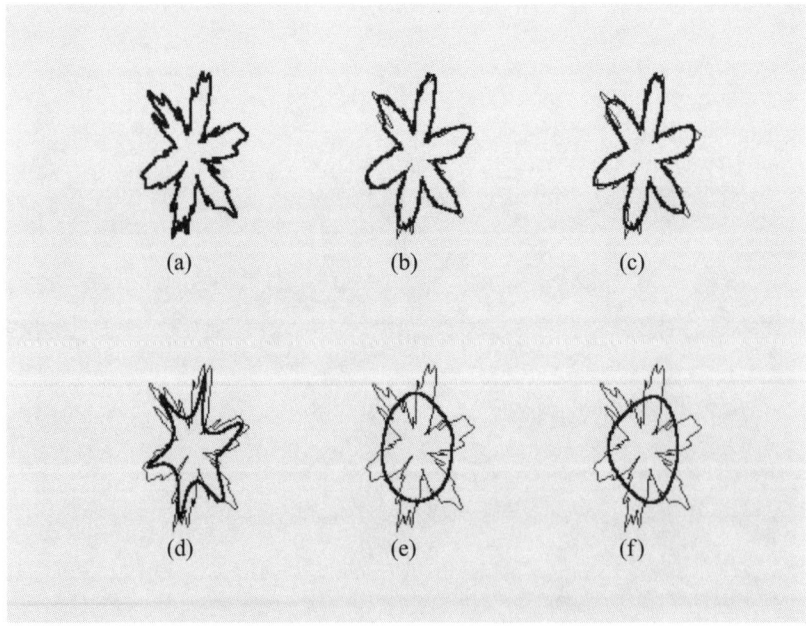

Figure 17.4   Multiresolution decomposition of the curve, in Figure 17.3, with Knot Decimation.

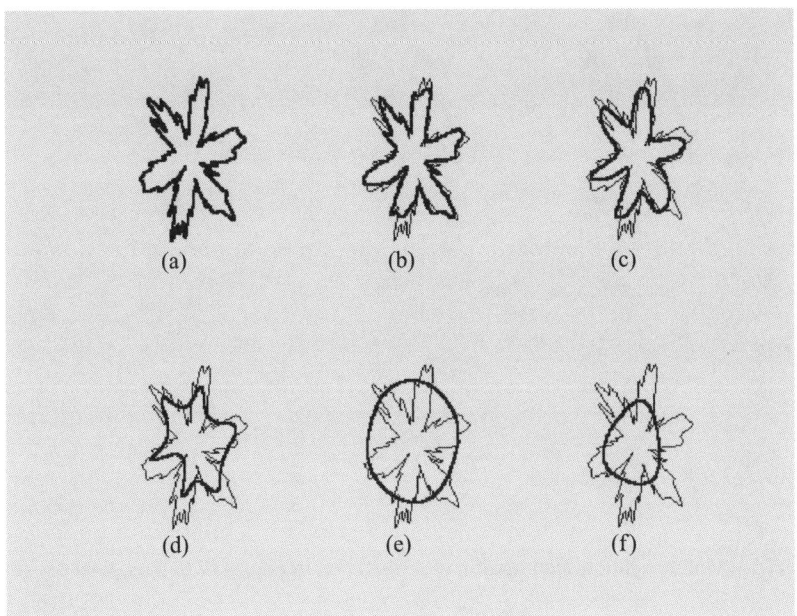

Figure 17.5   Multiresolution decomposition of the curve, in Figure 17.3, with point decimation.

The idea of curve multiresolution has been extended to rectangular surfaces too. The details of this method are not in the scope of this work and will be described in a later version. However, for the sake of sample demonstration, an example has been quoted here in Figures 17.10 and 17.11.

As part of muiltiresolution representation of NURBS surfaces, Figure 17.10 shows a NUBS surface drawn with 30×30 mesh of control points. This surface is decomposed by applying the multiresolution decomposition to obtain the low-resolution versions as shown in Figure 17.11.

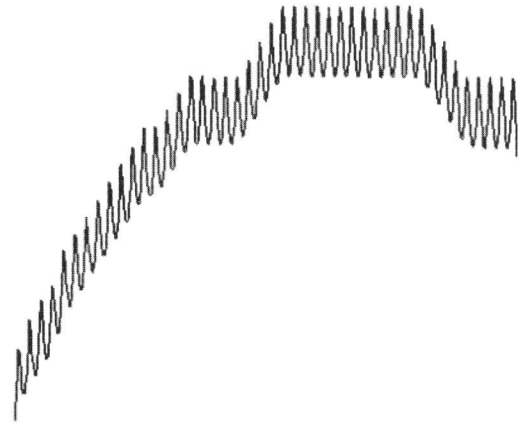

Figure 17.6   A NUBS Curve.

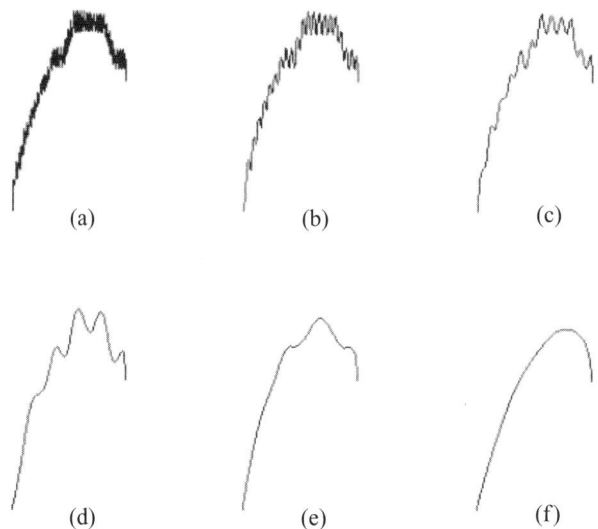

Figure 17.7   Multiresolution decomposition of the curve, in Figure 17.6, with point decimation.

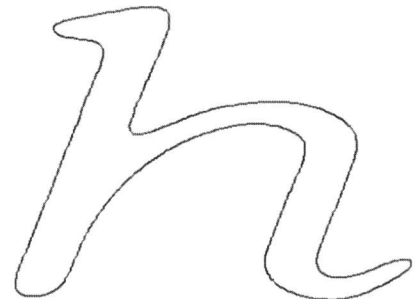

Figure 17.8   A NUBS Curve.

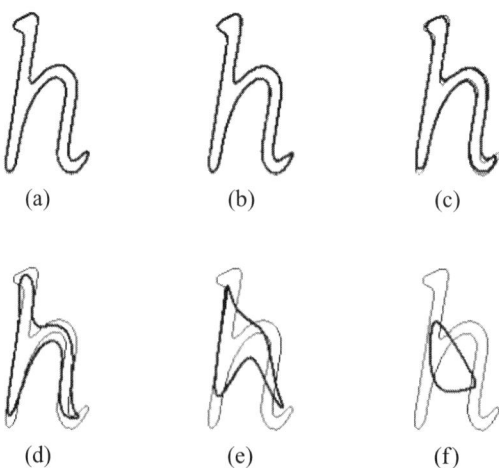

Figure 17.9   Multiresolution decomposition of the NUBS curve of Figure 17.8.

Figure 17.10   A NUBS Surface.

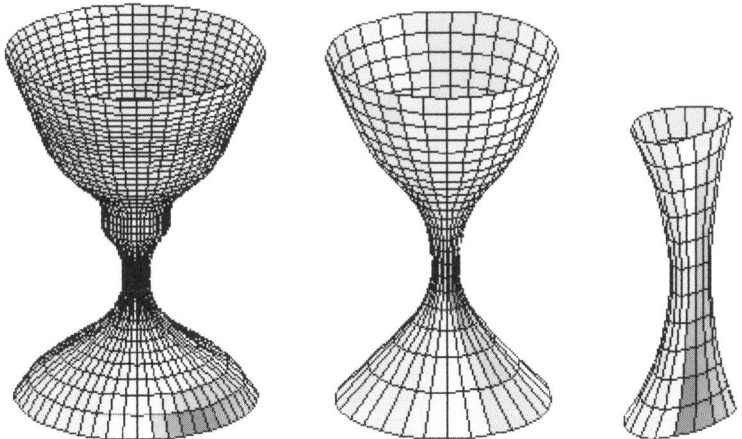

Figure 17.11   Multiresolution decomposition levels of the surface in Figure 17.10.

6. Summary

A framework for multiresolution representation of NUBS is developed for use in various computer graphics applications which require both local as well as global operations to be performed on B-splines. The developed method of multiresolution can be used for the purpose of performing editing on the B-splines. The proposed method is very efficient with respect to execution time as it uses a very simple technique for the decomposition, which does not require extensive calculations. As a future work, the authors think that a continuous multiresolution control would add a significant functionality to this method; investigation of any such method would be a major addition to the proposed model. The work can also be extended by using some other spline models like in [12–14]. This work is under consideration of the authors.

Acknowledgments

This work has been supported by the King Fahd University of Petroleum and Minerals under Project No. FT/2001-18.

References

[1] Celniker G. and Gossard D. (1991), Deformable Curve and Surface Finite Elements for Freeform Shape Design, *Computer Graphics,* 25(4).

[2] Chughtai M. S. A. (1999), ANURBS: An Alternative to the NURBS of Degree Three, *MS Thesis*, King Fahd University of Petroleum & Minerals, Dhahran Saudi Arabia.

[3] Cobb E. (1984), Design of Sculptured Surfaces using the B-spline Representation," *Ph. D. Thesis*, University of Utah, USA.

[4] Conner D., Snibble S., Herndon K., Robins D., Zeleznic R. and Van-Dam A. (1992), Three Dimensional Widgets, *In the Proceedings of the Symposium on Interactive 3D Graphics*.

[5] Gershon E. and Craig G. (1995), Multi-resolution Control for Nonuniform B-spline Curve Editing, In *Pacific graphics '95*.

[6] Gerald F. (1996), *Curves and Surfaces for Computer Aided Geometric Design: A practical Guide*, Academic Press Inc.

[7] Adam F. and David S. (1994), Multi-resolution Curves, In *proceedings of SIGGRAPH*, ACM, New York, 261–268.

[8] Foley T., Van Dam A., Feiner S., Hughes J. and Phillips R. (1994), *Computer Graphics*, Prentice Hall International.

[9] Gregory J. A., Sarfraz M. and Yuen P. K. (1994), Interactive Curve Design using C^2 Rational Splines, *Computers and Graphics*, 18(2), 153–159.

[10] Laurent G., Christophe S. and Carole B. (1997), An Hermitian Approach for Multi-resolution Splines, *Technical report no. 1192-97, LaBRI*.

[11] Rogers D. F. and Adams A. J. (1990), *Mathematical Elements for Computer Graphics*, 2nd Edition, McGraw-Hill International.

[12] Sarfraz M. (1995), Curves and Surfaces for CAD Using C^2 Rational Cubic Splines, *Engineering with Computers*, 11(2), 94–102.

[13] Sarfraz M. (1994), Cubic Spline Curves with Shape Control, *Computers & Graphics*, 18(5), 707–713.

[14] Sarfraz M. (1994), Generalized Geometric Interpolation for Rational Cubic Splines, *Computers and Graphics*, 18(1), 61–72.

[15] Stollnitz E. J., DeRose D. T. and Salesin H. D. (1995), Wavelets for Computer Graphics: A Primer, Part-1, *IEEE Computer Graphics and Applications*, 15(3), 76–84.

[16] Stollnitz J. E., DeRose D. T. and Salesin H. D. (1995), Wavelets for Computer Graphics: A Primer, Part-2, *IEEE Computer Graphics and Applications*, 15(4), 75–85.

[17] Stollnitz J. E., DeRose D. T. and Salesin H. D. (1996), *Wavelets for Computer Graphics: Theory and Applications*, Morgan Kaufman Publishers, San Francisco, USA.

[18] Welch W. and Witkin A. (1992), Variational Surface Modeling, *Computer Graphics*, 26(2).

18

Irregular Topology Spline Surfaces and Texture Mapping

Jin J. Zheng
Jian J. Zhang

National Centre for Computer Animation, Bournemouth University
Poole, Dorset BH12 5BB, United Kingdom

It is easy to texture map a surface model with triangular or rectangular patches. However, it is not so for irregular topology models with irregular patches (n-sided patches where n > 4) due to lack of proper global parameters for these irregular patches. In this chapter a texture mapping function capable of texturing an irregular surface model is proposed based on a spline surface model of irregular topology.

1. Introduction

Texture mapping is a powerful technique for adding realism to a computer-generated scene with less computational cost [1], [2]. It is widely used in computer graphics [3]. Basically, a texture map lays a texture pattern (image) onto an object in a scene, i.e., a geometric point on an object is associated with a pair of texture coordinates. For a parametric surface, texture mapping is to define a mapping function, which links the geometric parameters of a point on the surface to its texture coordinates. The quality of the final image is dependent on the quality of the mapping function. Due to the variety of curved surfaces and their geometric characteristics such as curvature, twist and topology, much effort in texture mapping has been expended to the development of proper mapping functions.

Texture mapping for surfaces of triangular and rectangular patches is relatively easy to perform. This is because the parameters of these surfaces can serve as the texture coordinates. A standard method is to treat the parametric coordinates of a surface as the texture coordinates. The texture values can then be interpolated across the surface.

Advances in Geometric Modeling. Edited by M. Sarfraz
© 2003 John Wiley & Sons, Ltd   ISBN: 0-470-85937-7

However, problems arise if we want to map a piece of texture onto an object with holes and branches. This often comes with an irregular topology model, i.e., a model contains n-sided surface patches, where $n > 4$, such as pentagons and hexagons. One understands that such an irregular surface patch does not have global parameters which can be used to interpolate the textures across the surface patch [4]. This problem also arises with the popular subdivision surfaces [5], and polygonal models, which are often used in building animated characters.

To overcome this limitation, in this chapter we first introduce a C^1 smooth spline surface over an irregular topology mesh by using Zheng-Ball surface [6]. Based on this spline surface, we then construct an analytical mapping function for texture mapping irregular surfaces. This mapping function provides a C^1 continuous interpolation scheme, which is used to index the texture map in the texture space. It is able to texture any n-sided surface area and can be incorporated into subdivision surface models to texture the neighborhood of the extraordinary points.

This proposed method is based on the use of Zheng-Ball surface patches [6] and Catmull-Clark subdivision rules [5]. Instead of keeping subdividing an irregular mesh to infinity, our method only subdivides the mesh once. A mapping function is then developed over the irregular surface. The steps of generating the mapping function can be summarized as follows:

- Subdividing an initial surface mesh in the 3D space R^3 using Catmull-Clark subdivision rules, and subdividing the corresponding initial single texture image once within the texture space using the same rules.
- Constructing a C^1 smooth spline surface, which consists of regular rectangular sub-patches and irregular n-sided sub-patches.
- Generating a mapping function for each regular or irregular sub-patch. Applying texture to each sub-patch has then become trivial.

The rest of this chapter is organised as follows: Section 2 gives a survey of the related work in texture mapping. Section 3 introduces a spline surface over an irregular topology mesh. In Section 4, a mapping function is developed for texture mapping of irregular patches and the procedure of texture mapping of irregular objects is explaind, followed by the results and examples in Section 5. Section 6 concludes this chapter.

2. Previous Work

Traditionally texture mapping of irregular surfaces required either a high amount of human intervention or degrading of visual quality. This applies to both smooth surfaces and polygonal modeling.

One method used quite often by animators is to split an irregular surface model into a number of triangular or quadrilateral pieces. Each of these pieces can be individually texture mapped using its texture coordinates. The problem of this approach, as can be easily understood, is the discontinuity between the seams. This problem further exacerbates when the model is animated, due to a lack of coherence of the moving images.

Two-part texture mapping, one of the early popular techniques proposed by Bier and Sloan [7] is another method employed in irregular surface texture mapping. Despite it effectiveness in many applications, it is not very easy to use, especially when the surface is arbitrarily concave.

Solid texturing [8] is theoretically applicable to both solids and surfaces of any complexity. In practice, however, it is tedious if one is to render a surface using solid texturing. A large amount of human intervention is required.

Subdivision surfaces have in recent years become a popular surface-modelling scheme in computer animation due to their capability of irregular topology. Several texture mapping methods have been proposed for such surface models. Since a large part of a subdivision surface model is usually covered by ordinary surface patches, such as B-Spline and Bézier patches, the focal point is on the texture mapping of the irregular regions. A method proposed by DeRose *et al.* [4] interpolates the texture coordinates with the surface subdivision rules, which are used to compute the texture values. It is proved that the interpolation scheme using the same subdivision rules as for the surface mesh produces a smooth mapping over the surface. Effective as it is, it heavily relies on the use of the subdivision procedure, which in theory requires an infinite number of iteration. In other words, this is not an analytical method. As a result, it makes any further operations, such as differentiation a difficult job, although not entirely impossible [9]. Another recent texture mapping method [10] was proposed by Piponi and Borshukov where a subdivision surface model is divided into a number of regions. Using a simplistic mass-spring system to even out the surface deformation, this method, called pelting, is able to produce a continuous color map by combining together the separate regions. Although it is not aiming for arbitrarily n-sided surface types, it is effective for subdivision surfaces.

3. C^1 **Spline Surfaces**

To texture map an irregular topology model, a C^1 smooth spline surface is introduced in this section. The spline surface model is based on the following n-sided quadratic Zheng-Ball [6] patch:

$$\mathbf{r}(\mathbf{u}) = \sum_{j=0}^{1} \sum_{\min \lambda = j} B_\lambda(\mathbf{u}) \mathbf{r}_\lambda, \tag{18.1}$$

where $n \geq 3$, $\lambda = (\lambda_1, \lambda_2, \ldots, \lambda_n)$ represents the n-ple subscripts. $\min \lambda = \min_{i=1}^{n} \{\lambda_i\}$. $\mathbf{u} = (u_1, u_2, \ldots, u_n)$ represents n parameters of which only two are independent. $\mathbf{r}_\lambda$ stands for the control points in $\mathbb{R}^3$, as shown in Figure 18.1, where we use a hexagon as an example.

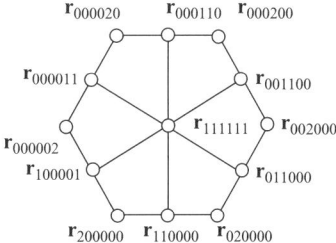

Figure 18.1   Control points for a 6-sided Zheng-Ball Patch.

$B_\lambda(u)$ are the associated basis functions whose expressions are given as:

$$B_\lambda(u) = \begin{cases} \binom{2}{\lambda_{i-1}}\binom{2}{\lambda_i}\prod_{j=1}^{n} u_j^{\lambda_j}(1 + f_\lambda(u)) + S, & \lambda = (1, \ldots, 1) \\ \binom{2}{\lambda_{i-1}}\binom{2}{\lambda_i}\prod_{j=1}^{n} u_j^{\lambda_j}(1 + f_\lambda(u)), & \text{for other } \lambda\text{'s} \end{cases}$$ (18.2)

where,

$$S(u) = 1 - \sum_\lambda \binom{2}{\lambda_{i-1}}\binom{2}{\lambda_i}\prod_{j=1}^{n} u_j^{\lambda_j}(1 + f_\lambda(u)),$$ (18.3)

and the functions $f_\lambda(u)$ and the parameters are given by:

- $n = 3$.

$$f_{2\delta_i}(u) = -2u_{i+1}u_{i+2},$$ (18.4)
$$f_{\delta_i+\delta_{i+1}}(u) = -u_{i+2},$$ (18.5)
$$f_\lambda(u) = 0 \quad \text{for} \quad \lambda = (1, 1, 1),$$ (18.6)
$$u_1 + u_2 + u_3 = 2u_1u_2u_3 + 1,$$ (18.7)

- $n \geq 5$.

$$f_\lambda(u) = 0,$$ (18.8)
$$u_i = 1 - \frac{\prod_{j\neq i-1,i}d_j + \prod_{j\neq i,i+1}d_j}{\sum_{k=1}^{n}\prod_{j\neq k-1,k}d_j},$$ (18.9)

in which d_i are auxiliary variables satisfying:

$$d_{i-1} + d_{i+1} = 1 + 2\cos\left(\frac{2\pi}{n}\right)d_i, \quad i = 1, 2, \ldots, n.$$ (18.10)

The starting point of this method is a user-defined irregular mesh M^0, which is a collection of vertices, edges and faces.

There are two steps in the construction of the spline surface model.

Step 1. Applying Catmull-Clark subdivision once over the initial mesh M^0.

The purpose of the first step is to sort out the mesh irregularity so that all faces of the mesh have exactly four edges. Given a user-defined irregular mesh M^0, a new refined mesh M^1 can be created by carrying out Catmull-Clark subdivision once. However, if all its faces are already 4-sided, the user may use the mesh directly. This will create a spline surface closest to the initial mesh M^0.

For a given mesh, three types of points are identified when applying Catmull and Clark [1] subdivision. They are the face points **f**, edge points **e** and vertex points **v**. All these points may be either ordinary or extraordinary depending on their valence. Vertices of valence 4 are called ordinary, and the others are called extraordinary.

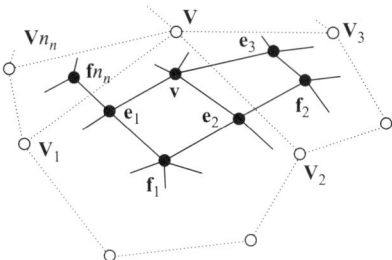

Figure 18.2 Applying Catmull-Clark subdivision once to vertex **V** whose valence is n.

Clearly, each vertex **V** of valence n of mesh M^0 is incident to n faces and n edges. In the refined mesh M^1, the new vertices associated with vertex **V** are computed by:

$\mathbf{f}_i$ = centroid of the surrounding vertices of the ith face incident to vertex

$$\mathbf{V}(i = 1, \ldots, n), \tag{18.11}$$

$$\mathbf{e}_i = \frac{1}{4}(\mathbf{V} + \mathbf{f}_{i-1} + \mathbf{f}_i + \mathbf{V}_i) \quad i = 1, \ldots, n, \tag{18.12}$$

where subscripts are taken modulo the valence n of vertex **V**, and $\mathbf{V}_i$ is the end point of the ith edge emanating from **V**.

$$\mathbf{v} = \frac{1}{4}\left(2\mathbf{V} + \frac{1}{n}\sum_{i=1}^{n}\mathbf{f}_i + \frac{1}{n}\sum_{i=1}^{n}\mathbf{e}_i\right). \tag{18.13}$$

Note that all faces of the new mesh M^1 are now 4-sided. The valence of the new vertex point **v** remains n. The valence of a new edge point is 4. The valence of a new face point is the number of edges of the corresponding face of mesh M^0.

Step 2. Constructing one sub-patch for each vertex in the resulting mesh.
The second step is to construct one surface patch for each vertex of mesh M^1 ensuring all such patches join together smoothly. For an ordinary vertex, as it is surrounded by four 4-sided faces, a bi-quadratic Bézier patch is used. For an extraordinary vertex, an n-sided quadratic Zheng-Ball patch will be generated where n is the valence of the vertex. Every edge of a patch of either type is a quadratic Bézier curve defined by three control points, two end-points and one mid-point. Understandably, the end-points are also the corner points of the corresponding surface patch, such as $\mathbf{c}_1, \ldots, \mathbf{c}_5$ in Figure 18.3.

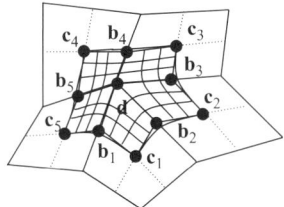

Figure 18.3 Control points generated corresponding to a vertex of valence 5.

Suppose that $\mathbf{d}$ is a vertex in the resulting mesh around which a surface patch is to be constructed. The control points for a corresponding patch can be generated using the following rules:

- A corner control point is defined as the centroid of the corresponding surrounding face.
- A mid-edge control point is found as the midpoint of the edge emanating from vertex $\mathbf{d}$.
- The central control point is identical to vertex $\mathbf{d}$.

The output of the above two steps is a collection of sub-patches, which are either tensor product Bézier patches or quadratic Zheng-Ball patches with an overall C^1 smoothness.

4. Mapping Functions

4.1. Mapping Functions for Irregular Patches

A texture mapping function can be viewed as one that maps a scalar value (color in this case) from a planar 2D space to a curved 2D space (the curved surface). In other words, it is to construct functions $f(u, v)$ and $g(u,v)$ for each pair of parameters s and t, such that:

$$s = f(u, v),$$
$$t = g(u, v),$$

where u, v are the parameters of the target surface; s, t are the original texture coordinates which are used to index the texture map to color a point on the surface. So, at each geometric point on the surface with parameters (u, v), there is correspondingly a pair of texture coordinates (s, t), which will index a unique color for the surface point.

To arrive at a mapping function for irregular patches, we propose that both $f()$ and $g()$ also be constructed by formula (18.1), under the condition that both the texture map and the surface to be textured have the same topology.

In practice, the 2D texture coordinate associated with a geometric control point $\mathbf{r}_\lambda$ of an n-sided surface is considered as the fourth and the fifth components of the control point. So, at each parameter vector $\mathbf{u}$ of the surface (Equation 18.1), not only a 3D geometric surface point but its associated 2D texture coordinates will be obtained as well. The 2D texture coordinate is then used to index a color value on the texture map. In this way, the original texture pattern is mapped across all the surface points.

A mapping function so constructed provides a smooth transformation from the original texture map to the mapping on the surface. To prove the smoothness is not difficult. Suppose the original scalar texture function $T = T(s, t)$ is differentiable. Since formula (18.1) is differentiable, so are the mapping functions f and g. It follows that the new texture function $T = T(f(u, v), g(u, v))$ is differentiable.

4.2. Mapping Functions for Irregular Spline Surfaces

Once all the sub-patches are identified, their texture mapping becomes straightforward using the proposed mapping function (18.1).

Suppose there is a pair of texture coordinates (s_i, t_i) for each initial mesh vertex $\mathbf{P}_i$. With the aforementioned subdivision procedure, these two texture coordinates are considered as the fourth and fifth coordinates of each vertex undergoing the same subdivision operation. As a result, a new refined mesh M^1 is produced, so are a set of new vertices, each associated with a pair of new texture coordinates.

In the process of constructing the sub-patches, the control points of each sub-patch are generated. The fourth and fifth coordinates of each control point are its associated texture coordinates. The texture coordinates of these vertices are then used to define the texture vertices used in mapping function (Equation (18.1)). Now we are ready to texture map the sub-patches.

To texture map a regular Bézier or B-spline patch is quite easy. As mentioned above, there are many methods available such as linear or bilinear interpolation. It is a challenge if irregular n-sided patches are involved. The method provided in this chapter is applicable to both regular and irregular patches.

As mentioned in the last section, we represent both the geometric control points and the texture control points by r_λ in Equation (18.1). Since they all employ the same function, it is intuitive to combine both the geometric and texture information together for each point on the spline surface. This means that we represent a surface patch in a 5D space. The first three coordinates, x, y, z are its real coordinates in 3D and the last two coordinates s and t are its texture coordinates. Hence, for each parameter vector u a vector $r(u)$ in 5D space results. Substituting the texture coordinates into a given texture function gives the texture color of a point on the patch.

5. Implementation

The texturing method presented in this chapter is implemented in a Pentium III 850 dual CPU PC using VC++ and OpenGL.

Figure 18.4a shows a textured regular pentagon consisting of 5 triangles. Figure 18.4b shows a deformed pentagon model, where discontinuity of the circled pattern is clearly visible. By using the texture mapping method provided in this chapter, no discontinuity occurs, as shown in Figure 18.4c.

Figure 18.5a shows a regular pentagonal texture. The texture is mapped onto a geometric model in Figure 18.5b. In Figure 18.5c, the model is concavely deformed, which is then smoothly textured with the same pattern (Figure 18.5d).

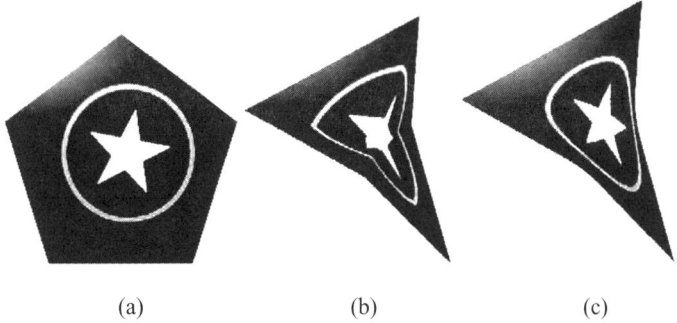

(a)   (b)   (c)

Figure 18.4   (a) Textured regular pentagon. (b) Texture mapped with five triangles. (c) Texture mapped using the method provided in this chapter.

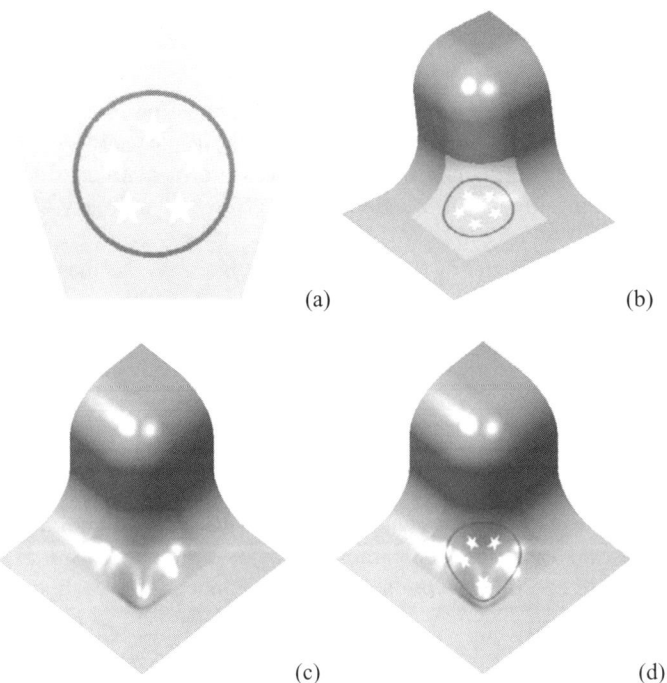

Figure 18.5 (a) Textured regular pentagon. (b) Texture mapped onto a model. (c) Deformed model.
(d) Deformed model with smoothly mapped texture.

6. Summary

Texture mapping a regular surface has been well studied and many effective methods are developed based on the use of their natural parametric coordinates. It represents more of a challenge, however, for irregular surfaces due to lack of such global parameters.

In this chapter a C^1 smooth spline surface is presented. This spline surface can be used to represent irregular topology models.

Based on the C^1 smooth spline surface, a mapping function is developed to texture map a model of irregular topology. To represent both the geometric and texture information concisely, a point on a curved surface is regarded as a 5-dimensional vector, where the first three components are the geometric coordinates and the last two are the texture coordinates. This method consists of three steps:

- Apply the Catmull-Clark subdivision rules once to a single texture map that corresponds to the initial mesh, resulting in one pair of new texture coordinates for each new mesh point.
- The texture coordinates are considered as the fourth and fifth components of new mesh points. Generate the control points of a sub-patch around each mesh point so that the fourth and fifth components of each control point are its texture coordinates.
- Substitute these new 5D control points into Equation (18.1) to create the texture coordinates of any point on the sub-patch. Surface texturing is then undertaken using these texture coordinates.

It is worth noting that texturing an n-sided surface has been made amiable to production only with recent popular development of subdivision surfaces. But these subdivision surface based texture mapping methods rely on the use of the subdivision rules. Because it is procedural, analytical properties may not be easily derived. The proposed method is actually also applicable to subdivision surfaces, as can be seen from the construction process. There may be a small amount of geometric discrepancy between the spline surfaces proposed here and the subdivision surfaces in the neighborhood of the extraordinary points. But the discrepancy has negligible effects on the texture mapping results. The fact that the mapping function is differentiable will make further operations much easier.

References

[1] Catmull, E.D. (1974), A subdivision algorithm for computer display of curved surfaces, *PhD Thesis*, University of Utah.

[2] Heckbert, P.S. (1986), Survey of texture mapping, *IEEE Computer Graphics and Applications* 6(11), 56–67.

[3] Zhang, J.J. (1998), Lease distorted bump mapping onto surface patches. *Computer and Graphics* 22(2–3). 233–242.

[4] DeRose, T, Kass, M. and Trong, T. (1998), Subdivision surfaces in character animation. *SIGGRAPH'98*. 85–94.

[5] Catmull, E. and Clark J. (1978), Recursively generated B-spline surfaces on arbitrary topological meshes. *Computer Aided Design* 10(6), 350–355.

[6] Zheng, J.J. and Ball A.A. (1997), Control point surfaces over non-four-sided areas. *Computer Aided Geometric Design* 14(9). 807–820.

[7] Bier, E.A. and Sloan, K.R. (1986), Two-part texture mapping. *IEEE Computer Graphics and Application* 6(9). 40–53.

[8] Peachey, D.R. (1985), Solid texturing of complex surfaces. *SIGGRAPH'85*. 279–286.

[9] Stam, J. (1998), Exact evaluation of Catmull-Clark subdivision surface at arbitrary parameter values. *SIGGRAPH'98*. 395–404.

[10] Piponi, D. and Borshukov, G. (2000). Seamless texture mapping of subdivision surfaces by model pelting and texture blending. *SIGGRAPH 2000*. 471–478.

19

Segmentation of Scanned Surfaces: Improved Extraction of Planes

R. Sacchi
J.F. Poliakoff
P.D. Thomas

Department of Computing and Mathematics, The Nottingham Trent University,
Burton St., Nottingham, NG1 4BU, England.

K.-H. Häfele

Forschungszentrum Karlsruhe, Institut für Angewandte Informatik,
Postfach 3640, D-76021 Karlsruhe, Germany.

Reverse engineering often involves the production of a digital representation of a physical object in order to copy or modify the object. The object surface is scanned to produce a large number of points, or point cloud. A triangulated surface can always be generated from an unstructured point cloud, so in all cases adjacency information can be obtained. The original object is often made up of a number of simple geometric components and could therefore be represented much more simply. We aim to segment a discretely represented surface into a small number of such simple geometric components using a 'region growing' approach. This chapter describes our improved algorithm for planar extraction using 'super triangles'. The algorithm requires only the data point coordinates together with some form of adjacency information. A new idea is presented for improving extraction using 'pseudo-randomization' of the data.

1. Introduction

Computer Aided Design (CAD) packages are used by engineers to design objects for production by Computer Aided Manufacturing (CAM) systems [9, 15, 16]. Such an object

---

Advances in Geometric Modeling. Edited by M. Sarfraz
© 2003 John Wiley & Sons, Ltd   ISBN: 0-470-85937-7

often consists of a fairly small number of parts of simple geometric shapes, which, when represented appropriately, can be relatively easy to manipulate or modify before manufacture. In some cases a physical prototype is produced and then modified directly, whereas in others the initial design consists of a physical object, which has to be copied. In many cases, a process of reverse engineering is required in order to create a digital representation of the object [18]. During reverse engineering a large number of surface points are measured to form what is known as a point cloud. Our aim is to develop algorithms for the automated segmentation of such a digitally represented surface, or point cloud, where possible, into simple geometric parts.

The measurement of the data points for reverse engineering may involve a tactile method, such as a contact probe, or the process can be nontactile, for example laser scanning. The measured data points may form what is known as a *structured* point cloud, i.e. it includes adjacency information derived from the scanning process; very often the points are collected in order along straight lines as the scanning device moves over the object. In some cases, the points may project onto a grid in, say, the x-y plane, which means that they can be easily parametrized. In the most general case, however, such a $2^1/_2$D property cannot be assumed. Previous methods for segmentation have often relied on the fact that the data points are related to such a grid, for example so-called range data [2] or image data based on pixels.

In cases when the data points form an unstructured point cloud, it is still possible to derive a structured representation. This can be done by generating a triangulated surface from the original points [6, 8, 19]. Our aim is to take such a general surface containing some form of adjacency information and to segment it into a relatively small number of simple geometric components. Such a segmented surface can then be manipulated easily by a CAD system. Each component must have all its data points lying, within a given tolerance, on part of a simple geometric shape or primitive. We have assumed that the geometric shape is to be one of the following: a plane, a sphere, a cylinder, cone or a torus. In many cases it is relatively easy for a human operator to classify regions into one of these types. For example, the surface shown in Figure 19.1 can be seen to contain seven planar segments. Merely looking at this one view of it allows others to be recognized as parts of cylinders, cones and, probably, spheres and tori. In the following sections we give a brief survey of segmentation methods and explain our algorithm for fast extraction of planar segments, which uses only data points and simple adjacency information.

The POMOS (POint-based MOdelling System) system has been developed at the Research Centre Karlsruhe to handle large sets of digitized data. It is able to triangulate completely unstructured $2^1/_2$D data [6], as well as to take manually segmented triangulated surfaces and approximate them with free-form surfaces [5]. The surfaces generated can then be further analyzed. We have used this system as a platform for the development and testing of our algorithms.

2. Segmentation of Surfaces

Previous work in segmentation of surfaces has mainly involved range data or image data [3, 11]. Split-and-merge is a top-down method for which the splitting process has usually relied on having parametrized data. Another approach involves starting bottom-up with a seed point and continuing to add further data points until no more suitable ones can be found [13]. Sometimes

POMOS FZK/IAI

Figure 19.1 An example of a triangulated surface of a technical object which has been rendered.
Components of many simple geometric shapes can be seen, including planes, cylinders
and cones.

the regions grown can be merged subsequently whenever there is a close match between
parameters. During the growing process, surface shape parameters may need to be adjusted.
Curvature has been used to provide preliminary information about surface quality [2]. One
approach to segmentation is to attempt to join up points where curvature is high in order to
identify ridge lines which form the boundaries of surface segments [12]. Some boundaries can
be found in this way. However, when there is a smooth join between two segments, such as
a plane and a cylinder joined tangentially, it cannot be identified in this way because there is
no ridge. There will often be a small change in the curvature itself but this cannot be detected
reliably in the presence of noisy data. Clustering methods, such as the Hough Transform, have
been applied successfully to range data, but many rely on having parametrized data [7]. Genetic
algorithms for primitive extraction are computationally intensive [14], so they are unsuitable
for many of the complex objects which occur in reverse engineering.

We have found that the region growing approach is the most appropriate for general triangulated surfaces [15, 16]. Suitable 'seed' points are chosen and then all adjacent points are
tested and added whenever they approximate to the required geometric shape. Readjustment of
shape parameters is usually needed during this growing process. We have developed a method
for fast planar extraction based on the region growing approach. This method can be applied
to general surface data, only provided that it contains some sort of adjacency information.

Planar patches are very important, because many mechanical objects are made up largely
of planes, as explained by Ashbrook *et al.* [1]. In 1982 Hebert *et al.* [7] described a method
for extracting geometric primitives using the Hough transform but it is time-consuming and
memory-intensive. Since then other methods have been proposed for the extraction of planar
segments from range images, but they rely on having a parametrized surface [4, 10, 17].

Our approach to segmentation of triangulated surfaces using region growing has been previously described in detail [15, 16]. The work has been based on our curvature estimation
algorithm for triangulated surfaces. Our region growing algorithm starts with an initial region
consisting of a triangle of low estimated mean curvature. The idea of this is to avoid false starts
by choosing only triangles that have a good chance of belonging to a planar segment.

We have found, however, that in the case of planar segments the time taken for curvature
estimation is longer than the time saved in avoiding false starts. We have therefore proposed a
modification of the approach to the segmentation problem by first extracting planar segments
without using curvature estimation. Subsequent curvature estimation will then take less time,
because it needs to be applied only to points remaining after the planar segments have been
extracted. Without the curvature estimation (which relies on the triangulation of the surface)
the method can be adapted immediately to more general surfaces with some sort of adjacency
information. Our new algorithm, described in Section 3.2, was based on an idea proposed
by Roth *et al.* [14] for representing a geometric primitive by an appropriate minimal set of
points.

3. Extraction of Simple Geometric Segments

We initially developed a region growing algorithm for triangulated surfaces [15, 16]. We
describe these algorithms briefly in Section 3.1. The simplest type of geometric primitive is
the plane and in many cases a considerable proportion of the surface area of an object is made
up of planar parts. Therefore a reduction in the time taken for the extraction of planes would
make a considerable difference to the total extraction time for the objects. For the extraction
of planes is not essential to have curvature estimation but it is needed for other geometric
primitives. We have found that one factor which was particularly time-consuming during the
whole process was the time for adjustment of the geometric primitive whenever a new point
was added to the region. We have therefore modified the region growing for planes by making
the adjustment less frequently and only allowing certain planes to be used for the adjustment, as
described in Section 3.2. The method involves 'representative triangles' and 'super triangles',
based on the idea of Roth *et al.*

Finally Section 3.3 presents a method for improving the selection of seed regions for the
many surfaces where the points have been stored systematically. With this improvement the
'tilt' problem is much less likely to occur.

3.1. Simple Region Growing

The first stage of simple region growing for a given geometric primitive involves finding a
'seed' region of a number of adjacent points which therefore define an initial region for the
geometric primitive. In the case of a plane, three non-collinear points are sufficient to define a
seed plane. For the other types of geometric primitive appropriate initial parameters need to be
calculated using at least four non-coplanar points. The growing stage consists of an attempt to
add a new point to the current region. To be added the new point must be able to satisfy certain
tolerance criteria. Each candidate point must be adjacent to a point which is already part of the
region. The growing process then continues until no more such points can be added. When a
point is considered for addition, one of two cases will occur:

(i) The point lies within the given tolerance of the geometric primitive associated with the current region;

(ii) it is not within the given tolerance of the geometric primitive.

In case (i) the point is added immediately to the region and the parameters of the geometric primitive are adjusted to the new region. In case (ii) it is considered for addition to the region. An attempt is made to adjust the geometric primitive so that all the points of the region *together with* the new point lie within the required tolerance of it. When this can be done the candidate point is added to the region and the plane is updated accordingly.

If the above approach is used, many very small regions will be extracted. Therefore, in order to prevent this from happening, an additional parameter is needed. For example we could set minimum number of points allowed for a segment, so that a region will not be extracted as a segment if it contains fewer points than that minimum. (We have used a minimum number of triangles in our implementation.)

The geometric primitive is adjusted every time a point is added in order to achieve a new geometric primitive which best fits the new set of data points. However, this adjustment adds considerably to the time taken for the region growing process. Our new algorithm speeds up the process in the case of planes by adjusting less frequently and reducing the number of possible planes.

3.2. Planar Region Growing with Super Triangles

In the case of planes the simple algorithm from Section 3.1 has been modified to be faster, as described below.

We have found that planar region growing can be made even faster by restricting the possible planes to those defined by any set of three points within the current region. We call the triangle formed by such a set of points a 'representative triangle' for the plane. We attempt to grow this representative triangle by making its area as large as possible. The idea is that for reasonably large regions there will only be a small 'tilt' to the plane caused by adjusting the plane when another point is added. This restriction could cause failure of some seeds to grow successfully. However, for any region which is approximately planar there will be many possible seed regions available.

At each stage of the region growing process, therefore, the current region is associated with a current plane defined by a representative triangle (with vertices among the points in the current region). There is also associated a 'super triangle' of area at least as large as that of the representative triangle. Initially the two triangles are both defined by the triangle consisting of the three non-collinear points of the seed region.

In case (ii), we consider the plane defined by the super triangle. If *all* the points (including the candidate point) lie within an 'adjustment tolerance' of this plane, it becomes the new plane and the candidate point is added to the region. The current super triangle thus becomes the new representative triangle and the super triangle is modified (see below). Therefore the area of the representative triangle will never decrease and will tend to increase, because of the way the super triangle is found. Figure 19.2 shows an example where a single planar segment has been extracted together with the final super triangle.

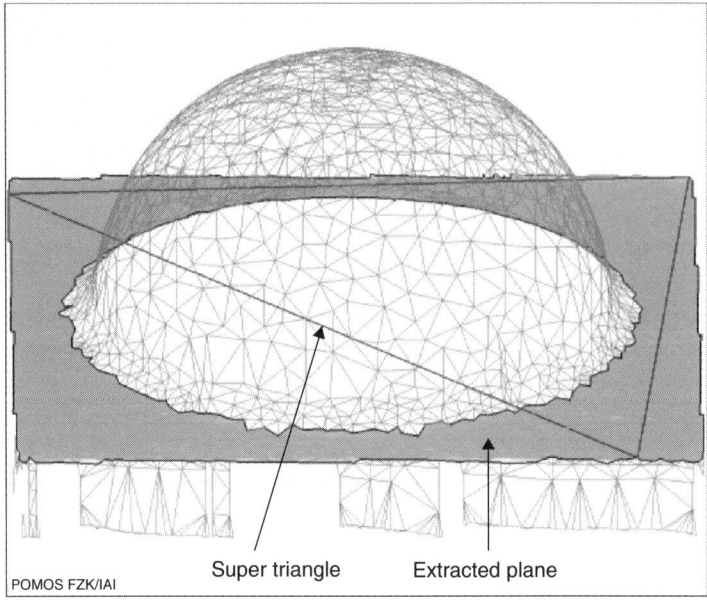

Figure 19.2   An example of a triangulated surface consisting of a hemisphere on a plane. As expected,
a single planar segment (shaded) has been found in 2 sec. The final super triangle has
been superimposed. (3473 points, tolerance 0.0005, adjustment tolerance 30%, minimum
triangles 150.)

Now, if case (i) occurs, the new point is added to the region and the super triangle is modified
using the new point, as described below. However, the representative triangle (and therefore
the current plane) is left unchanged.

We now describe how a new super triangle is found (in such a way that the area will tend to
increase). The new triangle is generated from the current one using both the new point and the
last point added previously. An attempt is made to replace the current triangle by one of larger
area by checking all possible triangles with vertices chosen from the above two points together
with the current triangle's three vertices. If an increase in area is possible, the replacement is
chosen as a triangle with the largest area. Otherwise, the current triangle is retained.

We have found that, if the adjustment tolerance has the *same* value as the given tolerance
for fitting, then case (ii) can lead to a considerable tilting of the super triangle, as shown
in Figure 19.3. This can prevent further region growth and result in the 'splitting' of one
approximately planar part into several planar segments. Figure 19.4 shows the extraction results
for the data from Figure 19.3 viewed from above so that the splitting into three segments can
be seen. Therefore we have used a smaller value than the given tolerance, which reduces the
tilting without having much effect on the size of region grown. In Figures 19.3 and 19.4 three
planar segments were extracted using adjustment tolerance 100%. However, when the same
data is used but with adjustment tolerance of 30% a single planar segment is obtained, as shown
in Figure 19.2 and is found more quickly. For the examples in Figures 19.5, 19.6 and 19.7 the
adjustment tolerance was set to 50% of the given tolerance.

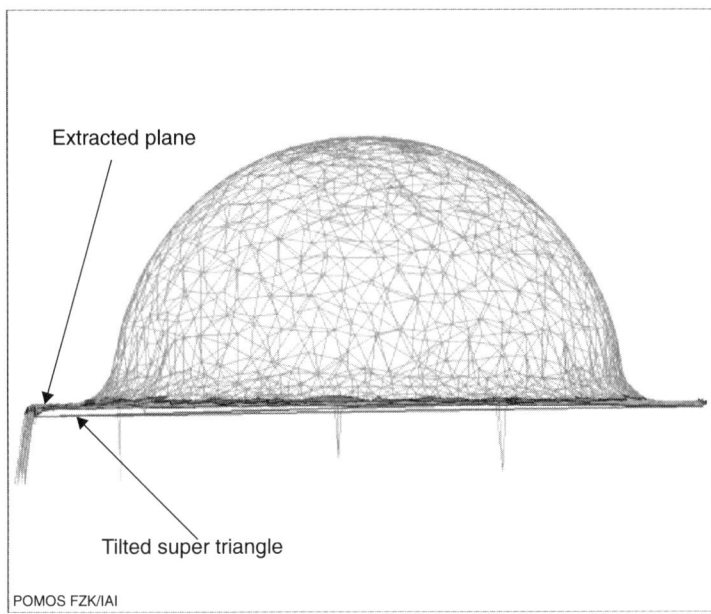

Figure 19.3   The data from Figure 19.2 has been processed again but with the adjustment tolerance
equal to the given tolerance. In the side view it can be seen that the final super triangle is
now tilted. Three planar segments are found instead of one and the time taken was 13 sec.

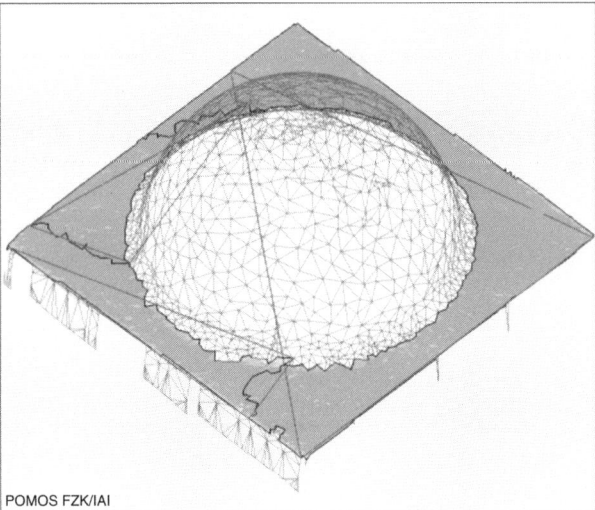

Figure 19.4   The extraction results from Figure 19.3 viewed from above. The tilting has caused splitting
of the planar component into three segments which can be seen, together with the three
super triangles.

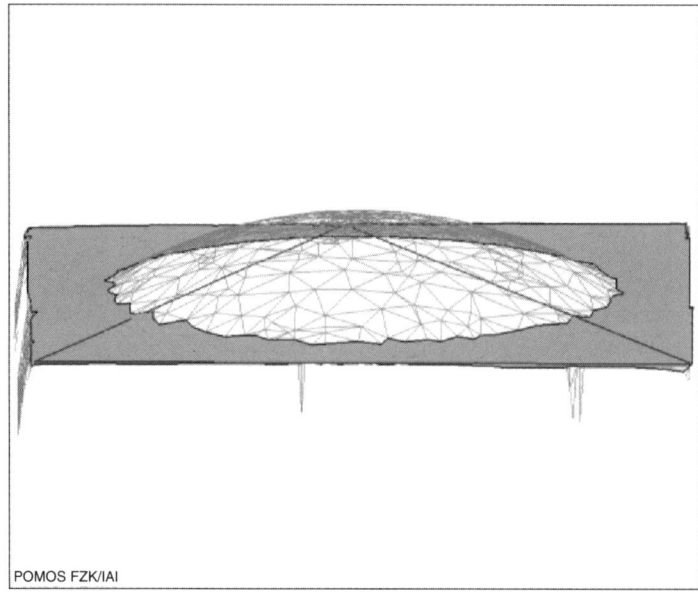

Figure 19.5 An example of a plane intersected by part of an ellipsoid. As expected, a single planar segment has been found in 10 sec. (2750 points, tolerance 0.00068, adjustment tolerance 50%, minimum triangles 150.)

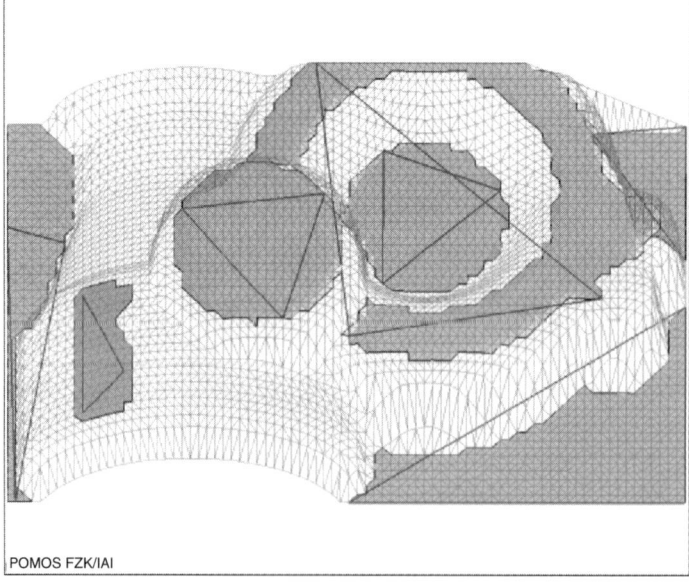

Figure 19.6 Seven planar segments have been found in 25 sec. for the surface from Figure 19.1. Six of them resemble planes but one segment appears to be part of a cylinder of large radius. (4018 points, tolerance 0.279, adjustment tolerance 50%, minimum triangles 100.)

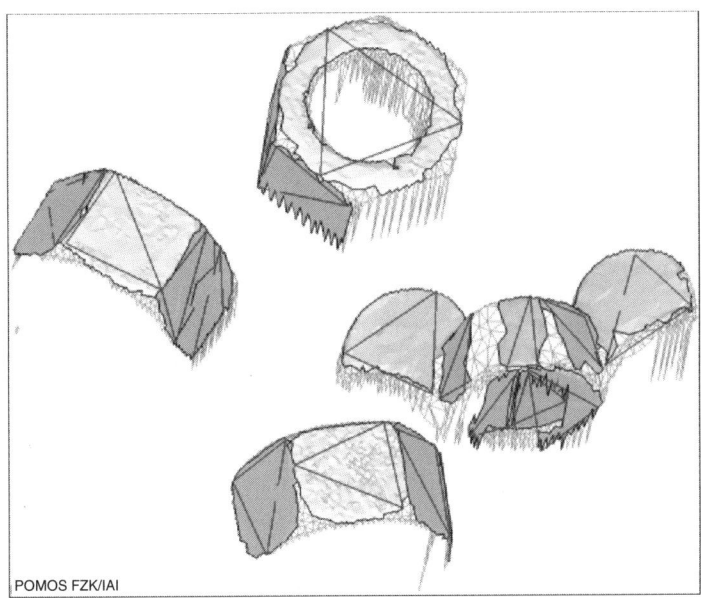

POMOS FZK/IAI

Figure 19.7   Twenty planar segments have been extracted in 23 sec. for this scanned surface consisting
of several hexagonal nuts arranged in different positions. (3666 points, tolerance 0.0005,
adjustment tolerance 50%, minimum triangles 20.)

3.3. Pseudo-Randomized Selection of Seed Regions

The results from region growing are always likely to be affected significantly by the order in
which the points or triangles are stored. We have found that many data sets are derived from
surfaces which have been scanned and stored in a systematic way. Sometimes the data points
are roughly equally spaced on a grid relative to a particular plane, as is the case for the data for
the surface in Figures 19.1 and 19.6. Typically a sensor will travel over the surface in straight
lines, passing backwards and forwards traversing the object until the complete surface has been
scanned.

For such data sets the first seed points selected from a known geometric component are
nearly always close to the boundary of that component. In the planar case the initial plane is
then sometimes tilted, giving a bad approximation to the actual planar component of the
surface. Adjustment may not be able to compensate for the tilt and the region will be unable
to grow very far. There may be many attempts at region growing with different seeds before
a segment can be extracted. The first successful segment will be the one nearest the boundary
that manages to achieve the given minimum number of triangles for a region and therefore
may still be somewhat tilted. In such cases the region is unlikely to be able to grow sufficiently
to extract the entire planar component. Thus, the extracted segments may still be small, again
causing splitting of planar components.

If this 'boundary effect' can be avoided, the results are likely to be improved, reducing both
time taken and splitting. We have attempted to do this by altering the order in which the points
are chosen as seed points using a 'pseudo-randomization' technique. This does not prevent a
point close to a boundary from being chosen but it makes it very much less likely. We have

found that using this technique has both increased the speed of the process and reduced splitting of planar components, as expected.

3.4. Automated Estimation of Tolerance

Another problem with region growing is when the value of tolerance is set inappropriately by the user. If it is too low, the segments extracted will tend to be small, because the user's requirement is too strict, and splitting will often be the result. On the other hand, if the tolerance chosen is too large, the likelihood of tilting will be increased, again causing splitting. Therefore we have developed algorithms for estimation of tolerance, whereby a more appropriate tolerance value is derived from the data itself. The idea is to obtain an estimate of the measurement errors present in the data based a small region of low curvature.

Automatic estimation of tolerance is performed during the extraction of the first planar segment. For this case alone the tolerance restriction is not applied until the region reaches a certain size. (We have used a fixed number of triangles for the size, 10 triangles in the examples shown here.) This 'test region' of the given size is then assessed for low curvature as follows. A 'minimum separation' is calculated as the length of the shortest edge among the triangles in the test region. Then a 'curvature tolerance' is found as half of this minimum separation. If all the points in the test region are within half this curvature tolerance of the current super triangle, then the estimated tolerance is calculated from it, as described below. If some points are not within the curvature tolerance, the test region is discarded and the process continues until a suitable one is found. Provided that the surface contains some parts of low curvature,

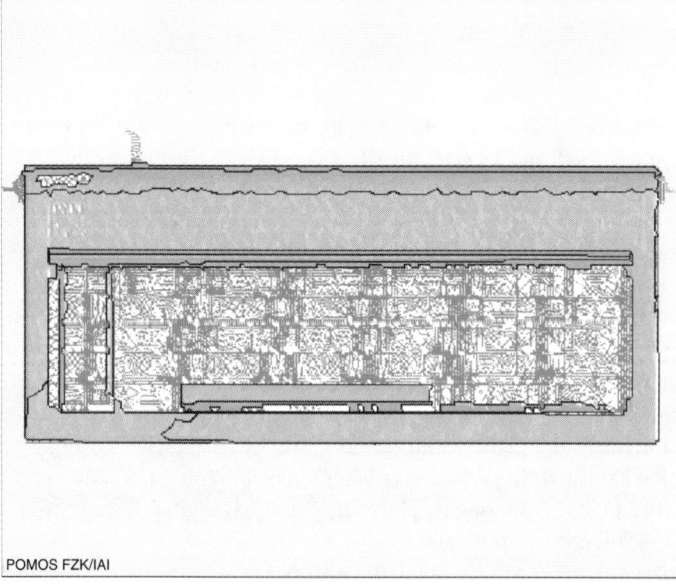

POMOS FZK/IAI

Figure 19.8   Thirteen segments have been extracted in about 11 secs. from the scanned surface of a keyboard but many small ones have been missed altogether. (Approximately 1100 points, adjustment tolerance 50%, minimum triangles 150.)

from which planar segments can be extracted, it is likely that a successful test region will be found.

Once a successful test region has been found, the distance of each point in the region from the super triangle is found and the average distance $\bar{a}$ is calculated. By assuming that the measurement errors have a normal distribution and demanding that about 99% of points are fitted within the tolerance of the plane, the standard deviation can be calculated as $\sigma = 2.6\bar{a}$. Leaving a small additional margin, we have used an estimated tolerance value of $\tau_e = 3\bar{a}$, slightly larger than σ itself. We have found that this value gives satisfactory results when it is used for subsequent extraction of planer segments and it has been used for our examples in Figures 19.2, 19.5–19.8.

4. Results

Fast planar segment extraction using both pseudo-randomization of seeds and estimated tolerance has been implemented within the POMOS system on a Silicon Graphics O2 workstation. Figures 19.2, 19.5–19.8 show the results of planar segment extraction for various triangulated surfaces using adjacency information alone. It can be seen that the planar regions found generally correspond well to what was expected. In both Figures 19.2 and 19.5 one planar segment has been extracted surrounding the domed part of the object. In Figure 19.2, where there is a large angle between the hemisphere and the plane, region growing stops appropriately at the join between hemisphere and plane. The surface in Figure 19.5 presents a greater challenge to the process, because the angle between ellipsoid and plane is much smaller. Again the region growing process has successfully distinguished between the two parts.

In Figure 19.6 it can be seen that seven planar segments have been extracted from the surface from Figure 19.1. Of these, six were as expected but one additional region has been grown on a part resembling a cylinder of large radius. This is not a fault in the implementation of the process but demonstrates the fact that this part of the cylinder does indeed lie within the tolerance of a plane. Such an effect will become more likely as the radius of the cylinder increases. It could be avoided by increasing the value of the minimum triangles parameter but then some small, genuinely planar, segments may be missed. In Figures 19.7 and 19.8 most of the planar parts have been extracted as single planar segments, although there is some splitting and some very small planar parts have been missed, probably because they do not have the minimum number of triangles demanded for a region.

5. Summary

Preliminary results have shown that the new algorithm is successful in achieving planar segment extraction when a suitable value of tolerance is provided.

Pseudo-randomization allows successful seeds to be found more quickly and the chance of splitting is reduced. The pseudo-randomization technique has the advantage that it is equally applicable to any region growing process. Whenever the points are stored systematically based on adjacency, pseudo-randomization is likely to both speed up region growing and reduce splitting. For surfaces with data already effectively randomized, however, little difference could be expected in the results.

The automatic estimation of tolerance has given promising results, as shown here, avoiding the need for the user to provide a tolerance value. Further investigation is needed to evaluate its performance for a larger sample of surfaces. The current algorithmic performance is by no means optimal, because of the data structure used in POMOS. On a dedicated system we expect that the computational time would be noticeably lower.

References

[1] Ashbrook A.P., Fisher R.B. *et al.* (1997) Segmentation of Range Data into Rigid Subsets using Planar Surface Patches. *Proc. British Machine Vision Conference*, Essex, 530–539.

[2] Besl P.J., Jain R.C. (1988) Segmentation Through Variable-Order Surface Fitting. *IEEE Transactions on Pattern Analysis & Machine Intelligence*, **10**, No. 2, 167–192.

[3] Biswas P.K., Biswas S.S. *et al.* (1995) An SIMD Algorithm for Range Image Segmentation. *Pattern Recognition*, **28**, No.2, 255–267.

[4] Faugeras O.D., Hebert M. *et al.* (1983) Segmentation of Range Data into Planar and Quadric Patches. *Proc. 3rd Conference on Computer Vision and Pattern Recognition*, Arlington, VA, 8–13.

[5] Häfele K.-H. (1996) POMOS – POint Based MOdelling System in: Hoschek, J., Dankwort, W. (Eds.) *Reverse Engineering*, Verlag B.G. Teubner, Stuttgart, Germany.

[6] Häfele K.-H., Hellmann M. (1996) 3D-Meßdatenaufbereitung und Weiterverarbeitung mit POMOS. 2. *Workshop Optische 3D-Formerfassung Großer Objekte*, Esslingen, Germany.

[7] Hebert M., Ponce J. (1982) A New Method for Segmenting 3-D Scenes into Primitives. *Proc. International Conference on Pattern Recognition*, Munich, 836–838.

[8] Hoppe H. (1994) Surface Reconstruction from Unorganised Points, *PhD Thesis*, University of Washington.

[9] Hoschek J., Lasser D. (1993) *Fundamentals of Computer Aided Geometric Design*. A.K. Peters, Wellesley, Massachusetts.

[10] Jiang X.Y., Bunke H. (1994) Fast Segmentation of Range Images into Planar Regions by Scan Line Grouping. *Machine Vision & Applications*, **7**, No. 2, 115–122.

[11] Leonardis A., Gupta A. *et al.* (1995) Segmentation of Range Images as the Search for Geometric Parametric Models. *Int. Journal of Computer Vision*, **14**, No. 3, 253–277.

[12] Lukács G., Andor L. (1998) Computing Natural Division Lines on Free-Form Surfaces Based on Measured Data. in: Daehlen M., Lyche T., Schumaker L.L. (Eds.) *Mathematical Methods for Curves and Surfaces II*, 319–326.

[13] Maître G., Hügli H. *et al.* (1990) Range Image Segmentation Based on Function Approximation. in: Gruen, A. and Baltsavias, E. (Eds.) *Close-Range Photogrammetry Meets Machine Vision*, SPIE **1395**, 275–282.

[14] Roth G., Levine M.D. (1994) Geometric Primitive Extraction using a Genetic Algorithm. *IEEE Transactions on Pattern Analysis and Machine Intelligence*, **16**, No. 9, 901–905.

[15] Sacchi R., Poliakoff J.F. (1999) Curvature Estimation for Segmentation of Triangulated Surfaces. *Proc. 2nd Int. Conference on 3-D Digital Imaging and Modeling*, Ottawa, Canada, 536–544.

[16] Sacchi R. (2001) Primitive-Based Segmentation for Triangulated Surfaces, *PhD Thesis*, the Nottingham Trent University.

[17] Taylor R.W., Savini M. (1989) Fast Segmentation of Range Imagery into Planar Regions. *J. Computer Vision, Graphics and Image Processing*, **45**, No. 1, 42–60.

[18] Vàrady T., Martin R.R. *et al.* (1997) Reverse Engineering of Geometric Models – An Introduction. *Computer-Aided Design*, **29** No. 4, 255–268.

[19] Yemez Y., Schmitt F. (1999) Progressive Multilevel Meshes from Octree Particles. *Proc. 2nd Int. Conference on 3D Digital Imaging and Modeling*, Ottawa, 290–299.

20

Constraint-Based Visualization of Spatiotemporal Databases

Peter Revesz
Lixin Li
Computer Science and Engineering Department, University of Nebraska-Lincoln
Lincoln, NE 68588, U.S.A.

We propose using a constraint relational representation for spatial and spatiotemporal data derived using an inverse distance weighting interpolation method. The advantage of our approach is that many queries that could not be done in traditional GIS systems can now be easily expressed and evaluated in constraint database systems. The data visualization can also be based on constraint techniques.

1. Introduction

To visualize and query a set of spatial data in GIS (Geographic Information Systems) applications, we often need *spatial interpolation*, that is, to estimate the unknown values at unsampled locations with a satisfying level of accuracy. For example, suppose we have the following two sets of sensory data in our database:

1. *Incoming* (y, t, u) records the amount of incoming ultraviolet radiation u for each pair of latitude degree y and time t, where time is measured in days.
2. *Filter* (x, y, r) records the ratio r of ultraviolet radiation that is usually filtered out by the atmosphere above location (x, y) before reaching the earth.

Figures 20.1 and 20.2 illustrate the locations of the (y, t) and (x, y) pairs where the measurements for u and r are recorded. Tables 20.1 and 20.2 show the corresponding instances of these two relations.

Advances in Geometric Modeling. Edited by M. Sarfraz
© 2003 John Wiley & Sons, Ltd   ISBN: 0-470-85937-7

Table 20.1 Incoming.

ID	Y	T	U
1	0	1	60
2	13	22	20
3	33	18	70
4	29	0	40

Table 20.2 Filter.

ID	X	Y	R
1	2	1	0.9
2	2	14	0.5
3	25	14	0.3
4	25	1	0.8

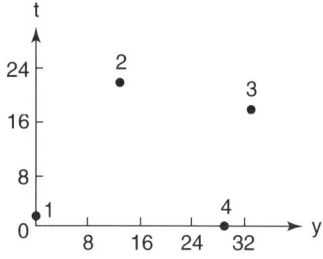

Figure 20.1 The sample points in Incoming.

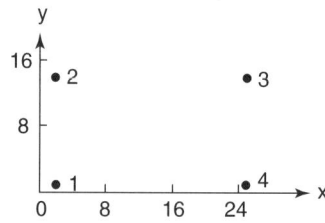

Figure 20.2 The sample points in Filter.

Since *Incoming* (y, t, u) and *Filter* (x, y, r) only record incoming ultraviolet u and filter ratio r at a few sampled locations, they cannot be displayed directly. Some spatial interpolation is needed to estimate u and r for all the locations in the domain. The spatial interpolation is usually used to calculate the interpolation values at each pixel to be displayed. This will result in pixel-based data.

Pixel-based data are of great use for GIS applications, where the basic idea is to map each data value to a pixel in display. Many algorithms developed for pixel-based data stem from the graphics and image processing areas, such as the algorithms for planar transformation, shape filling, and clipping [7]. An overview of the pixel-based visualization technique is given in [13]. However, the resulting pixel-based data file of interpolation has some potential problems. For example, the number of pixels in the display is limited. In some applications, the number of data values may exceed the number of available pixels. In this case, the pixel file will not have complete information. Therefore, it is difficult to use to answer many queries. For example, consider the following query:

Query 1.1 Find the amount of ultraviolet radiation for each ground location (x, y) at time t.

Let *INCOMING* (y, t, u) and *FILTER* (x, y, r) be the relations that represent the interpolations of *Incoming* (y, t, u) and *Filter* (x, y, r), respectively. Then the above query can be expressed in Datalog as follows:

$$
\begin{aligned}
GROUND\,(x, y, t, i) :- \; & INCOMING\,(y, t, u), \\
& FILTER\,(x, y, r), \\
& i = u(1 - r).
\end{aligned}
\tag{20.1}
$$

The above query could be also expressed in SQL style or relational algebra. Whatever language is used, it is clear that the evaluation of the above query requires a **join** of the *INCOMING* and

FILTER relations. Unfortunately, **join** operations are difficult to perform on pixel-based files and are not supported by most GIS systems, including the ArcGIS systems.

Several authors noted that interpolation constraints can be stored in *constraint relations*, which can be easily joined together, making the evaluation of queries like Query 1.1 feasible. (The textbook [20] discusses the relationship of constraint databases and GIS data models.) Also, in contrast to the pixel data representation, the constraint representation is capable of an arbitrary precision.

Chen *et al.* [3] and Revesz *et al.* [21] considered piecewise linear interpolation of time series data. Grumbach *et al.* [10] considered linear interpolation between snapshots of moving points and the interpolation of a landscape surface based on TIN (Triangular Irregular Network) elevation data. Chen and Revesz [4] used a similar linear interpolation for landscape elevation, aspect, slope, and related data. All of these interpolations are represented in linear constraint databases.

Cai *et al.* [2], and Tossebro and Güting [24] considered the interpolation of snapshots of moving regions. Cai *et al.* [2] represent the interpolation by sets of *parametric rectangles*, and Tossebro and Güting [24] represent the interpolation by a *sliced representation* that was introduced by Forlizzi *et al.* [8]. Both parametric rectangles and sliced representations can be translated into linear constraint relations.

However, many practical spatial interpolations, such as inverse distance weighting [5, 23], Kriging [6, 18], splines [9], trend surfaces [26], and Fourier series [11], require non-linear constraints. In this paper, we focus on the Inverse Distance Weighting (IDW) interpolation, which is non-linear, relatively easy, and gives good results in practice [14]. We also look at visualization, which is generally ignored in the earlier papers. This chapter also sees beyond [22] by applying IDW to real spatiotemporal data.

The rest of this chapter is organized as follows. Section 2 discusses how to represent IDW interpolation in polynomial constraint databases. Section 3 describes the application of IDW in constraint databases to the example in this section. Section 4 gives some visualization results. Section 5 discusses using IDW for spatiotemporal interpolation. Finally, in Section 6, we present some ideas for future work.

2. Representation of IDW in Constraint Databases

The rationale for IDW is consistent with most natural properties of spatial data, in particular, that their values vary continuously and tend to be similar at closer than at further locations. In IDW, the measured values (known values) closer to a prediction location will have more influence on the predicted value (unknown value) than those farther away. More specifically, IDW assumes that each measured point has a local influence that diminishes with distance. Thus, points in the near neighborhood are given high weights, whereas points at a far distance are given small weights.

According to reference [12], the general formula of IDW interpolation is the following:

$$w(x, y) = \sum_{i=1}^{N} \lambda_i \omega_i, \qquad \lambda_i = \frac{\left(\dfrac{1}{d_i}\right)^p}{\sum_{k=1}^{N} \left(\dfrac{1}{d_k}\right)^p}. \tag{20.2}$$

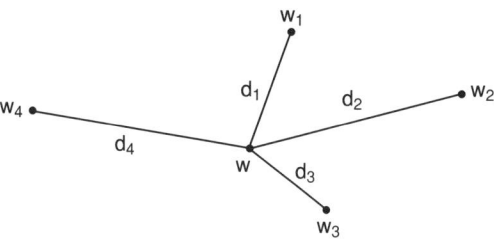

Figure 20.3 IDW interpolation.

As shown in Figure 20.3, w is the predicted value for location (x, y), N is the number of nearest known points surrounding (x, y), λ_i are the weights assigned to each known point value w_i at location (x_i, y_i), d_i are the distances between each (x_i, y_i) and (x, y), and p is the exponent, which influences the weighting of w_i on w. The optimal value of the exponent is dependent on the statistical characteristics of the data set. Please note that in Equation (20.2), $\sum_{i=1}^{N} \lambda_i = 1$.

Query 2.1 Suppose $(x_1, y_1) = (0, 0)$, $(x_2, y_2) = (10, 0)$ and $(x_3, y_3) = (10, 5)$ are the three closest sampled locations to the location $(x, y) = (8, 2)$, as shown in Figure 20.4. Let $w_1 = 1$, $w_2 = 2$ and $w_3 = 3$ be the values of the three sampled locations and $d_i = \sqrt{(x_i - x)^2 + (y_i - y)^2}$. We can interpolate the unknown value w at location (x, y) by IDW with $N = 3$ and $p = 2$ as:

$$w = \sum_{i=1}^{3} \lambda_i w_i$$

$$= \frac{\left(\frac{1}{d_1}\right)^2}{\left(\frac{1}{d_1}\right)^2 + \left(\frac{1}{d_2}\right)^2 + \left(\frac{1}{d_3}\right)^2} w_1 + \frac{\left(\frac{1}{d_2}\right)^2}{\left(\frac{1}{d_1}\right)^2 + \left(\frac{1}{d_2}\right)^2 + \left(\frac{1}{d_3}\right)^2} w_2$$

$$+ \frac{\left(\frac{1}{d_3}\right)^2}{\left(\frac{1}{d_1}\right)^2 + \left(\frac{1}{d_2}\right)^2 + \left(\frac{1}{d_3}\right)^2} w_3$$

$$= 0.07 \times 1 + 0.35 \times 2 + 0.58 \times 3 = 2.51.$$

2.1. High-Order Voronoi Diagrams

To represent the IDW interpolation, we first need to find the nearest neighbors for a given point. Therefore, we borrow the idea of higher-order Voronoi diagrams (or k-th order Voronoi diagrams) from computational geometry. Higher-order Voronoi diagrams generalize ordinary Voronoi diagrams by dealing with k closest points. The ordinary Voronoi diagram of a finite set S of points in the plane is a partition of the plane so that each region of the partition is the locus of points which are closer to *one member* of S than to any other member [19]. The higher-order Voronoi diagram of a finite set S of points in the plane is a partition of the plane into regions

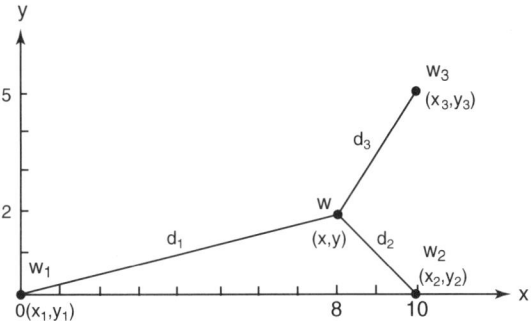

Figure 20.4   IDW example with three neighbors.

such that points in each region have the *same closest members* of S. As in an ordinary Voronoi diagram, each Voronoi region is still convex in a higher-order Voronoi diagram.

From the definition of higher-order Voronoi diagrams, it is obvious that the problem of finding the k closest neighbors for a given point in the whole domain, which is closely related to the IDW interpolation method with $N = k$, is equivalent to constructing k-th order Voronoi diagrams.

Although higher-order Voronoi diagrams are very difficult to create by imperative languages, such as C, C++, and Java, they can be easily constructed by declarative languages, such as Datalog. For example, we can express a 2nd order Voronoi region for points (x_1, y_1), (x_2, y_2) in Datalog as follows.

At first, let $P(x, y)$ be a relation that stores all the points in the whole domain. Also let $Dist (x, y, x_1, y_1, d_1)$ be a Euclidean distance relation where d_1 is the distance between (x, y) and (x_1, y_1). It can be expressed in Datalog as $Dist (x, y, x_1, y_1, d_1) :- d_1 = \sqrt{(x - x_1)^2 + (y - y_1)^2}$.

Note that any point (x, y) in the plane does *not* belong to the 2nd order Voronoi region of the sample points (x_1, y_1) and (x_2, y_2) if there exists another sample point (x_3, y_3) such that (x, y) is closer to (x_3, y_3) than to either (x_1, y_1) or (x_2, y_2). Using this idea, the complement can be expressed as:

$$Not\_2Vor (x, y, x_1, y_1, x_2, y_2) :- P(x_3, y_3), Dist (x, y, x_1, y_1, d_1),$$
$$Dist (x, y, x_3, y_3, d_3), d_1 > d_3.$$
$$Not\_2Vor (x, y, x_1, y_1, x_2, y_2) :- P(x_3, y_3), Dist (x, y, x_2, y_2, d_2),$$
$$Dist (x, y, x_3, y_3, d_3), d_1 > d_3.$$

Finally, we take the negation of the above to get the 2nd order Voronoi region as: $2Vor (x, y, x_1, y_1, x_2, y_2) :- not\ Not\_2Vor (x, y, x_1, y_1, x_2, y_2)$.

The 2nd order Voronoi diagram will be the union of all the nonempty 2nd order Voronoi regions. Similarly to the 2nd order, we can also construct any kth-order Voronoi diagram.

2.2. IDW in Constraint Databases

After finding the closest neighbors for each point by constructing higher-order Voronoi diagrams, we can represent IDW interpolation in constraint databases. In this section, we describe how to represent the IDW interpolation with $N = 2$ and $p = 2$. The representation of other

IDW interpolations in constraint databases is straightforward. The representation is obtained by constructing the appropriate Nth-order Voronoi diagram (where $N \geq 2$) and using Equation (20.2) with the proper p.

Based on the previous section, assume that the 2nd order Voronoi region for points (x_1, y_1), (x_2, y_2) is stored by the relation $Vor\_2nd\ (x, y, x_1, y_1, x_2, y_2)$, which is a conjunction C of some linear inequalities corresponding to the edges of the Voronoi region. Then, the value w of any point (x, y) inside the Voronoi region can be expressed by the cubic constraint tuple as follows:

$$R(x, y, w) :- ((x - x_2)^2 + (y - y_2)^2 + (x - x_1)^2 + (y - y_1)^2)w$$
$$= ((x - x_2)^2 + (y - y_2)^2)w_1 + ((x - x_1)^2 + (y - y_1)^2)w_2 \quad (20.3)$$
$$+ Vor\_2nd x, y, x_1, y_1, x_2, y_2).$$

or equivalently as,

$$R(x, y, w) :- ((x - x_2)^2 + (y - y_2)^2 + (x - x_1)^2 + (y - y_1)^2)w$$
$$= ((x - x_2)^2 + (y - y_2)^2)w_1 + ((x - x_1)^2 + (y - y_1)^2)w_2 + C. \quad (20.4)$$

In the above polynomial constraint relation, there are three variables x, y and w. The highest-order terms in the relation are $2x^2w$ and $2y^2w$, which are both cubic. Therefore, this is a cubic constraint tuple.

3. Application

Let us return now to the example in Section 1. Figures 20.5 and 20.6 show the 2nd order Voronoi diagrams for the sample points in *Incoming* (y, t, u) and *Filter* (x, y, r), respectively. Please note that some 2nd order Voronoi regions are empty. For example, there is no (1, 3) region in Figure 20.5, and there are no (1, 3) and (2, 4) regions in Figure 20.6.

Based on Equation (20.4), *INCOMING* (y, t, u) and *FILTER* (x, y, r), which are the IDW interpolation for *Incoming* (y, t, u) and *Filter* (x, y, r), can be represented in constraint databases as shown in Tables 20.3 and 20.4. Note that the five tuples in Table 20.3 represent the five 2nd

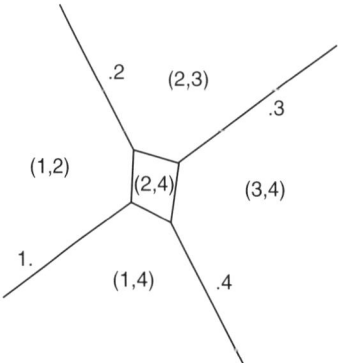

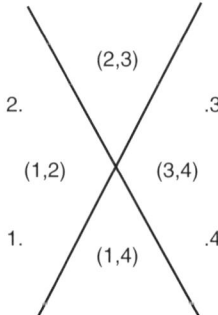

Figure 20.5 The 2nd order Voronoi diagram for incoming.

Figure 20.6 The 2nd order Voronoi diagram for filter.

Table 20.3 INCOMING (y, t, u).

Y	T	U	
y	t	u	$13y + 7t - 286 \leq 0, 2y - 3t - 12 \leq 0, y \leq 15,$ $((y - 13)^2 + (t - 22)^2)60 + (y^2 + (t - 1)^2)20$ $= ((y - 13)^2 + (t - 22)^2 + y^2 + (t - 1)^2)u$
y	t	u	$2y - 3t - 12 \geq 0, 2y + 5t - 60 \leq 0, 2y + t - 44 \leq 0,$ $((y - 29)^2 + t^2)60 + (y^2 + (t - 1)^2)40$ $= ((y - 29)^2 + t^2 + y^2 + (t - 1)^2)u$
y	t	u	$2y + t - 44 \geq 0, 7y - t - 136 \geq 0, 8y - 11t - 47 \geq 0,$ $((y - 29)^2 + t^2)70 + ((y - 33)^2 + (t - 18)^2)40$ $= ((y - 29)^2 + t^2 + (y - 33)^2 + (t - 18)^2)u$
y	t	u	$8y - 11t - 47 \leq 0, y + 3t - 54 \geq 0, 13y + 7t - 286 \geq 0,$ $((y - 33)^2 + (t - 18)^2)20 + ((y - 13)^2 + (t - 22)^2)70$ $= ((y - 33)^2 + (t - 18)^2 + (y - 13)^2 + (t - 22)^2)u$
y	t	u	$y \geq 15, y + 3t - 54 \leq 0, 7y - t - 136 \leq 0, 2y + 5t - 60 \geq 0,$ $((y - 29)^2 + t^2)20 + ((y - 13)^2 + (t - 22)^2)40$ $= ((y - 29)^2 + t^2 + (y - 13)^2 + (t - 22)^2)u$

order Voronoi regions in Figure 20.5. These five regions are (1, 2), (1, 4), (3, 4), (2, 3) and (2, 4). Similarly, the four tuples in Table 20.4 represent the four 2nd order Voronoi regions in Figure 20.6. These four regions are (1, 2), (1, 4), (3, 4) and (2, 3).

The final result of the Datalog query, *GROUND* (x, y, t, i), can be represent by Table 20.5. Since there are five tuples in *INCOMING* (y, t, u) and four tuples in *FILTER* (x, y, r), there should be twenty tuples in *GROUND* (x, y, t, i). Note that the constraint relations can be easily joined by taking the conjunction of the constraints from each pair tuples of the two input

Table 20.4 FILTER (x, y, r).

X	Y	R	
x	y	r	$2x - y - 20 \leq 0, 12x + 7y - 216 \leq 0,$ $((x - 2)^2 + (y - 14)^2)0.9 + ((x - 2)^2 + (y - 1)^2)0.5$ $= (2(x - 2)^2 + (y - 14)^2 + (y - 1)^2)r$
x	y	r	$2x - y - 20 \geq 0, 12x + 7y - 216 \leq 0,$ $((x - 25)^2 + (y - 1)^2)0.9 + ((x - 2)^2 + (y - 1)^2)0.8$ $= (2(y - 1)^2 + (x - 25)^2 + (x - 2)^2)r$
x	y	r	$2x - y - 20 \geq 0, 12x + 7y - 216 \geq 0,$
			$((x - 25)^2 + (y - 14)^2)0.8 + ((x - 25)^2 + (y - 1)^2)0.3$ $= (2(x - 25)^2 + (y - 14)^2 + (y - 1)^2)r$
x	y	r	$2x - y - 20 \leq 0, 12x + 7y - 216 \geq 0,$ $((x - 25)^2 + (y - 14)^2)0.5 + ((x - 2)^2 + (y - 14)^2)0.3$ $= (2(y - 14)^2 + (x - 25)^2 + (x - 2)^2)r$

Table 20.5 GROUND (x, y, t, i).

X	Y	T	I
x	y	t	i	$2x - y - 20 \leq 0, 12x + 7y - 216 \leq 0, 13y + 7t - 286 \leq 0,$ $2y - 3t - 12 \leq 0, y \leq 15, i = u(1-r),$ $((x-2)^2 + (y-14)^2)0.9 + ((x-2)^2 + (y-1)^2)0.5$ $= (2(x-2)^2 + (y-14)^2 + (y-1)^2)r,$ $((y-13)^2 + (t-22)^2)60 + (y^2 + (t-1)^2)20$ $= ((y-13)^2 + (t-22)^2 + y^2 + (t-1)^2)u$
x	y	t	i	$2x - y - 20 \geq 0, 12x + 7y - 216 \leq 0, 13y + 7t - 286 \leq 0,$ $2y - 3t - 12 \leq 0, y \leq 15, i = u(1-r),$ $((x-25)^2 + (y-1)^2)0.9 + ((x-2)^2 + (y-1)^2)0.8$ $= (2(y-1)^2 + (x-25)^2 + (x-2)^2)r,$ $((y-13)^2 + (t-22)^2)60 + (y^2 + (t-1)^2)20$ $= ((y-13)^2 + (t-22)^2 + y^2 + (t-1)^2)u$
x	y	t	i	$\vdots$

relations. Finally, in a constraint database system the constraint in each tuple are automatically simplified by eliminating the unnecessary variables u and r. We do not show the result of the simplification step.

4. Visualization

In Section 2, we have described how to represent IDW interpolation in constraint databases. In Section 1, we have seen it is very easy to express queries (such as join operation) in constraint database on interpolation data. In this section, we will discuss how to visualize interpolation data and give some analysis on the quality of IDW interpolation.

For visualization, six basic colors are chosen: red, yellow, green, turquoise, blue, and purple. The 24 bits RGB values for these colors are the following: red = (255, 0, 0), yellow = (255, 255, 0), green = (0, 255, 0), turquoise = (0, 255, 255), blue = (0, 0, 255), purple = (255, 0, 255). 400 smoothly changing colors have been used for the color plot. These 400 colors are created by a linear interpolation scheme that is used between each of the following pair of the basic colors:

- red and yellow,
- yellow and green,
- green and turquoise,
- turquoise and blue,
- blue and purple.

This color rendering yields a smooth change of colors in the visualization.

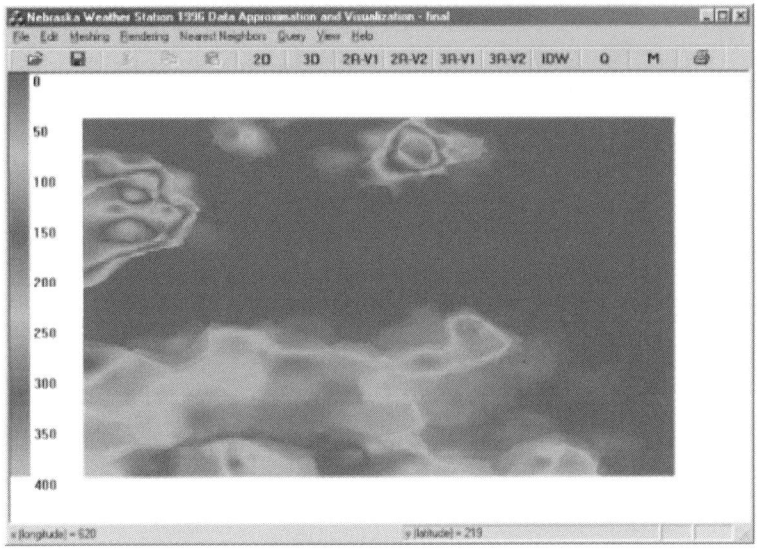

Figure 20.7   IDW ($n = 3$, $p = 2$) on 255 points.

In Figures 20.7 and 20.8, the graphical interface for the presentation of IDW interpolation data is illustrated. Specifically, these two figures illustrate IDW interpolation with $n = 3$ and $p = 2$ on randomly selected DEM (Digital Elevation Model) data over the same area. Figure 20.7 visualizes the interpolation data based on 255 input points, while Figure 20.8 visualizes the interpolation data based on 1271 input points.

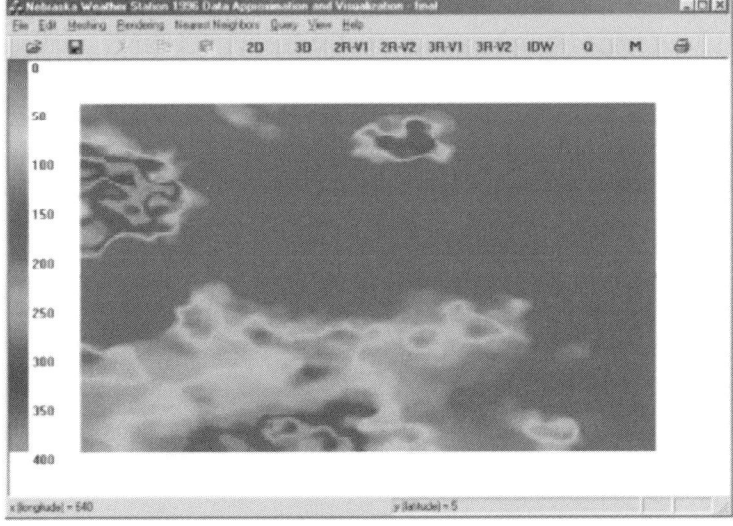

Figure 20.8   IDW ($n = 3$, $p = 2$) on 1271 points.

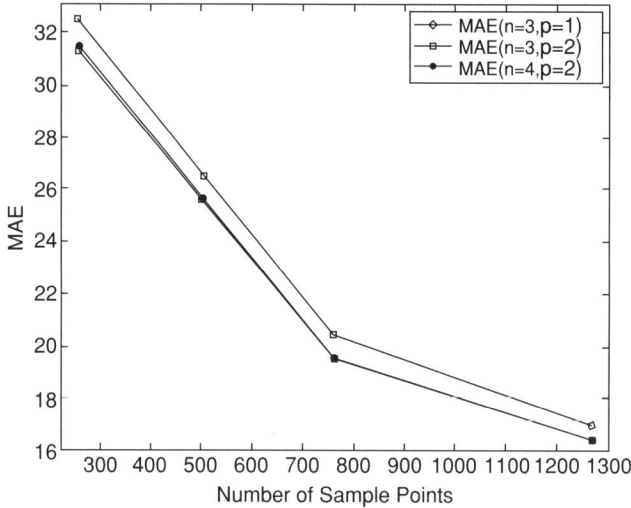

Figure 20.9   MAE result.

Experiments have been conducted to analyze the quality of IDW interpolation according to Mean Absolute Error (MAE) and Root Mean Square Error (RMSE). The definition of MAE and RMSE is as follows:

$$MAE = \frac{\sum_{i=1}^{N} |I_i - O_i|}{N}, \quad RMSE = \sqrt{\frac{\sum_{i=1}^{N}(I_i - O_i)^2}{N}},$$

where N is the number of test houses, I_i is the interpolated house price, and O_i is the original house price. A set of sample points have been selected from a DEM surface in the northern part of San Francisco, which has 1525991 original points. The numbers of randomly selected points are 255, 509, 763 and 1271. For each dataset, three kinds of IDW interpolation methods with different n (the number of neighbors) and p (exponent) have been experimented: (i) $n = 3$, $p = 1$; (ii) $n = 3$, $p = 2$; (iii) $n = 4$, $p = 2$. The number of pixels in the display is between 214775 and 215380.

Figure 20.9 and illustrates the quality analysis of IDW interpolation based on different sets of randomly selected points from the DEM data. We can see that under the condition of randomly selecting points, MAE almost decreases to half when the number of sample points increases from 255 to 1271. In particular, when $n = 4$, $p = 2$, and the dataset contains 1271 points, the MAE is 16.34, which is approximately 17.3% of 94.55, the original average elevation value. This is a very good result, considering that the size of input points is condensed over 1200 times, that is from 1525991 to 1271. RMSE result is very similar as MAE result shown in Figure 20.9.

Although we only discuss the visualization of interpolation data in this section, the same visualization technique can apply to animating, that is, visualizing for each time instance, a query result, such as *GROUND* (x, y, t, i) in the Datalog query in Equation (20.1).

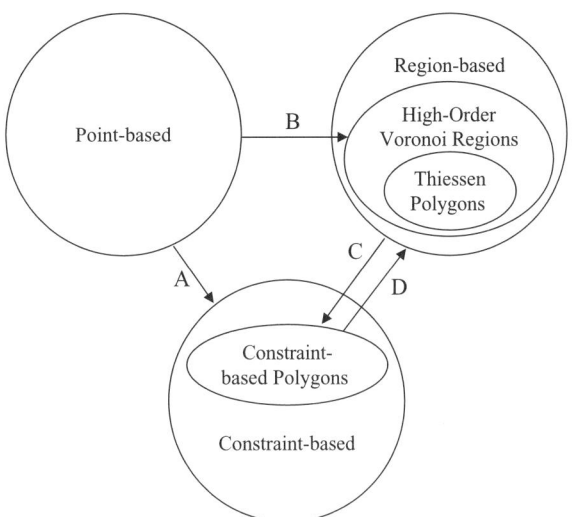

Figure 20.10   The relationship among spatiotemporal databases.

5. Spatiotemporal Interpolation Using IDW

IDW can also be used for spatiotemporal interpolation. There exist different types of spatiotemporal databases (STDBs), such as point-based, region-based, and constraint-based. The relationship among these STDBs is illustrated in Figure 20.10.

Edge A in Figure 20.10 represents the conversion from point-based STDBs to constraint-based STDBs. The conversion can be achieved by multiple methods, such as shape functions [1, 25], spline functions, and Kriging and IDW. We are interested in using the IDW method for the conversion indicated by Edge A. The other edges in Figure 20.10 between STDBs can be found in [17]. There are two fundamentally different ways to approach spatiotemporal interpolation. These methods can be described briefly as follows [15]:

Reduction. This way is to try to reduce the spatiotemporal interpolation problem to the 2-dimensional case. First, for each sample point separately we interpolate (using any 1-dimensional interpolation) all the measured values. This yields a function of time for the measured value at each sample point. Then by substituting the desired time instant into these functions, any of the 2-dimensional spatial interpolation methods can be used. Assume we are interested in the value of the unsampled point at location (x, y) and time t. This approach first finds the nearest neighbors of for each unsampled point and calculates the corresponding weights λ_i. Then, it calculates the value at time t by some time interpolation method for each neighbor. Assume the value at sampled point i at time t_1 is w_{i1}, and at time t_2 the value is w_{i2}. If we use a simple linear interpolation in time, the formula of this approach can be expressed as [16]:

$$w(x, y, t) = \sum_{i=1}^{N} \lambda_i \omega_i(t), \quad \lambda_i = \frac{\left(\frac{1}{d_i}\right)^p}{\sum_{k=1}^{N} \left(\frac{1}{d_k}\right)^p}. \tag{20.5}$$

Table 20.6 Sample (x, y, t, p).

X	Y	T	P(price/ft^2)
888	115	4	56.14
888	115	76	76.02
1630	115	118	86.02
1630	115	123	83.87
2215	110	27	60.57
2215	110	77	69.11
2215	110	114	75.20
...	...	...	...
2650	1190	78	70.34
1950	1760	33	65.44
2240	2380	51	91.87
2650	1190	43	63.27

Table 20.7 Test (x, y, t).

X	Y	T
115	1525	16
115	1525	58
115	1525	81
115	1610	63
115	1610	119
890	1880	36
890	1880	75
...	...	...
1065	680	91
930	785	115
120	1110	30
615	780	59

where,

$$w_i(t) = \frac{t_{i2} - t}{t_{i2} - t_{i1}} w_{i1} + \frac{t - t_{i1}}{t_{i2} - t_{i1}} w_{i2}. \tag{20.6}$$

Each neighbor may have different beginning and ending times t_{i1} and t_{i2} in Equation (20.6) if each points are sampled at different times.

Extension. Several of the interpolation methods can be extended from 2- to 3-dimensional space. Hence for these methods, we can treat time as the third dimension. Since this method treats time as a third dimension, the IDW based spatiotemporal formula is of the form of Equation (20.2) with $d_i = \sqrt{(x_i - x)^2 + (y_i - y)^2 + (t_i - t)^2}$.

To test IDW spatiotemporal interpolation methods, we randomly select 126 residential houses from a quarter of a section of a township, which covers an area of 160 acres. Furthermore, from these 126 houses, we randomly select 76 houses as sample data, and the other 50 houses as test data. Tables 20.6 and 20.7 show instances of these two data-sets. Based on the fact that the earliest sale of the houses in this neighborhood is in 1990, we encode the time in such a way that 1 represents January 1990, 2 represents February 1990, ..., 148 represents April 2002. Note that some houses are sold more than once in the past, so they have more than one tuple. For example, the house at the location (2215, 110) was sold three time in the time at time 27, 77, and 114 (which represent 3/1992, 5/1996, and 6/1999).

We use the 76 house sale history information to estimate the remaining 50 test house sale histories, using both the reduction and the extension methods based on IDW with $n = 3$ and $p = 1$. We compare the estimated values of price per square foot with the true values for each sale instance of the 50 test houses according to MAE and RMSE. The result is summarized in Table 20.8.

Table 20.8 IDW spatiotemporal interpolation
MAE and RMSE result.

Method	MAE	RMSE
Reduction	10.05	11.96
Extension	11.14	13.63

6. Summary

This paper discusses the representation, querying and visualization of IDW interpolation in polynomial constraint databases. In constraint databases, the details of the interpolation are at a lower level, which is transparent for the users. This property makes querying and visualization easy in constraint databases [20].

Kriging [18] is similar to IDW but the weights are derived using error statistics of the data. Beside IDW, it is easy to see that Kriging is also representable by constraint relations if the variogram, or the statistically derived function of weight and distance, is representable using constraints. If we take some (distance, weight) samples from the variogram, then we get a time series data, which can be interpolated and translated into a linear constraint relation using the algorithm in [21].

We are currently extending our work to animation using a 3D spatial interpolation that is the combination of a 2D spatial interpolation and a function of time. That is, at each sample point we would no longer have a constant value measured, but we would have a time series of the measurements. If the time series is itself interpolated, then we get a function of time that can be combined with the spatial interpolation to get a 3D spatiotemporal interpolation that is also representable by constraint relations.

References

[1] Buchanan G.R. (1995) *Finite Element Analysis*. McGraw-Hill, New York

[2] Cai M. Keshwani D. and Revesz P. (2000) Parametric rectangles: A model for querying and animating spatiotemporal databases. *Proc. 7th International Conference on Extending Database Technology*, volume 1777 of *Lecture Notes in Computer Science*, 430–444. Springer-Verlag

[3] Chen R. Ouyang M. and Revesz, P. (2000) Approximating data in constraint databases. *Proc. Symposium on Abstraction, Reformulation and Approximation*, volume 1864 of *Lecture Notes in Computer Science*, 124–143. Springer-Verlag.

[4] Chen R. and Revesz P. (2000) Geo-temporal data transformation and visualization. *Proc. 1st International Conference on Geographic Information Science*, 240–242, Savannah, Georgia, USA

[5] Demers M.N. (2000) *Fundamentals of Geographic Information Systems*, 2nd edition. John Wiley & Sons, New York.

[6] Deutsch C.V. and Journel A.G. (1998) *GSLIB: Geostatistical Software Library and User's Guide*, 2nd edition. Oxford University Press, New York

[7] Foley J.D. Dam A.V. Feiner S.K. and Hughes J.F. (1996) *Computer Graphics: Principles and Practice*, Second Edition in C. Addison-Wesley

[8] Forlizzi L. Güting R.H. Nardelli E. and Schneider M. (2000) A data model and data structure for moving object databases. *Proc. ACM SIGMOD International Conference on Management of Data*, 319–330

[9] Goodman J.E. and O'Rourke J. eds (1997) *Handbook of Discrete and Computational Geometry*. CRC Press, Boca Raton, New York

[10] Grumbach S. Rigaux P. and Segoufin L. (2000) Manipulating interpolated data is easier than you thought. *Proc. IEEE International Conference on Very Large Databases*, 156–165

[11] Harbaugh J.W. and Preston F.W. (1968) *Fourier analysis in geology. Spatial Analysis: A Reader in Statistical Geography*, 218–238. Prentice-Hall, Englewood Cliffs

[12] Johnston K. Ver Hoef J.M. and Krivoruchko K. and Lucas N. (2001) *Using ArcGIS Geostatistical Analyst*. ERSI Press

[13] Keim D.A. (1996) Pixel-oriented database visualizations. *SIGMOD Record* (ACM Special Interest Group on Management of Data), 25(4):35–39

[14] Lam N.S. (1983) Spatial interpolation methods: A review. *The American Cartographer*, 10(2):129–149

[15] Li L. and Revesz P. (2002) A comparison of spatio-temporal interpolation methods. *Proc. of the Second International Conference on GIScience* 2002, 145–160, Egenhofer M. and Mark D. eds, Vol. 2478 of *Lecture Notes in Computer Science*, Springer-Verlag

[16] Li L. and Revesz P. (in press, 2003) Interpolation methods for spatio-temporal geographic data. *Computers, Environment and Urban Systems*, Elsevier

[17] Li L. (2003) Spatiotemporal interpolation methods in GIS. *Doctoral Dissertation*, University of Nebraska-Lincoln

[18] Oliver M.A. and Webster R. (1990) Kriging: A method of interpolation for geographical information systems. *International Journal of Geographical Information Systems*, 4(3):313–332

[19] Preparata F.P. and Shamos M.I. (1985) *Computational Geometry: An Introduction*. Springer-Verlag

[20] Revesz P. (2002) *Introduction to Constraint Databases*. Springer-Verlag

[21] Revesz P. Chen R. and Ouyang M. (2001) Approximate query evaluation using linear constraint databases. *Proc. Symposium on Temporal Representation and Reasoning*, 170–175, Cividale del Friuli, Italy

[22] Revesz P. and Li Li. (2002) Constraint-based visualization of spatial interpolation data. *Proc. of the Sixth International Conference on Information Visualization*, 563–569, IEEE Press, London, England

[23] Shepard D. (1968) A two-dimensional interpolation function for irregularly spaced data. *Proc. 23nd National Conference ACM*, 517–524, ACM Press

[24] Tossebro E. and Güting R.H. (2001) Creating representation for continuously moving regions from observations. *Proc. 7th International Symposium on Spatial and Temporal Databases*, 321–344, Redondo Beach, CA

[25] Zienkiewics O.C. and Taylor R.L. (2000) *Finite Element Method*, Vol. 1, The Basis. Butterworth Heinemann, London

[26] Zurflueh E.G. (1967) Applications of two-dimensional linear wavelength filtering. *Geophysics*, 32:1015–1035

21

Surface Oriented Triangulation of Unorganized 3D Points Based On Laszlo's Algorithm

Thomas Schädlich
Guido Brunnett
Marek Vanco

Computer Graphics and Visualization, Faculty of Computer Science, Technical University of Chemnitz, D-09107 Chemnitz, Germany.

This chapter is concerned with the triangulation of unorganized 3D points in the context of reverse engineering. A new method is presented that extends Laszlo's edge based triangulation algorithm to 3D. The approach is surface oriented, thus it is very fast and provides excellent reconstruction results.

1. Introduction

This chapter in concerned with the triangulation of unstructured sets of 3D points in the context of reverse engineering. Reverse engineering addresses the problem of automatic reconstruction of CAD models from digitized 3D objects. For the digitization a fast and reliable laser range scanner can be used these have become widely available over the past few years. Difficulties arise when objects are too complex to be scanned in a single pass and multiple viewpoints must be used to get a representative sample of the surface. As a result the point set representing the objects boundary will be completely unstructured.

For the reconstruction of the object a triangulation of the point set is often required as an intermediate step. Due to the importance of this problem several algorithms for triangulation of unorganized point sets have been presented. For a good overview of these methods see [9]. Existing methods for 3D triangulation can be broadly classified into *volumetric oriented methods* and *surface oriented methods*.

Advances in Geometric Modeling. Edited by M. Sarfraz
© 2003 John Wiley & Sons, Ltd   ISBN: 0-470-85937-7

Most of the algorithms for 3D triangulation follow the volumetric approach which is based on the idea of creating a volumetric reconstruction of the whole object. Very often the Delaunay tetrahedrization is used as an initial structure for the solid. The triangulation is then extracted from the boundary of the reconstructed solid. An early influential paper was published by Boissonnat [3] in 1984. Three dimensional α-shapes, carefully chosen subsets of the Delaunay triangulation, were introduced by Edelsbrunner and Mücke [6]. Recently, Amenta *et al.* [1, 2] published a volumetric method called the crust method.

The volumetric approach is useful if an object with a closed boundary has to be reconstructed since holes in the boundary do not occur. However, if the original part contains holes this effect may be unwanted. Furthermore, the volumetric method is very time and memory consuming due to the computation of the Delaunay tetrahedrization. This is especially so for the crust algorithm developed by Amenta *et al.* There, the point set is extended by points in the interior of the object before the tetrahedrization is applied. Since the quadratic worst case behavior of the tetrahedrization usually does not occur if all data points lie on the object's boundary this extension has an unfavorable effect on the performance of the method.

In contrast to volumetric methods, surface oriented methods intend to triangulate the 3D points directly. Unfortunately, existing methods of this category cannot be applied to unorganized point sets since they exploit an underlying structure of the gathered sites or do not provide satisfactory results with respect to the quality of the reconstruction. The triangulation method presented in this chapter is surface oriented but avoids the above mentioned problems, i.e. it is a general purpose method that provides very good results for most data sets available.

The layout of this chapter is as follows. In Section 2 we describe Laszlo's algorithm for the computation of 2D Delaunay triangulations. In our presentation we already point out problems that may arise in a 3D generalization. In Section 3 the general outline of the 3D triangulation method is given which consists of the following consecutive steps: neighborhood search, tangent plane estimation, triangulation and postprocessing. In this chapter we omit the details of how the neighborhood search and the tangent plane estimation is actually performed. Only the triangulation is described in more detail in Section 4. In the final section we demonstrate how the algorithm works in practice, evaluate its time and space efficiency and show a few reconstructed objects.

2. Laszlo's Edge-Based Triangulation Algorithm

Let S be a set of $n \geq 3$ sites in general position in $\mathbf{R}^2$. In the following we describe an algorithm for computing the Delaunay triangulation of S, which was first suggested by Laszlo [8]. In this algorithm the triangulation grows face by face in contrast to, for example, divide-and-conquer methods. The algorithm's time complexity is $O(n^2)$.

First, it is important to realize that each edge in a two-dimensional triangulation has exactly two adjacent triangles (or *faces*), one of which may be the unbounded plane in case the edge connects two sites that lie on the convex hull of S. As a vital data structure, the algorithm maintains a set of oriented edges which form the current triangulation's 'frontier'. Edges can thus be classified as 'dormant' (undiscovered), 'live' (edge is in the frontier, one of its adjacent triangles is known), and 'dead' (edge has been removed from the frontier, both adjacent triangles have been found). At the beginning, the frontier consists of a single edge between an arbitrary site and its closest neighbor. This edge is part of the Euclidean minimum spanning tree of

S and thus also of its Delaunay triangulation. Sites are then incrementally (i.e. one by one) inserted by arbitrarily choosing an edge from the frontier and searching a third vertex (called the 'mate' of the edge). If a mate has been found, a new triangle is created, the chosen edge dies and new edges may be inserted into the frontier.

The notion of *orientation* will become important in later chapters. In $\mathbf{R}^3(\mathbf{p}_1, \mathbf{p}_2, \mathbf{p}_3, \mathbf{p}_4)$ have positive orientation, if looking from $\mathbf{p}_1$ to plane spanned by $\mathbf{p}_2, \mathbf{p}_3, \mathbf{p}_4$ ($\mathbf{p}_2, \mathbf{p}_3, \mathbf{p}_4$) are counterclockwise in order. In $\mathbf{R}^2$ ($\mathbf{p}_1, \mathbf{p}_2, \mathbf{p}_3$) are positively oriented if $\mathbf{p}_1$ lies to the left of the line directed from $\mathbf{p}_2$ to $\mathbf{p}_3$. All edges in the frontier are oriented such that their unknown face lies on the right hand side. Correct edge orientation is extremely important for the algorithm and must be maintained throughout the entire computation.

The pseudo-code for Laszlo's randomized incremental Delaunay triangulation algorithm can be given as follows.

```
Algorithm TRIANGULATE
Input: a set S of sites, n = |S|
       Output: the Delaunay triangulation D (S)
 1  D = { }
 2  s'= closest neighbor of arbitrary site s ∈ S
 3  initialize frontier to contain edges (s, s') and (s', s)
 4  while frontier ≠ Ø do
 5    remove some edge e = (s, s') from frontier
 6    if there exists a mate m of e then
 7       add triangle (s, s', m) to D
 8       update the frontier
 9    end if
10  end while
11  return D
```

Finding the Mate of an Edge. Recall that S is a finite set of sites in $\mathbf{R}^2$. For any edge $\mathbf{e} = (\mathbf{s}, \mathbf{s}')$ contained in the frontier, a third site $\mathbf{m}$ is sought such that $\mathbf{s}$, $\mathbf{s}'$, and $\mathbf{m}$ form $\mathbf{e}$'s yet unknown face in the Delaunay triangulation of S. $\mathbf{m}$ is called the *mate* of $\mathbf{e}$. To find $\mathbf{m}$, observe that edge $\mathbf{e}$ defines an infinite family of circles whose centers lie along $\mathbf{e}$'s perpendicular bisector (Figure 21.1). The bisector can be parameterized such that each parametric value α corresponds to the center C_α of a circle (denoted by C_α). Assuming the current frontier contains $\mathbf{e}$, one of $\mathbf{e}$'s faces is already known. Let C_μ denote the circumcircle of this face and let $\mu \geq -\infty$ be the corresponding parametric value; $\mu = -\infty$ if and only if the known face of $\mathbf{e}$ is the unbounded plane. Now, the smallest parametric value $v > \mu$ is sought such that some site $\mathbf{m}$ ($\mathbf{m} \neq \mathbf{s}, \mathbf{m} \neq \mathbf{s}'$) lies on the boundary of the circle C_v associated with v. To find v, the algorithm selects only sites $R \subset S$ that lie to the right of $\mathbf{e}$ thus forcing $v > \mu$. If there are none ($R = \emptyset$) then $\mathbf{e}$ has no mate. Otherwise, a parametric value v_r is computed for each site $\mathbf{r} \in R$ as the intersection of the perpendicular bisectors of $(\mathbf{s}, \mathbf{s}')$ and $(\mathbf{s}', \mathbf{r})$. The mate $\mathbf{m}$ of $\mathbf{e}$ is the site $\mathbf{r}$ with the smallest value of v_r: let $v = \min_{r \in R} v_r$, then $\mathbf{m} = \mathbf{r}_v$. Figure 21.2 depicts, how the mate of an edge $\mathbf{e} = (\mathbf{s}, \mathbf{s}')$ is chosen from the sites lying to the right of $\mathbf{e}$.

Note that C_v is site-free on the left hand side of e, because C_v is contained in C_μ and C_μ is site-free. Furthermore, C_v is also site-free on the right hand side of $\mathbf{e}$ because if there was any site in the interior of C_v, the parametric value associated with its circumcircle would have been smaller than v, contradicting to the choice of v. Therefore it follows by

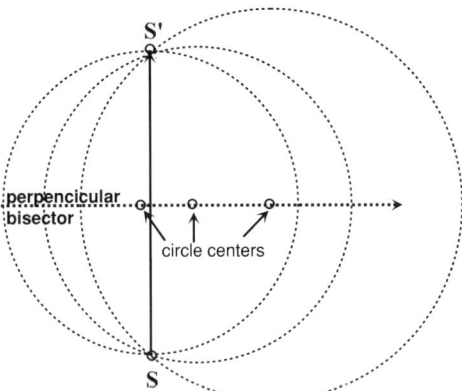

Figure 21.1   Family of circles.

induction over all edges that the triangulation created by the presented algorithm is the Delaunay triangulation.

Updating the Frontier. If in step 6 of the algorithm a mate **m** for edge **e** = (**s**, **s**′) was found, as explained in the previous paragraph, the frontier must be updated. Recall that **e** had already been removed from the frontier in step 5. To bring the frontier up to date, the following three cases have to be checked in the order given.

1.1 *Edge* (**s**′, **m**) *is already in the frontier.* Remove edges (**s**′, **m**), insert edge (**s**, **m**) into frontier.

1.2 *Edge* (**m**, **s**) *is already in the frontier.* Remove edge (**m**, **s**) and insert only edge (**m**, **s**′); this case is depicted in Figure 21.3.

2.1 **m** *is the vertex of exactly one edge, not equal to* (**s**′, **m**) *and* (**m**, **s**) *in the frontier.* If **m** is an edge source then insert merely edge (**s**′, **m**), else if **m** is an edge destination, insert merely edge (**m**, **s**). This case only applies in $\mathbf{R}^3$.

2.2 **m** *is the vertex of exactly two edges, not equal to* (**s**′, **m**) *and* (**m**, **s**), *in the frontier.* This brings about a potentially dangerous situation, as depicted in the left image of Figure 21.4,

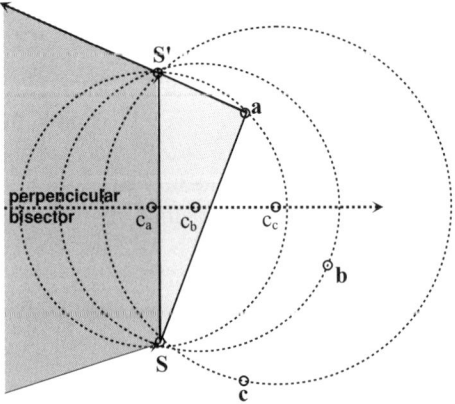

Figure 21.2   Finding the mate.

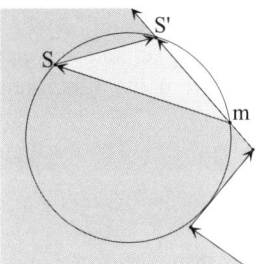

Figure 21.3   Case 1.2.

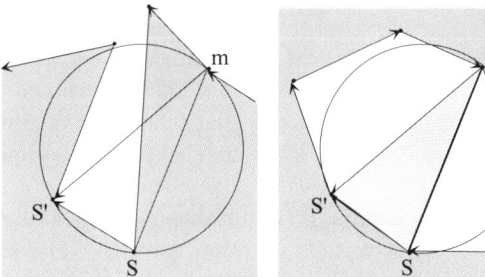

Figure 21.4   Case 2.2.

which was not considered by Laszlo's original algorithm. It frequently occurs when reconstructing surfaces in three dimensions as well as in two-dimensional site sets which are not in a general position. If this phenomenon is encountered for the first time, edge (**s**, **s**′) is re-inserted into the frontier and the frontier update subroutine reports a failure, causing the triangle (**s**, **s**′, **m**) to be removed from D(S). The idea behind this strategy is simply the hope that this situation will be resolved elsewhere. If it is not, and the same configuration occurs again, it is safe to assume that (**s**, **s**′, **m**) must indeed be a part of the Delaunay triangulation, e.g. like the triangle in the right image of Figure 21.4. Then edges (**s**, **m**) and (**m**, **s**′) are inserted into the frontier.

3.1 **m** *is not a vertex of any edge in the frontier but occurs in the triangulation.* To detect this case, each site has a flag associated with it that determines whether the site has already been discovered by the algorithm. If it has, (**s**′, **m**) or (**m**, **s**) potentially cross the frontier which is strictly prohibited (Figure 21.5). Edge (**s**, **s**′) is *not* re-inserted into the frontier; it simply dies. This case only applies when surfaces are reconstructed in three dimensions.

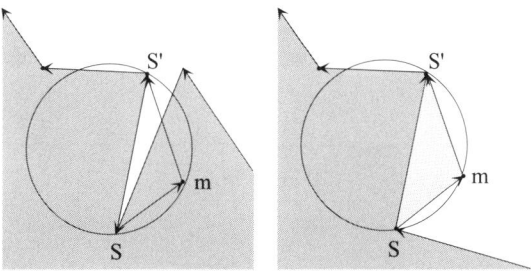

Figure 21.5   Cases 3.1 and 3.2.

3.2 **m** *is not a vertex of any edge in the frontier and does not occur in the triangulation.* This is the simplest case of all. **m** is actually an undiscovered site, therefore (**s**, **m**) and (**m**, **s**′) are inserted into the frontier, as shown in the right image of Figure 21.5, and m is marked 'traversed'.

Naturally, the frontier should be represented by a search tree supporting fast site and edge retrieval.

3. From Two to Three Dimensions

In this section, we are going to reveal the fundamental concepts behind our triangulation method and explain in detail how Laszlo's two-dimensional algorithm is generalized to three dimensions.

An obvious approach to triangulating S is consequently to reduce the three-dimensional problem to two dimensions. In particular, a plane π must be chosen which the sites can be projected onto. A two-dimensional Delaunay triangulation of the projected sites is then computed and 'lifted' back to $\mathbf{R}^3$. 'Lifting the triangulation' implies that $\mathbf{s}, \mathbf{s}' \in S \subset \mathbf{R}^3$ are connected by an edge, if and only if there is an edge connecting the projections of $\mathbf{s}$ and $\mathbf{s}'$ in the corresponding planar triangulation. It is immediately apparent that a single projection plane π is insufficient to reliably recover M. Therefore we will choose an entire *set* of planes $\{\pi(\mathbf{e})\}$, one for each edge $\mathbf{e} = (\mathbf{s}, \mathbf{s}')$. Recall that a particular iteration of the incremental triangulation algorithm described in Section 2 starts by selecting an edge $\mathbf{e}$ from the frontier. Then, as a new feature of our three-dimensional reconstruction algorithm, a plane $\pi(\mathbf{e})$ and a set of candidate sites for the edge's mate $C(\mathbf{e})$ are computed. Next, the algorithm projects $\mathbf{e}$ and the candidate sites onto the plane; it finds $\mathbf{e}$'s mate m in two dimensions and lifts the result back to three dimensional space, i.e. (**s**, **s**′, **m**) becomes a triangle in the triangulation of S. Finally, the frontier of oriented edges has to be updated. This process is repeated, until the frontier is empty.

It remains to clarify how the plane $\pi(\mathbf{e})$ and the candidate sites $C(\mathbf{e})$ should be computed. Informally speaking, the *neighborhood* of a site **s** is a set of sites N(s) geometrically close to s. A single site in $N(\mathbf{s})$ called *neighbor* of **s**. Candidates for the mate **m** of a given edge $\mathbf{e} = (\mathbf{s}, \mathbf{s}')$ are all sites which lie in the intersection of the neighborhoods of **s** and **s**′:

$$C(\mathbf{e}) = N(\mathbf{s}) \cap N(\mathbf{s}'). \tag{21.1}$$

Finally, a proper plane $\pi(\mathbf{e})$ has to be found for each edge $\mathbf{e}$. $\pi(\mathbf{e})$ must represent the local geometry of M well enough to assure that lifting the newly created triangle (**s**, **s**′, **m**) to three dimensions is topologically correct. An obvious choice for $\pi(\mathbf{e})$ is therefore an approximate tangent plane. Let $\tau(\mathbf{s}) : \mathbf{n}(\mathbf{s}) \cdot (\mathbf{p} - \mathbf{c}(\mathbf{s})) = 0$ be the tangent plane of site **s**, estimated in a least-squares sense from $N(\mathbf{s})$, $\mathbf{n}(\mathbf{s})$ is the surface normal and $\mathbf{c}(\mathbf{s})$ some point on the plane. Then,

$$\pi(e) : \frac{n(s) + n(s')}{\mid n(s) \mid n(s') \mid} \cdot \left(p - \frac{c(s) + c(s')}{2} \right) = 0. \tag{21.2}$$

The fact that **e**'s tangent plane $\pi(\mathbf{e})$ is estimated from local site neighborhoods $N(\mathbf{s}) \cup N(\mathbf{s}')$ supports the fundamental assumption on which the entire algorithm is based – lifting

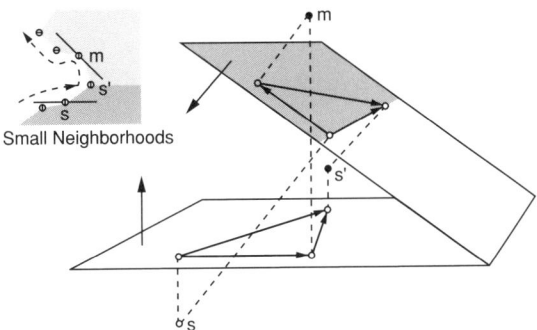

Figure 21.6   Orientation fault in small neighborhoods.

$(\mathbf{s}, \mathbf{s}', \mathbf{m})$, where $\mathbf{s}, \mathbf{s}', \mathbf{m} \in C(\mathbf{e})$, $C(\mathbf{e}) = N(\mathbf{s}) \cap N(\mathbf{s}')$, always produces a valid triangle. Indeed, this prerequisite is met in practice, as demonstrated by our results, provided site neighborhoods are carefully selected.

Recall that Laszlo's algorithm requires a frontier consisting of properly oriented edges. In three dimensions, however, sites can only be classified as lying to the left or right of a *plane*; directed edges are insufficient to specify an orientation. Therefore, in $\mathbf{R}^3$ the term 'oriented edges' refers to 'projected oriented edges'. Specifically, if a directed edge, contained in the frontier, is projected onto its tangent plane $\pi(\mathbf{e})$, the unknown face of $\mathbf{e}$ must lie to the right of $\mathbf{e}$. Only sites that project to the right of $\mathbf{e}$ are considered when computing the mate. Now suppose tangent planes $\tau(\mathbf{s})$ have been estimated from comparatively small neighborhoods, as shown in the left image of Figure 21.6. The frontier is forging ahead from left to right as indicated by the arrows. Figure 21.6 depicts the puzzling situation that evolves. Note how the orientation reverses: first, $\mathbf{m}$ lies to the right of $(\mathbf{s}, \mathbf{s}')$, later it lies to its left, causing the frontier to advance in the wrong direction! This is because the tangent planes subtend an angle of more than $\pi/2$. Therefore, neighborhoods chosen must be large enough to assure 'smooth' transitions between tangent planes of geometrically close sites. This not only helps to maintain the frontier orientation; it is also a prerequisite for consistent normal alignment, which is required for properly building a local coordinate system. Naturally, tangent planes estimated from large neighborhoods may not provide a perfectly close approximation to the actual situation on the manifold M. However, here we are only interested in planes that allow lifting the mate to three dimensions, and that change smoothly across the surface, i.e. close subtend an angle $< \pi/2$.

The pseudo-code for the algorithm now looks as follows:

```
Algorithm RECONSTRUCTSURFACE
Input: a set S of sites on or near M, n = |S|
Output: a triangulation of S
1 D = { }
2 for each site s ∈ S do
3     determine neighborhood N(s)
4     estimate tangent plane τ(s)
5 end for
```

```
6 consistently orient all tangent planes
7 s'=closest neighbor of arbitrary site s ∈ S
8 initialize frontier to contain edges (s,s') and (s',s)
9 while frontier ≠ ∅ do
10    remove edge e = (s,s') from frontier in a FIFO manner
11    determine π(e) and build a local coordinate system
12    project e and candidate sites C(e) onto π(e)
13    if there exists a mate m of projected e then
14        add triangle (s,s',m) to D
15      update the frontier
16    end if
17 end while
18 return D
```

This algorithm reconstructs the oriented two-dimensional manifold M to the extent possible. It may, however, due to the lifting, produce triangles of arbitrary poor aspect ratio (especially in areas of high curvature).

4. Implementation Details

The complete surface reconstruction algorithm, fundamentals of which have been described in the previous section, can be split into four conceptually different phases. First, neighborhood search is carried out. During this important step, a neighborhood $N(s)$ is computed for each site $s \in S$. Phase one is indispensable for all subsequent computations and the obtained neighborhoods have great influence on the outcome of the entire reconstruction algorithm. During a second phase, surface tangent planes are estimated from the neighborhoods computed in step one. They are consistently oriented such that their normals vectors point towards the object's exterior. Third, the set of unorganized sites is triangulated using the ideas and concepts which were introduced in Section 3. Finally, the triangular mesh reconstructed in that way is rendered and displayed.

   This section elaborates on the different phases of the algorithm. Specifics of vital data structures are given; problems that occurred with inadequate approaches are described, several possible solutions introduced, and their chances of success are considered.

4.1. Neighborhood Search

The objective of the algorithm's first phase, neighborhood search, is to find an neighborhood $N(s)$ for each site $s \in S$. $N(s)$ contains sites, called *neighbors* of s, which are geometrically close to s. This can be either:

1. all sites within a given Euclidean distance r from s, or
2. the k sites which are closest to s,

depending on the approach that was taken. Other metrics for deciding which sites are 'geometrically close' to a given site may exist, but may not be as straightforward to compute and have therefore not been explored.

As mentioned before, finding neighbors is the crucial first step of the algorithm. All succeeding computations build on the results acquired here. Due to its significance, various approaches for neighborhood search have been implemented and investigated. They differ in quality of the resulting neighborhoods for reconstruction purposes; in performance, coging complexity and userfriendliness. A common feature of all such algorithms is some sort of spatial subdivision structure to avoid the worst-case of having to compute all inter-site distances which would result in a time complexity of $O(n^2)$ for a set S of n sites. Formally, *spatial subdivision* dicomposes some spatial domain D into a set of smaller pieces. Problems involving D can in that way be reduced to smaller (and simpler) subproblems. For example, the volume of a polyhedron may be computed as the sum of the volumes of the tetrahedra of its Delaunay triangulation. In practice, *hierarchical subdivision* schemes are popular, which recursively partition D. The subdivision structure that was tested first is, however, nonhierarchical and probably the simplest of all – a *grid*.

The Grid Method. Our goal is to find all sites $N(\mathbf{s})$ that lie within a given distance r of $\mathbf{s} \in S$. Problems of this type are called spherical *range queries* and the sphere with radius r centered at $\mathbf{s}$ is termed a *range*. r will be referred to as the *search radius*. The problem can be generalized to other types of ranges (such as cubes, cones etc.) but in our case a sphere will suffice.

A series of range queries, one for each site $\mathbf{s} \in S$, must be performed because neighborhoods are required for all sites in the set. It is thus advantageous to organize the sites in a spatial data structure which efficiently supports range queries. For now, a *grid* is the structure of choice. It divides some domain D into a regular lattice of small *cells*. For spherical range queries, a grid can be constructed as follows. First, compute the axis-parallel bounding box of S, represented by two points $\mathbf{b}_{\min}$ and $\mathbf{b}_{\max}$. For simplicity, cubical cells with an edge length equal to the search radius r will be used. The number of grid cells in each dimension is computed as:

$$n_x = \left\lceil \frac{b_{\max,x} - b_{\min,x}}{r} \right\rceil, \; n_y = \left\lceil \frac{b_{\max,y} - b_{\min,y}}{r} \right\rceil, \; n_z = \left\lceil \frac{b_{\max,z} - b_{\min,z}}{r} \right\rceil. \quad (21.3)$$

Next, the bounding box is adjusted to contain an integer number of cells. For its x-coordinates, the formulas are:

$$b'_{\min,x} = b_{\min,x} - (n_x r - (b_{\max,x} - b_{\min,x}))/2$$
$$b'_{\max,x} = b_{\max,x} + (n_x r - (b_{\max,x} - b_{\min,x}))/2 \quad (21.4)$$

Similarly, the bounding box is extended in y and z direction. Each cell C in the grid is associated with three coordinates $C(c_x, c_y, c_z)$. To determine the cell in which a given site $s = (s_x, s_y, s_z)$ is contained, the cell's coordinates are computed as follows:

$$c_x = \left\lceil \frac{s_x - b'_{\min,x}}{r} \right\rceil, \; c_y = \left\lceil \frac{s_y - b'_{\min,y}}{r} \right\rceil, \; c_z = \left\lceil \frac{s_z - b'_{\min,z}}{r} \right\rceil. \quad (21.5)$$

A grid can thus be constructed in $O(n)$ linear time from a set of sites. To use it for efficient range querying, recall that the cell edge length is equal to the search radius r. When determining the neighborhoods $N(\mathbf{s})$ for all sites s in an arbitrary cell C, the distances between these sites and those contained in any cell C' adjacent to C (including C itself) have to be computed. As depicted in Figure 21.7, the spherical range centered about some $\mathbf{s} \in C$ may

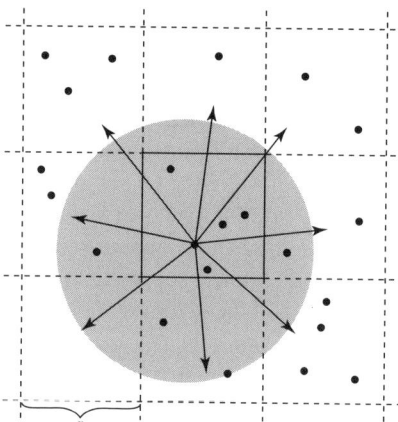

Figure 21.7 Range covers adjacent cells.

cover all adjacent grid cells, validating the just stated rule. In three dimensions, up to 27 such cell-to-cell comparisons would be necessary to determine neighborhoods for all sites in a given (central) grid cell. A single such comparison consequently involves computing the intersite distances between every two sites $\mathbf{s} \in C$, $\mathbf{s}' \in C'$ in cells C and C' and adding sites that are within range r to each others neighborhood. Note that for range queries, 'neighborhood' is a symmetric relationship: $\mathbf{s} \in N(\mathbf{s}') \Leftrightarrow \mathbf{s}' \in N(\mathbf{s})$. We can use this observation to limit the number of cell-to-cell comparisons by traversing the grid in a specific, e.g. lexicographic, order. The core for a spherical range query could then look as follows:

```
1 for z = 0 to nz do
2   for y = 0 to ny do
3     for x = 0 to nx do
4       cell-to-cell compare
            (x, y, z) to (x,   y,   z),
            (x, y, z) to (x+1, y,   z),
            (x, y, z) to (x-1, y+1,z),
            (x, y, z) to (x,   y+1, z),
            (x, y, z) to (x+1, y+1, z),
            (x, y, z) to (..., ..., z+1) [9x]
```

Figure 21.8 illustrates how range queries are efficiently performed on a grid consisting of square cells. The total asymptotic run time can be derived as follows: Let m be the maximum number of sites in a cell of the grid. Then the number of site-to-site comparisons for any single cell-to-cell comparison is bounded by $O(m^2)$. Now let c denote the total number of cells, $c = n_x n_y n_z$. Because each cell is compared to at most fourteen others, as shown in the pseudo-code, there are total $O(c)$ cell-to-cell comparisons and the overall number of individual site-to-site comparisons is consequently $O(cm^2)$. Because the total number of sites n is bounded by $O(cm)$, and building the grid takes $O(n)$ time, the entire algorithm has a complexity of $O(nm)$. The $O(n^2)$ worst case occurs if there is only a single cell, namely the bounding box, and thus $m = n$. In practice, $m << n$. The grid method performs pretty well, also because it introduces little overhead into the computation.

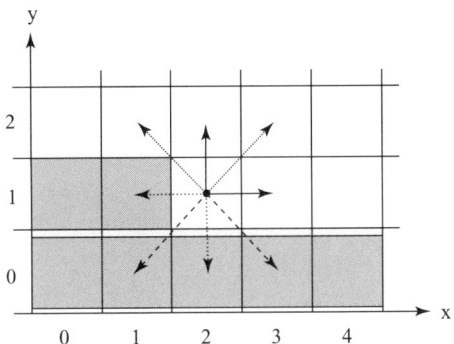

Figure 21.8 Efficient grid range querying.

One disadvantage of the grid method is its memory inefficiency. If the search radius r is chosen too small, the number of cells in the grid easily exceeds 2^{32} and thus the address space of a 32-bit CPU. The user should therefore be provided with an estimate of r to avoid such situations. Assume the site set is bounded by a box with diagonal length $d = |\mathbf{b}_{max} - \mathbf{b}_{min}|$. This bounding box is regarded as a cube with the same diagonal length d and an edge length $a = d/\sqrt{3}$. The number of cells in the grid is then given as $c = \lceil a/r \rceil^3$. Assuming a uniform site distribution within the bounding cube, the number m of sites per cell is approximately:

$$m = \frac{c}{n} \approx \frac{1}{n}\left(\frac{a}{r}\right)^3 = \frac{1}{n}\left(\frac{d}{\sqrt{3}r}\right)^3 \Rightarrow r \approx \frac{d}{\sqrt{3^3}\sqrt{nm}}. \tag{21.6}$$

Empirically, $m = 27$ gives acceptable results for many site sets. Thus, r can be approximated as:

$$r \approx 0.19245\frac{d}{\sqrt[3]{n}}. \tag{21.7}$$

Though the user is given an idea of r's magnitude, the above formula still only provides an estimate of the search radius. For large site sets, relatively small changes to r may cause $c > 2^{32}$ and the application to crash. Furthermore, the grid method is inefficient if the sites are unevenly distributed. Many cells remain empty while others contain a large number of sites, resulting in the performance being pushed towards the $O(n^2)$ worst case. In conclusion, the grid method is easy to code, but not flexible enough for the purpose of surface reconstruction.

The kD-Tree. To overcome the grid's disadvantages, in particular its inefficient memory usage and lack of flexibility, an adaptive hierarchical subdivision scheme is now introduced. Instead of creating cells of constant size, each containing a variable number of sites, adaptive subdivision generates cells of variable size, each enclosing approximately the same amount of sites. In other words, the cell size varies in response to the distribution of the sites. Subdivision is fine where the sites are dense and coarse. Examples of data structures that support adaptive spatial subdivision are quad-trees and oct-trees, respectively.

For spherical range queries, a k-dimensional binary search tree, or *kD-tree*, is practical. This is a binary tree that recursively subdivides k-dimensional space by alternating cut planes which

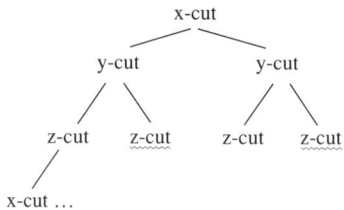

Figure 21.9   Alternating tree cuts.

are parallel to the coordinate planes. The three-dimensional case, i.e. building a 3D-tree, is naturally of particular interest to surface reconstruction. For $k = 3$, cut planes parallel to the yz coordinate plane will be referred to as x-cut. Equivalently, y-cuts and z-cuts are defined. Cut directions alternate when descending down the tree, as shown in Figure 21.9.

Briefly explained, a 3D-tree can be constructed as follows. First, the sites in S are presorted in x, y, and z direction, respectively. During each step of the recursive construction, the present set of n sites is split into two roughly equal parts along the current cut direction. The median **m** is taken from position $n/2$ of the set that was presorted by the split coordinate. This median is stored in the tree node just created; it uniquely determines the cut plane associated with it.

Now, the two remaining presorted site sets are each partitioned into two halves; one half contains all sites smaller than m, the other all sites greater than m. The split is carried out, such that ordering in the two newly created subsets is preserved. Essentially, this operation performs the opposite of a merge in merge sort. Tree construction then continues recursively on both subsets of the three presorted sets, generating the left and right son of the current tree node, until a set consists of only a single site. Figure 21.10 depicts how the initial set is split into two subsets. A complete 2D-tree and the subdivision structure associated with it are shown in Figure 21.11.

If, during the tree's construction, a set contains an even number of sites, the median is chosen such that the larger subset is always associated with the left son of the current tree node. Then, the size, or number of nodes in the kD-tree is bounded by $2^{\lceil \log_2 n \rceil}$. Just like a heap, the kD-tree can therefore be very memory efficiently implemented as an array.

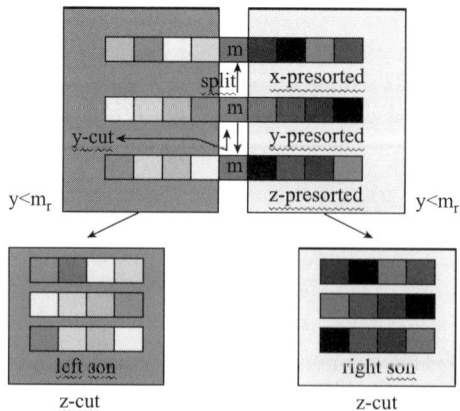

Figure 21.10   Splitting the presorted sets.

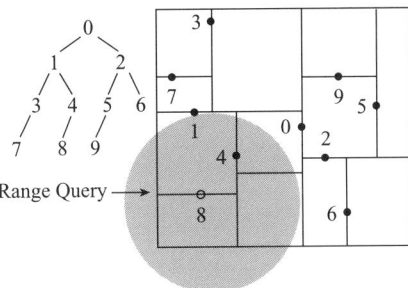

Figure 21.11   Querying the 2D-tree.

kD-trees allow fast spherical range queries for any query center **s** and search radius r by recursing through the tree. For the current tree node, we first check whether site **m** associated with this node lies within the range. If this is the case, **m** is added to $N(\mathbf{s})$. Next, if a part of the spherical range lies to the left, top, or front respectively, of the current cut plane through the median, **m**'s left son is queried recursively. Symmetrically, if a part of the spherical range lies to the right, bottom, or back, respectively, of the current cut plane, **m**'s right son is queried. This way it is assured that the query is restricted to one son whenever possible. Figure 21.11 illustrates a two-dimensional range search.

A kD-tree can be constructed in $O(n \log n)$ time. For $k = 3$, a single range query that reports r sites takes $O(n^{2/3} + r)$. The total time complexity for a complete neighborhood search is thus $O(n^{5/3} + nr)$. The kD-tree performs almost as well as a grid; it is much more flexible and consumes considerably less memory. The implementation is, however, slightly more complicated. kD-trees can be efficiently computed for large sets of sites.

One problem still remains: determining a suitable search radius r requires some knowledge about the distribution of the sites within the bounding box. Though one can use the same approximation for r as the grid method uses to get an idea of r's magnitude, r still largely depends on properties of the site set. This disadvantage is common to all methods that determine neighborhoods using range search.

K Nearest Neighbors. We are now going to present the technique that is currently used in our software to find site neighborhoods for surface reconstruction. In it, instead of specifying a search radius r, the user provides an integer constant k. For each site $\mathbf{s} \in S$, the application searches the k sites closest to **s** which then form $N(\mathbf{s})$. This computational task is also called the *k-nearest neighbors problem.* Obviously, k can be chosen independent of the object's scale, which is a huge advantage over the range search approaches. The application can now be used more conveniently and intuitively. Neighborhood sizes $|N(\mathbf{s})|$ may be directly specified, the radius of a neighborhood (i.e. the distance to the farthest neighbor) adapts to the local site density.

All the advantages just mentioned are bought at the expense of performance, however. Solving the k-nearest neighbors problem is more complex than carrying out a series of range searches. The best known theoretical time bound for k-nearest neighbors search is due to Callahan [4]. He computes the Well Separated Pairs Decomposition (WSPD) for a site set based on its fair split tree in $O(n \log n)$. Geometric properties of the WSPD allow for solving the k-nearest neighbors problem in additional $O(kn)$ steps. We implemented the algorithm and it proved

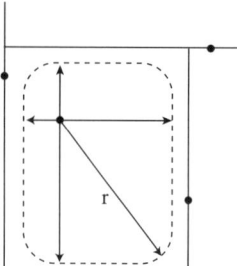

Figure 21.12   Finding r.

not to be useful in practice. Though the fair split tree and the WSPD of the site set can be computed efficiently, obtaining the k nearest neighbors for each site takes too much time and space. There is a large constant, about 50, hidden in the $O(kn)$ asymptotic time which accounts for the overhead of the method and makes it unsuitable.

Other approaches to the k-nearest neighbors problem yield better performance, although their theoretical bounds are worse. We have developed our own method. We use a kD-tree to determine a search radius for each site which guarantees to return at least k neighbors. The algorithm can be described as follows. If, during the construction of a kD-tree, there are $n \geq k$ sites in the current set, only $n/2 - 1 < k$ of which will be associated with the right son, a bounding box is computed for the current set (Figure 21.12). For each coordinate of any sites **s** in the present set, the two distances to its corresponding bounding box planes are derived. Let **r** be the vector, the components of which are the larger of the two distances. Then the length of **r** is the search radius **r** for site **s**. After the kD-tree has been constructed, search radiuses for most sites have been obtained that way.

Finally, r needs to be determined for sites s representing high-level internal tree nodes. This is done by searching an adjacent kD-tree node which contains $\geq k$ sites and the median of which has already been assigned a search radius. Within this node, the site closest to **s** is found. The search radius of **s** is then the sum of the distance to the closest site and the closest site's search radius (Figure 21.13). This computation is carried out top-down for each node, the median of which has not yet been assigned a search radius, starting at the root node of the kD-tree. After having assigned some radius r to each site $\mathbf{s} \in S$, a range search is carried out to obtain $\geq k$ sites in the vicinity of **s**. Then, the k nearest sites are found in linear time by selecting the k-th

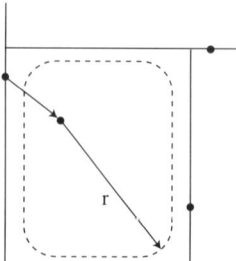

Figure 21.13   Finding r.

median and retrieving all closer sites. Note that the search radius r is always a conservative estimate. In many cases, especially if the data is scattered, r can be chosen smaller. Over all neighborhoods $N(\mathbf{s})$, the application therefore determines a *confidence factor* < 1 by which the estimated search radiuses can be diminished without losing the property of having at least k sites within range of each site $\mathbf{s} \in S$. The confidence factor can be used to speedup subsequent searches for the same number of neighbors in the same set of sites. The neighbors $N(\mathbf{s})$ are sorted by increasing distance from $\mathbf{s}$ and are represented as a fixed-size array of pointers.

Despite the fact that the k-nearest neighbors problem cannot be solved as fast as a series of individual range queries, it is very well suited for neighborhood search in surface reconstruction. This is mainly because the method guarantees a certain neighborhood size without having to worry about the local distribution and scale of the site set.

As an additional feature, the user can reduce site neighborhoods after they have been determined. The basic idea is that the surface triangulation algorithm requires large neighborhoods (i.e. large k) in coarsely sampled regions which in turn may be oversize in high-curvature areas where tangent planes are inaccurately estimated as a result. *Principal component analysis* is used to compute the smallest eigenvalue of the covariance matrix of a site neighborhood. The user can specify an upper bound for this eigenvalue, i.e. a maximum value for variance in normal direction. The application successively removes neighbors, starting with the farthest, until the smallest eigenvalue drops below the given maximum. The smaller the smallest eigenvalue, the better the neighborhood can be approximated by plane. As three distinct points always lie on a common plane, the smallest eigenvalue is zero for neighborhoods of size three or less.

4.2. Tangent Plane Estimation

Given the neighborhoods $N(\mathbf{s})$, tangent planes $\tau\ (\mathbf{s}) : \mathbf{n}(\mathbf{s}) \cdot (\mathbf{p} - \mathbf{c}(\mathbf{s})) = 0$ can be estimated for each site $\mathbf{s} \in S$. The plane's normal $\mathbf{n}(\mathbf{s})$ and its centroid $\mathbf{c}(\mathbf{s})$ are computed. Denote the $k = |N(\mathbf{s})|$ neighbors of s as:

$$N(s) = \{n_i\}_{i=1}^k. \tag{21.8}$$

The centroid of $N(\mathbf{s})$, which corresponds to the neighborhood's center of gravity, is then simply obtained by averaging the neighbors of $\mathbf{s}$:

$$c(s) = \frac{1}{k} \sum_{i=1}^k n_i. \tag{21.9}$$

Principal component analysis can now be employed to estimate the tangent plane's normal vector. If $\mathbf{c}(\mathbf{s})$ is regarded as the sample mean vector, the sample variance-covariance matrix $\mathbf{S}$ may be computed by evaluating the following sum:

$$S = \frac{1}{k-1} \sum_{i=1}^k (n_i - c(s))(n_i - c(s))^T. \tag{21.10}$$

S is a 3×3, positive semi-definite matrix and can thus be decomposed into real eigenvalues $\lambda_1 \geq \lambda_2 \geq \lambda_3 \geq 0$ and the corresponding eigenvectors. $\tau(\mathbf{s})$'s normal vector $\mathbf{n}(\mathbf{s})$ is chosen to

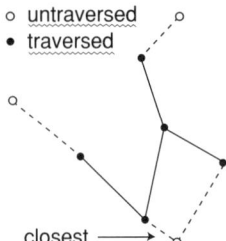

Figure 21.14   EMST.

be the eigenvector associated with the smallest eigenvalue λ_3. The tangent plane thus obtained is optimal in a least-squares sense in that it minimizes the sum of squared distances between the $\mathbf{n}_i$ and $\tau(\mathbf{s})$. The above eigenproblem is solved by utilizing an appropriate LAPACK (Linear Algebra Package) driver routine.

Having estimated tangent planes for all sites $\mathbf{s} \in S$, they need to be properly oriented, such that every normal $\mathbf{n}(\mathbf{s})$ points towards the objects exterior. Correct tangent plane orientation is crucial when local coordinate systems are constructed during surface triangulation. Erroneous planes may cause topologically incorrect triangles to be generated or, what is just as serious, the entire frontier could collapse.

To orient tangent planes, we have adopted the approach of Hoppe *et al.* [7]. The procedure starts at the site $\mathbf{s} \in S$ with the smallest z-coordinate, the normal vector of which is aligned such that its z-value is negative. Because the $\mathbf{n}_i$ are sorted by increasing distance to $\mathbf{s}$, the nearest neighbor $\mathbf{n}_1$ of $\mathbf{s}$ can be found in $O(1)$. Now, the dot product $\mathbf{n}(\mathbf{s}) \cdot \mathbf{n}(\mathbf{n}_1)$ of the two normal vectors is considered. Under the assumption made in Section 3, tangent planes of adjacent sites must enclose an angle of less than $\pi/2$. Consequently, if the above scalar product is less than zero, $\mathbf{n}(\mathbf{n}_1)$ is incorrectly oriented and must be flipped by multiplying it with -1. Next, $\mathbf{n}_1$ is marked 'traversed' to indicate that its tangent plane normal is now valid. In the i-th step of the algorithm, $i + 1$ sites have been traversed. Some of these sites possess a closest, yet untraversed neighbor. The algorithm continues checking the normal orientation of closest such neighbor among all traversed sites (Figure 21.14).

To do this efficiently, a heap of untraversed sites, sorted by distance to a traversed neighbor, is maintained. It allows the sought minimum to be retrieved in $O(1)$. Note that this algorithm practically constructs a Euclidean Minimum Spannning Tree (EMST) of the neighborhood graph. By propagating orientation information over the EMST, consistent tangent planes can be obtained, as geometrically close sites are likely to have similar tangent planes. Normal orientation fails if the neighborhood graph is not connected, i.e. if some sites remain untraversed. In this case, the user should increase the neighborhood size.

4.3. Triangulation and Mesh Post-Processing

During the algorithm's third phase, the site set S is triangulated based on the sites' neighbor hood and tangent plane information which have been computed before. The triangular mesh thus created undergoes a post-processing procedure which involves checking and probably correcting triangle orientations, filling holes and smoothing the mesh.

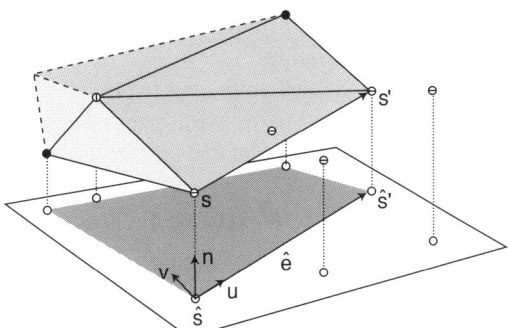

Figure 21.15   Local coordinate system.

In its entirety, the triangulation process was described in Sections 2 and 3. Recall that each step consists of selecting an edge $\mathbf{e} = (\mathbf{s}, \mathbf{s}')$ from the frontier, deriving a set of candidate sites $C(\mathbf{e})$ and a plane $\pi(\mathbf{e}) = \mathbf{n}(\mathbf{e}) \cdot (\mathbf{p} - \mathbf{c}(\mathbf{e}))$, projecting $\mathbf{e}$ and $C(\mathbf{e})$ onto $\pi(\mathbf{e})$, and finding the mate $\mathbf{m}$ of $\mathbf{e}$ in two dimensions. Then, the triangle $(\mathbf{s}, \mathbf{s}', \mathbf{m})$ is added to the mesh and the frontier of oriented edges is updated. This section elaborates on some aspects of the triangulation process that have not been covered in previous sections.

Assume, edge $\mathbf{e} = (\mathbf{s}, \mathbf{s}')$ has been removed from the frontier in the current triangulation step. The frontier is implemented as a randomized braided binary search tree. It combines two basic data structures, a tree and a list. Each tree node represents an edge of the frontier and contains pointers to its left and right child in the tree as well as one to the next node in the list, or *braid*, which is maintained in a FIFO manner. Thus, removing the braid's head yields the 'oldest' edge in the tree, namely $\mathbf{e}$. To assure efficient dictionary operations, the search tree should be balanced. So, each node is additionally assigned a random value, called *priority*. Besides the standard binary search tree rule, i.e. items in the left subtree are smaller than items in the right subtree, it must now also be assured that the priority of each node is smaller than the priority of every one of its descendants. This is done by applying appropriate rotations when inserting or deleting tree nodes. It can be proven that the expected depth of a tree node is $O(\log n)$ if n is the total number of edges currently stored in the frontier. Retrieving an edge involves finding the head of the braid in $O(1)$ and removing the corresponding node in expected $O(\log n)$ from the tree. Having computed $C(\mathbf{e})$ and $\pi(\mathbf{e})$, the candidates sites must be projected onto the plane to reduce the problem of finding $\mathbf{e}$'s mate to two dimensions. For this purpose, a local (u, v, n)-coordinate system has to be built. Its n-axis is simply given by $\pi(\mathbf{e})$'s normalized normal $\mathbf{n}(\mathbf{e})$. The uaxis corresponds to $\mathbf{e}$'s projection onto $\pi(\mathbf{e})$. Let $\hat{\mathbf{e}} = (\hat{\mathbf{s}}, \hat{\mathbf{s}}')$ denote this projected edge, where $\hat{\mathbf{s}} = \mathbf{s} + [\mathbf{n}(\mathbf{e}) \cdot \mathbf{c}(\mathbf{e}) - \mathbf{n}(\mathbf{e}) \cdot \mathbf{s}]\mathbf{n}(\mathbf{e})$ (analog for $\hat{\mathbf{s}}'$). Normalizing $\hat{\mathbf{e}}$ yields the sought u-axis and $\hat{\mathbf{s}}$ is taken as the origin of the local coordinate system. The v-axis is chosen such that it is directed towards $\mathbf{e}$'s known face, $\mathbf{v} = \mathbf{u} \times \mathbf{n}$.

To find the mate of $\mathbf{e}$, each candidate site contained in $C(\mathbf{e})$ is transformed into local (u, v, n)-coordinates by means of a basis transformation. Disregarding the n-coordinate yields the desired two-dimensional, projected sites. Only candidates which have a negative v-coordinate, and thus lie to the right of $\hat{\mathbf{e}}$ (i.e. on the side of $\hat{\mathbf{e}}$'s unknown face), are considered for mate search. Having found a mate $\hat{\mathbf{m}}$ in two dimensions, the triangle $(\hat{\mathbf{s}}, \hat{\mathbf{s}}', \hat{\mathbf{m}})$ is 'lifted' merely by inserting $(\mathbf{s}, \mathbf{s}', \mathbf{m})$ into the mesh. Then, the frontier is updated and mate search starts all over again, until the frontier contains no more edges.

Mesh post-processing begins while the triangulation is in progress. When a triangle $(\mathbf{s}, \mathbf{s}', \mathbf{m})$ is created and added to the mesh, the algorithm investigates whether it is properly oriented. To that end, the post-processing procedure maintains a separate tree data structure which is similar to the binary search tree that represents the frontier of oriented edges. It checks if any of the triangle's three sides $(\mathbf{s}, \mathbf{s}')$, $(\mathbf{s}', \mathbf{m})$, or $(\mathbf{m}, \mathbf{s})$ occur in the tree. If this is not the case, then the triangle is consistent with the mesh and added to it. At the same time, all three edges of the triangle are inserted into the tree. If, on the other hand, any of the triangle's edges are found in the search tree, an orientation fault has occurred, e.g. due to improperly chosen neighborhoods. In this case, the triangle's orientation is flipped and an attempt is made to add it again. If it still cannot be inserted into the mesh, the triangle is ignored. This strategy guarantees to generate a consistent, though not necessarily connected, mesh. Correct triangle orientations are required for backface culling and lighting as the mesh is rendered.

The application is furthermore capable of filling holes that may remain in the final mesh. It first removes all edges from the edge tree which occur twice, in reverse orientation. Then, edges that cannot be deleted correspond to mesh boundaries. For each such edge $\mathbf{e} = (\mathbf{s}, \mathbf{s}')$, the algorithm checks whether a third site $\mathbf{m}$ exists which is a neighbor of both $\mathbf{s}$ and $\mathbf{s}'$ and also a part of at least one other boundary edge in the tree. If such a site $\mathbf{m}$ exists, a hole filling triangle $(\mathbf{s}, \mathbf{s}', \mathbf{m})$ can be added to the mesh.

A further feature of the application is mesh smoothing. Just like smoothing filters in image processing, mesh smoothing can be used to remove small wrinkles from the surface which occur, for example, due to noise in the scanning process. Though the result may be visually pleasing, subtle surface detail can be lost. Many approaches to mesh smoothing have been suggested. More complicated methods attempt to increase the quality of the mesh while keeping gentle features of the surface. We have used the simplest possible smoothing algorithm, however. Let the mesh *neighbors* of a site $\mathbf{s}$ in the set be all sites $\mathbf{s}'$ that share an edge $\mathbf{e} = (\mathbf{s}, \mathbf{s}')$ with $\mathbf{s}$ in the triangulation of S. Note that the *mesh neighborhood $M(\mathbf{s})$* can easily be obtained from the edge tree which is created during the triangulation as described previously. To smooth the mesh, a site $\mathbf{s}$ is repositioned to the coordinates of the centroid of $M(\mathbf{s})$. Then, the smoothing procedure is invoked recursively on all unsmoothed neighbors of $\mathbf{s}$. The effects of this approach will be evaluated in Section 5.

5. Results

Numerous, different kinds of objects can properly and efficiently be reconstructed and visualized from unorganized points using the method described in this chapter. Figure 21.16 shows nine reconstructions. Most of the three-dimensional data was taken from Hoppe *et al*. Table 21.1 contains some statistical information about the time and space required for reconstructing each particular model. It has been obtained on a machine equipped with a 500 MHz AMD-K7 and 256 MB RAM running RedHat Linux. In this table, $n = |S|$ is the total number of sites in the set, k the neighborhood size (i.e. the number of neighbors), f the confidence factor, and $\lambda_3^{\max}$ an upper bound on the smallest eigenvalue of the local covariance matrix. '#Tri's' refers to the total number of triangles in the reconstructed mesh and is followed by

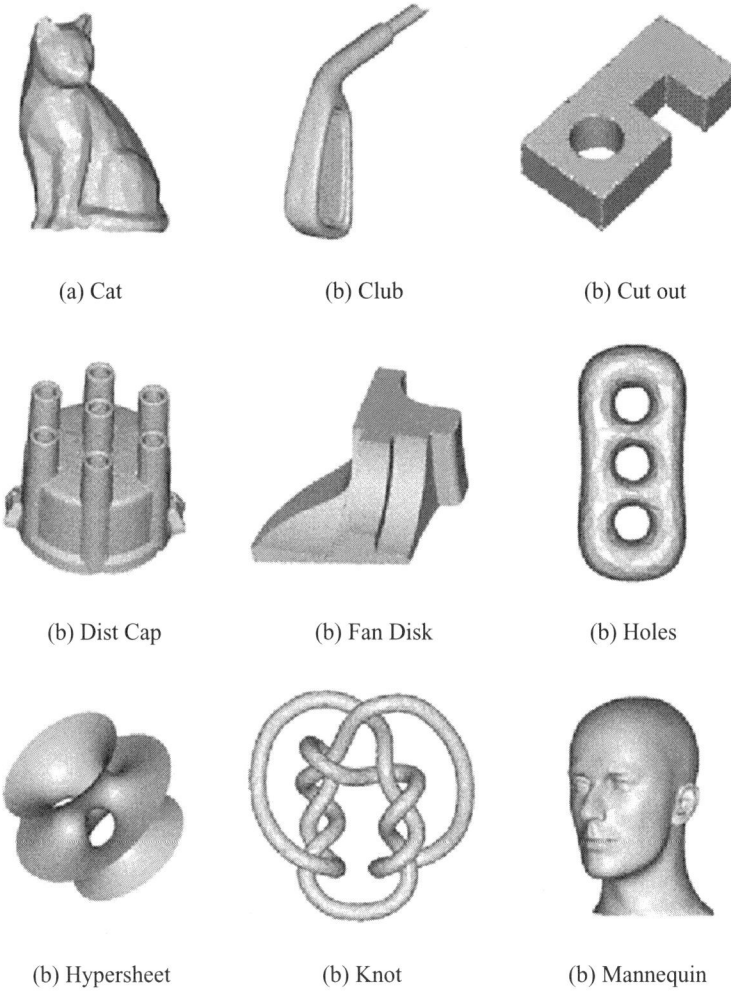

(a) Cat   (b) Club   (b) Cut out

(b) Dist Cap   (b) Fan Disk   (b) Holes

(b) Hypersheet   (b) Knot   (b) Mannequin

Figure 21.16   Reconstruction results.

'Time', indicating how many seconds the entire reconstruction process took. 'KNN' specifies what portion of the total time the k-nearest neighbors search accounted for. The penultimate column provides a quality measure for the sample (first number) and its reconstruction (second number). Quality can be either 1 (poor), 2 (fair), 3 (good), or 4 (perfect).

In evaluating the results, it can first be determined that sets consisting of up to several ten thousand sites may be reconstructed. Two facts have proven prohibitive to larger sets. First, as all sites are required simultaneously (the algorithm is *not* incremental in this sense), the amount of actually installed physical memory is a serious limiting factor. Moreover, neighborhood parameters, especially k, require careful tuning which is worn intensive if reconstruction times exceed a few minutes.

Table 21.1 Statistical data on reconstruction results.

Model Name	n	k	f	λ_3^{max}	#Tri's	Time[s]	KNN[%]	Quality
Cat	6779	32	0.64	1.0E+10	13501	9.58	70.8	3/4
Club	16864	28	0.77	1.0E+10	33695	22.53	74.9	4/4
Cutout	10639	38	0.78	1.0E+10	21294	16.49	70.3	3/3
DistCap	12745	40	0.80	3.8E−04	25224	23.50	57.0	1/2
FanDisk	16475	28	0.81	1.0E+10	32958	18.85	64.9	4/4
Hypersheet	6752	30	0.68	1.0E+10	13054	8.88	71.2	4/4
Holes	2650	31	0.74	1.0E+10	5308	3.41	72.1	3/4
Knot	10000	35	0.79	1.0E−03	19999	14.74	64.9	4/4
Mannequin	12772	48	0.88	1.0E+10	25504	33.70	81.4	4/4
MechPart	4102	34	0.69	2.0E−03	8228	6.38	69.4	2/3
Moeller	20021	26	0.71	1.0E+10	39947	17.67	67.9	4/4
Monkey	19208	28	0.86	1.0E+10	37783	25.76	67.0	4/4
Nascar	20621	25	0.70	1.0E+10	40890	24.93	73.8	4/4
OilPump	30933	10	0.70	1.0E+10	61205	27.75	72.0	2/2
Skidoo	37974	14	0.82	1.0E+10	75520	20.18	57.7	4/4
Sphere	4096	28	0.79	1.0E+10	8188	4.08	64.7	4/4
Teapot	25667	58	0.61	1.0E+10	51314	62.18	73.7	4/4
Thor	71292	35	0.66	5.5E+01	141927	110.16	76.0	3/2

Most objects can be reconstructed in a matter of seconds which is quite fast. For example, the recently published crust algorithm by Amenta et al. [1] requires 12 minutes to reconstruct, 'Club' almost 32 times as much as the method introduced in this thesis. On the other hand, our application may require several runs to fine-tune neighborhood parameters. But for all data sets considered thus far, less than ten tests were necessary to achieve a satisfactory reconstruction quality, which makes our algorithm comparatively efficient. The table of results further indicates that the total run time is dominated by k-nearest neighbors search. As the currently used algorithm is a straightforward approach to solving the problem, more sophisticated methods could yield even better performance for both, large k and n.

6. Summary

This chapter has successfully presented a new method concerned with the triangulation of unorganized 3d points. This method is meant for the reverse engineering solution. It is based on the surface oriented approach. This newly designed method has been constructed in such a way that it extends Laszlo's edge based triangulation alogrithm into 3D. The demonstration of the results has shown that the method is reasonably fast and provides excellent reconstruction results.

References

[1] Amenta, N., Bern, M., and Kamvysselis, M. (1998), A new Voronoi-Based Surface Reconstruction Algorithm. *SIGGRAPH '98 Proceedings*, 415–421.
[2] Amenta, N., and Bern, M. (1998), Surface Reconstruction by Voronoi Filtering. *Proceedings 14th ACM Symposium on Computer Geometry*, 39–48.

[3] Boissonnat, J-D. (1984), Geometric Structures for Three-Dimensional Shape Representation. *ACM Transactions on Graphics* 3:266–286.

[4] Callahan, P. (1995), Dealing with Higher Dimensions–The Well-Separated Pairs Decomposition and Its Applications. *PhD Thesis*, John Hopkins University, Baltimore.

[5] Curless, B., and Levoy, M. (1996), A Volumetric Method for Building Complex Models from Range Images. *SIGGRAPH '96 Proceedings*, 303–312.

[6] Edelsbrunner, H., and Mücke, E. (1994), Three-Dimensional Alpha Shapes. *ACM Transactions on Graphics* 13:43–72.

[7] Hoppe, H. *et al.* (1992), Surface Reconstruction from Unorganized Points. *SIGGRAPH '92 Proceedings*, 71–78.

[8] Laszlo, M. (1996), *Computation Geometry and Computer Graphics in C++*, Prentice-Hall, Upper Saddle River, New Jersey.

[9] Mencl, R., and Müller, H. (1998), Interpolation and Approximation of Surfaces from Three-Dimensional Scattered Data Points. In *Computer Graphics Forum (EUROGRAPHICS '96 Proceedings)*, State of the Art Report (STAR), 17.

[10] Turk, G., and Levoy, M. (1994), Zippered Polygon Meshes from Range Images. *SIGGRAPH '94 Proceedings*, 311–318.

22

Modifying the Shape of Cubic B-spline and NURBS Curves by Means of Knots

Imre Juhász

Department of Descriptive Geometry, University of Miskolc, Egyetemváros H-3515, Hungary

Miklós Hoffmann

Institute of Mathematics and Computer Science, Károly Eszterházy College Leányka str. 4-6. H-3300 Eger, Hungary

Shape control methods of cubic B-spline and NURBS curves by the modification of their knot values, and simultaneous modification of weights and knots are presented in this chapter. Theoretical aspects of knot modification concerning the paths of points of a curve, and the existence of an envelope for the family of curves obtained by the modification of a knot are also discussed for curves of order k.

1. Introduction

B-spline and NURBS curves are standard description methods of CAD systems and widely used in computer aided design today. There are several books and papers on these curves describing their properties, with the help of which one can apply them as powerful design tools.

A kth order B-spline curve is uniquely defined by its control points and knot values, while in terms of NURBS curves the weight vector has to be specified in addition. The shape modification of these curves plays central role in CAD, hence numerous methods have been published that discuss how to control the shape of a curve by modifying one of its data mentioned above. The most basic possibilities can be found in any book of the field, such as in [6]. Further control point-based shape modification is discussed in [2] and [5], weight-based modification is

Advances in Geometric Modeling. Edited by M. Sarfraz
© 2003 John Wiley & Sons, Ltd   ISBN: 0-470-85937-7

described, in [3], [5] and [8], while others present shape control by the simultaneous modification of control points and weights, cf. [1] and [7].

It is also well-known that the change of the knot vector affects the shape of the curve. The properties of this change however, have not been described yet. The aim of this chapter is to present the geometrical and mathematical representation of the effects of knot modification for B-spline curves. After the basic definitions some theoretical results are presented, by means of which one can describe the effects of the modification of a knot value on the shape of the curve. In the subsequent section constraint-based shape control possibilities are discussed that utilize the modification of knot values of non-rational B-spline curves, while the effect of the simultaneous modification of knots and weights is presented for the rational case. In the latter section we restrict our consideration to curves of degree 3, since this is the most widely used type of B-spline and NURBS curves.

2. Theoretical results

In this section the modification of a knot value of a kth order B-spline curve will be examined. We begin our discussion with the basic definitions.

Definition 1. The recursive function $N_j^k(u)$ given by the equations

$$N_j^1(u) = \begin{cases} 1 & if \ u \in [u_j, u_{j+1}), \\ 0 & otherwise, \end{cases}$$

$$N_j^k(u) = \frac{u - u_j}{u_{j+k-1} - u_j} N_j^{k-1}(u) + \frac{u_{j+k} - u}{u_{j+k} - u_{j+1}} N_{j+1}^{k-1}(u),$$

is called normalized B-spline basis function of order k (degree $k - 1$). The numbers $u_j \leq u_{j+1} \in \Re$ are called knot values or simply knots, and $0/0 = 0$ by definition.

Definition 2. The curve $\mathbf{s}(u)$ defined by

$$\mathbf{s}(u) = \sum_{l=0}^{n} N_l^k(u)\mathbf{d}_l, \quad u \in [u_{k-1}, u_{n+1}],$$

is called B-spline curve of order k (degree $k - 1$), where $N_l^k(u)$ is the lth normalized B-spline basis function, for the evaluation of which the knots $u_0, u_1, \ldots, u_{n+k}$ are necessary. The points $\mathbf{d}_i$ are called control points or de Boor-points, while the polygon formed by these points is called control polygon.

The jth span of the B-spline curve can be written in the form

$$\mathbf{s}_j(u) = \sum_{l=j-k+1}^{j} N_l^k(u)\mathbf{d}_l, \quad u \in [u_j, u_{j+1}).$$

Modifying the knot u_i, the point of this span associated with the fixed parameter value

$\tilde{u} \in [u_j, u_{j+1})$ will move along the curve

$$s_j(\tilde{u}, u_i) = \sum_{l=j-k+1}^{j} N_l^k(\tilde{u}, u_i)\mathbf{d}_l, \quad u_i \in [u_{i-1}, u_{i+1}].$$

Hereafter, we refer to this curve as the path of the point $s_j(\tilde{u})$. In [4] the authors proved the following basic properties of these paths:

Theorem 1. *Modifying the knot value $u_i \in [u_{i-1}, u_{i+1}]$ of a kth order B-spline curve, the points of the spans $s_{i-k+1}(u), \ldots, s_{i+k-2}(u)$ move along rational curves. The degree of these paths decreases symmetrically from $k - 1$ to 1 as the indices of the spans getting farther from i, i.e., the paths $s_{i-m}(\tilde{u}, u_i)$ and $s_{i+m-1}(\tilde{u}, u_i)$ are rational curves of degree $k - m$ with respect to u_i, ($m = 1, \ldots, k - 1$).*

The theorem states that modifying $u_i \in [u_{i-1}, u_{i+1}]$ the points of the spans $s_{i-k+1}(\tilde{u}, u_i)$ and $s_{i+k-2}(\tilde{u}, u_i)$ move along straight lines. One can easily prove the following corollary, which will be strongly used in the next section.

Corollary. *If u_i runs from u_{i-1} to u_{i+1}, then the points of the spans $s_{i-k+1}(\tilde{u}, u_i)$ and $s_{i+k-2}(\tilde{u}, u_i)$ move along straight lines parallel to the sides $\mathbf{d}_{i-k}, \mathbf{d}_{i-k+1}$ and $\mathbf{d}_{i-1}, \mathbf{d}_i$ of the control polygon, respectively.*

Beside these paths we can also consider the one-parameter family of B-spline curves with the family parameter u_i

$$s(u, u_i) = \sum_{l=0}^{n} N_l^k(u, u_i)\mathbf{d}_l, \quad u \in [u_{k-1}, u_{n+1}], \quad u_i \in [u_{i-1}, u_{i+1}),$$

which is resulted by the modification of the knot value u_i.

In case of $k = 3$ the spans of the curves are parabolic arcs. It is a well-known fact, that the tangent lines of these arcs at the knot values coincide with the corresponding sides of the control polygon. Modifying a knot value u_i the tangent line remains the same, which can be interpreted as the side of the control polygon is an envelope of the family of these quadratic B-spline curves. The generalization of this property has also been proved by the authors for arbitrary k, cf. [4].

Theorem 2. *The family of the kth order B-spline curves $s(u, u_i) = \sum_{l=0}^{n} N_l^k(u, u_i)$, $(k > 2)$ has an envelope which is also a B-spline curve of order $(k - 1)$, and can be written in the form*

$$\mathbf{b}(v) = \sum_{l=i-k+1}^{i-1} N_l^{k-1}(v)\mathbf{d}_l, \quad v \in [v_{i-1}, v_i],$$

where

$$v_j = \begin{cases} u_j & \text{if } j < i \\ u_{j+1} & \text{if } j \geq i \end{cases},$$

that is the ith knot value is removed from the knot vector of the original curves.

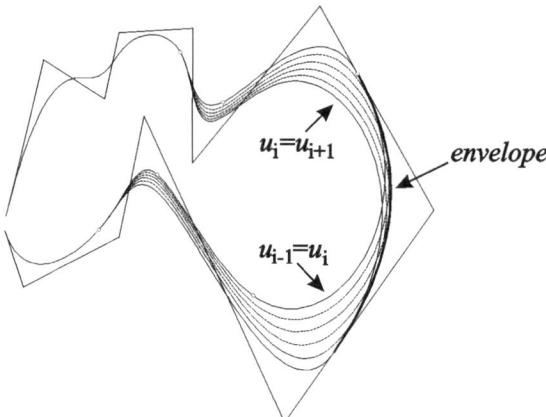

Figure 22.1   The envelope of the family of cubic B-spline curves is a quadratic B-spline curve with the same control polygon.

Until now, only non-rational B-spline curves have been examined, but similar results hold for the rational case. A rational B-spline curve can always be considered as a central projection of a non-rational B-spline curve. The degree of a curve cannot increase by a central projection, thus Theorem 1 and its corollary hold for paths of the points of a NURBS curve, except the property that parallel paths become concurrent, which will be discussed in the next section. Similarly, Theorem 2 holds for the rational case too, but the envelope will also be a NURBS curve. Figure 22.1 shows the envelope of the family of cubic B-spline curves resulted by the modification of one of the knot values.

These theoretical results help us to develop some interesting tools for shape control of B-spline and NURBS curves by the modification of their knot values, that will be examined in the next section.

3. Shape control

For the sake of simplicity we restrict our consideration for the case $k = 4$. Cubic curves are powerful design tools for most of the applications in the plane as well as in the 3D space. Some of the algorithms discussed below can be generalized for arbitrary k, while others use the specific properties of cubic curves.

3.1. Non-rational B-spline curve passing through a point

Let $s(u)$ be a non-rational cubic B-spline curve with control points d_i, $(i = 0, \ldots, n)$ and knot values u_j, $(j = 0, \ldots, n + 4)$. Until now, the only possibility for the modification of this curve has been the repositioning of its control points. Now, we give an algorithm for changing this curve by modifying its knot values in such a way that the curve will pass through a given point p at the given parameter value $\tilde{u}$. This point, of course cannot be anywhere: the algorithm works if this point is inside the region defined by the sides of the control polygon and the envelopes mentioned in Theorem 2, which are parabolic arcs in the cubic case.

Let the point $\mathbf{p}$ be in the region defined by the control points $\mathbf{d}_{j-2}, \mathbf{d}_{j-1}, \mathbf{d}_j$. Let a parameter value $\tilde{u} \in [u_j, u_{j+2})$ be also given. Consider a quadratic B-spline curve $\mathbf{b}(v)$ with the same control points and the knot values $v_0 = u_0, \ldots, v_{j-1} = u_{j-1}, v_j = u_j, v_{j+1} = u_{j+2}, \ldots, v_{n+3} = u_{n+4}$. Hence for the given value $\tilde{u} \in [v_j, v_{j+1})$ holds. Consider the jth span of the quadratic curve

$$\mathbf{b}_j(v) = \sum_{l=j-2}^{j} N_l^3(v)\mathbf{d}_l, \quad v \in [v_j, v_{j+1}).$$

Using the monotonicity of the knot values, one can write

$$v - v_{j-1} = (v_{j+1} - v_{j-1}) - (v_{j+1} - v)$$
$$v_{j+2} - v = (v_{j+2} - v_j) - (v - v_j).$$

Substituting these formulae to the original equation we obtain the form

$$\mathbf{b}_j(v) = \mathbf{d}_{j-1} + N_{j-2}^3(v)(\mathbf{d}_{j-2} - \mathbf{d}_{j-1}) + N_j^3(v)(\mathbf{d}_j - \mathbf{d}_{j-1}).$$

Let us consider the affine coordinate system the origin of which is $\mathbf{d}_{j-1}$ and the base vectors are $\mathbf{e}_1 = \mathbf{d}_{j-2} - \mathbf{d}_{j-1}$ and $\mathbf{e}_2 = \mathbf{d}_j - \mathbf{d}_{j-1}$. Let the coordinates of the given point $\mathbf{p}$ in this coordinate system be x and y. This yields the following system of equations:

$$x = \frac{(v_{j+1} - v)^2}{(v_{j+1} - v_{j-1})(v_{j+1} - v_j)},$$
$$y = \frac{(v - v_j)^2}{(v_{j+2} - v_j)(v_{j+1} - v_j)},$$

Since x, y and $v = \tilde{u}$ are given, one can choose two unknowns from the knot values $(v_{j-1}, v_j, v_{j+1}, v_{j+2})$. The system can be solved for any two of them, but to avoid the unnecessary changes of farther spans it is better to chose two neighboring values, thus 8 spans will be modified. Solving the system, e.g., for v_{j-1}, v_j and considering the quadratic curve $\bar{\mathbf{b}}(v)$ with these knot values, $\bar{\mathbf{b}}(\tilde{u}) = \mathbf{p}$ holds. Therefore, because of Theorem 2, the cubic curve $\bar{\mathbf{s}}(u)$ with the knot values $(\ldots u_{j-1} = v_{j-1}, u_j = v_j, u_{j+1} = \tilde{u}, u_{j+2} = v_{j+1}, \ldots)$ also passes through the point $\mathbf{p}$ at the parameter value $\tilde{u}$. As we have mentioned above, the point $\mathbf{p}$ cannot be chosen arbitrarily. The general permissible area of $\mathbf{p}$ can be seen in Figure 22.2 (the shaded area), but the choice of the parameter and the two unknown knot values enable us to specify further restrictions for the positions of $\mathbf{p}$. Fixing a parameter value $\tilde{u}$ these restricted areas are subsets of the general permissible area and differ from each other for every pair of unknowns. The boundaries of these areas are special rational curves which can be described as paths of certain points of the quadratic B-spline curve when one of its knot values is altered. Now, we will discuss these areas in detail.

3.1.1 The unknowns are v_{j-1} and v_j

In this case the boundaries of the area for a given parameter value $\tilde{u}$ can be described as follows. Considering the knot values $(v_{j-2}, v_{j-1}, v_j, v_{j+1})$ of the quadratic B-spline curve arc $\mathbf{b}_j(v)$

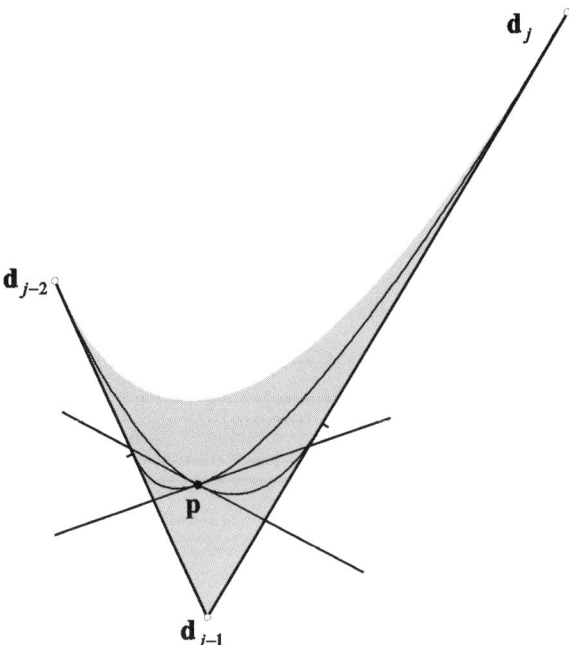

Figure 22.2   The permissible positions of the point **p** and the extreme tangent directions of the parabolic arc.

and the parameter value $\tilde{u} \in \lfloor v_j, v_{j+1})$, the two unknowns can be altered between the fixed values v_{j-2} and $\tilde{u}$ to keep the monotonicity. Hence the extreme situations are

$$(1) \;\; v_{j-2} = v_{j-1} = v_j < \tilde{u} < v_{j+1},$$
$$(2) \;\; v_{j-2} = v_{j-1} < v_j = \tilde{u} < v_{j+1},$$
$$(3) \;\; v_{j-2} < v_{j-1} = v_j = \tilde{u} < v_{j+1}.$$

The first two positions are connected by the path of the point $\mathbf{b}(\tilde{u})$ of the quadratic B-spline curve with the knot vector $\{\ldots v_{j-2} = v_{j-1} = v_j, v_{j+1} \ldots\}$ with the alteration of the knot value $v_j \in [v_{j-1}, \tilde{u}]$. Similarly, the positions (1) and (3) are connected by the path of the same point of the same B-spline curve altering the knot value $v_{j-1} = v_j \in [v_{j-2}, \tilde{u}]$ (here the moving knot value has the multiplicity 2). Finally, the positions (2) and (3) are connected by the path of the point $\mathbf{b}(\tilde{u})$ of the quadratic B-spline curve with the knot vector $\{\ldots v_{j-2} = v_{j-1}, \tilde{u}, v_{j+1} \ldots\}$ varying the knot value v_{j-1} between v_{j-2} and $\tilde{u}$. These boundaries can be seen in Figure 22.3a).

3.1.2 The unknowns are v_{j-1} and v_{j+1}

The boundaries of the permissible positions of the point **p** in this case are the paths connecting the following four extreme positions of a point of the quadratic B-spline curve arc $\mathbf{b}_j(v)$,

$(v_{j-1} \in [v_{j-2}, v_j]$ and $v_{j+1} \in [\tilde{u}, v_{j+2}])$:

$$(1)\ \ v_{j-2} = v_{j-1} < v_j < \tilde{u} = v_{j+1} < v_{j+2},$$
$$(2)\ \ v_{j-2} = v_{j-1} < v_j < \tilde{u} < v_{j+1} = v_{j+2},$$
$$(3)\ \ v_{j-2} < v_{j-1} = v_j < \tilde{u} = v_{j+1} < v_{j+2},$$
$$(4)\ \ v_{j-2} < v_{j-1} = v_j < \tilde{u} < v_{j+1} = v_{j+2}.$$

The paths can be described similarly to the preceding case, but only three of them form the actual boundaries, the other three paths run inside the region. The boundaries can be seen in Figure 22.3b).

3.1.3 The unknowns are v_{j-1} and v_{j+2}

The boundaries in this case are straight line segments due to the corollary of Theorem 1. The paths connect the following extreme positions:

$$(1)\ \ v_{j-2} = v_{j-1} < v_j < \tilde{u} < v_{j+1} = v_{j+2} < v_{j+3},$$
$$(2)\ \ v_{j-2} = v_{j-1} < v_j < \tilde{u} < v_{j+1} < v_{j+2} = v_{j+3},$$
$$(3)\ \ v_{j-2} < v_{j-1} = v_j < \tilde{u} < v_{j+1} = v_{j+2} < v_{j+3},$$
$$(4)\ \ v_{j-2} < v_{j-1} = v_j < \tilde{u} < v_{j+1} < v_{j+2} = v_{j+3}.$$

In this case only four of the six paths form the boundary of the permissible area that can be seen in Figure 22.3c).

3.1.4 The unknowns are v_j and v_{j+1}

With this choice of unknowns the four extreme cases are the following ($v_j \in [v_{j-1}, \tilde{u}]$ and $v_{j+1} \in [\tilde{u}, v_{j+2}]$):

$$(1)\ \ v_{j-1} = v_j < \tilde{u} = v_{j+1} < v_{j+2},$$
$$(2)\ \ v_{j-1} < v_j = \tilde{u} = v_{j+1} < v_{j+2},$$
$$(3)\ \ v_{j-1} = v_j < \tilde{u} < v_{j+1} = v_{j+2},$$
$$(4)\ \ v_{j-1} < v_j = \tilde{u} < v_{j+1} = v_{j+2}.$$

As one can see in Figure 22.3d), generally this case gives the largest permissible area, which also includes the control point $\mathbf{d}_{j-1}$. As a further advantage, the number of modified spans of the original curve is the least in this case as well, hence the two unknowns are neighboring knot values.

3.1.5 The unknowns are v_{j+1} and v_{j+2}

This case yields similar effect and area to that one, with the unknowns v_{j-1}, v_j due to the symmetry of the problem with respect to the parameter value $\tilde{u}$. Now, we have only three extreme cases which are the following ($v_{j+1}, v_{j+2} \in [\tilde{u}, v_{j+3}]$):

$$(1) \ \ v_j < \tilde{u} = v_{j+1} = v_{j+2} < v_{j+3},$$
$$(2) \ \ v_j < \tilde{u} = v_{j+1} < v_{j+2} = v_{j+3},$$
$$(3) \ \ v_j < \tilde{u} < v_{j+1} = v_{j+2} = v_{j+3}.$$

The permissible area can be seen in Figure 22.3e).

3.1.6 The unknowns are v_j and v_{j+2}

Due to the symmetry mentioned above, this final case is similar to that one with the unknowns v_{j-1}, v_{j+1}. The four extreme cases can be described as follows ($v_j \in [v_{j-1}, \tilde{u}]$ and $v_{j+2} \in [v_{j+1}, v_{j+3}]$):

$$(1) \ \ v_{j-1} = v_j < \tilde{u} < v_{j+1} = v_{j+2} < v_{j+3},$$
$$(2) \ \ v_{j-1} < v_j = \tilde{u} < v_{j+1} = v_{j+2} < v_{j+3},$$
$$(3) \ \ v_{j-1} = v_j < \tilde{u} < v_{j+1} < v_{j+2} = v_{j+3},$$
$$(4) \ \ v_{j-1} < v_j = \tilde{u} < v_{j+1} < v_{j+2} = v_{j+3}.$$

The resulted region can be seen in Figure 22.3f).

3.2. Tangential constraint for B-spline curve

Since we have four free parameters v_{j-1}, v_j, v_{j+1} and v_{j+2}, some additional constraints can be imposed on the quadratic curve $\mathbf{b}(v)$. Such a constraint can be the prescribed tangent direction at $\mathbf{p}$, but the initial position of $\mathbf{p}$ and the given direction cannot be arbitrary. To describe the permissible positions and directions, consider the parabolic arcs with the tangent lines $\mathbf{d}_{j-2}, \mathbf{d}_{j-1}$ and $\mathbf{d}_j, \mathbf{d}_{j-1}$. The points of contact are $\mathbf{b}(v_j)$ and $\mathbf{b}(v_{j+1})$, i.e., the points where the spline arcs are connected. The extreme positions of the points $\mathbf{b}(v_j)$ and $\mathbf{b}(v_{j+1})$ of the parabolic arc are $\mathbf{d}_{j-2}$ and $\mathbf{d}_j$, respectively. If the position of both end-points is extreme, then the parabolic arc (the Bézier curve) defined by the control points $\mathbf{d}_{j-2}, \mathbf{d}_{j-1}, \mathbf{d}_j$ will be obtained, hence the point $\mathbf{p}$ can be given in the region defined by this arc and the two legs ($\mathbf{d}_{j-2}, \mathbf{d}_{j-1}$ and $\mathbf{d}_{j-1}, \mathbf{d}_j$) of the control polygon (see the shaded area in Figure 22.2).

If the point $\mathbf{p}$ is on this arc, then the tangent direction cannot be prescribed, since a parabolic arc is uniquely defined by two of its points and the tangents at them. However, if the point $\mathbf{p}$ is an inner point of the area mentioned above, then the tangent line can be specified in addition.

The extreme positions of this tangent line are determined by the tangents of the extreme parabolic arcs passing through the point $\mathbf{p}$ and fulfilling the conditions. To obtain these extreme arcs, consider the following two situations: $\mathbf{b}(v_j) = \mathbf{d}_{j-2}$ and $\mathbf{b}(v_{j+1})$ is an inner point of the segment $\mathbf{d}_{j-1}, \mathbf{d}_j$, and the other, when $\mathbf{b}(v_{j+1}) = \mathbf{d}_j$ and $\mathbf{b}(v_j)$ is an inner point of the segment $\mathbf{d}_{j-2}, \mathbf{d}_{j-1}$. These parabolic arcs can easily be calculated by considering the affine coordinate system described at the beginning of this section, in which let the coordinates of the point $\mathbf{p}$ be (x, y). The control points of the first extreme arc are: $\mathbf{d}_{j-2}, \mathbf{d}_{j-1}, \mathbf{d}_{j-1} + \mu \mathbf{e}_2$, $\mu < 1$, and it can be written in the parametric form

$$\mathbf{c}(v) = \mathbf{d}_{j-1} + (1-v)^2 \mathbf{e}_1 + v^2 \mu \mathbf{e}_2.$$

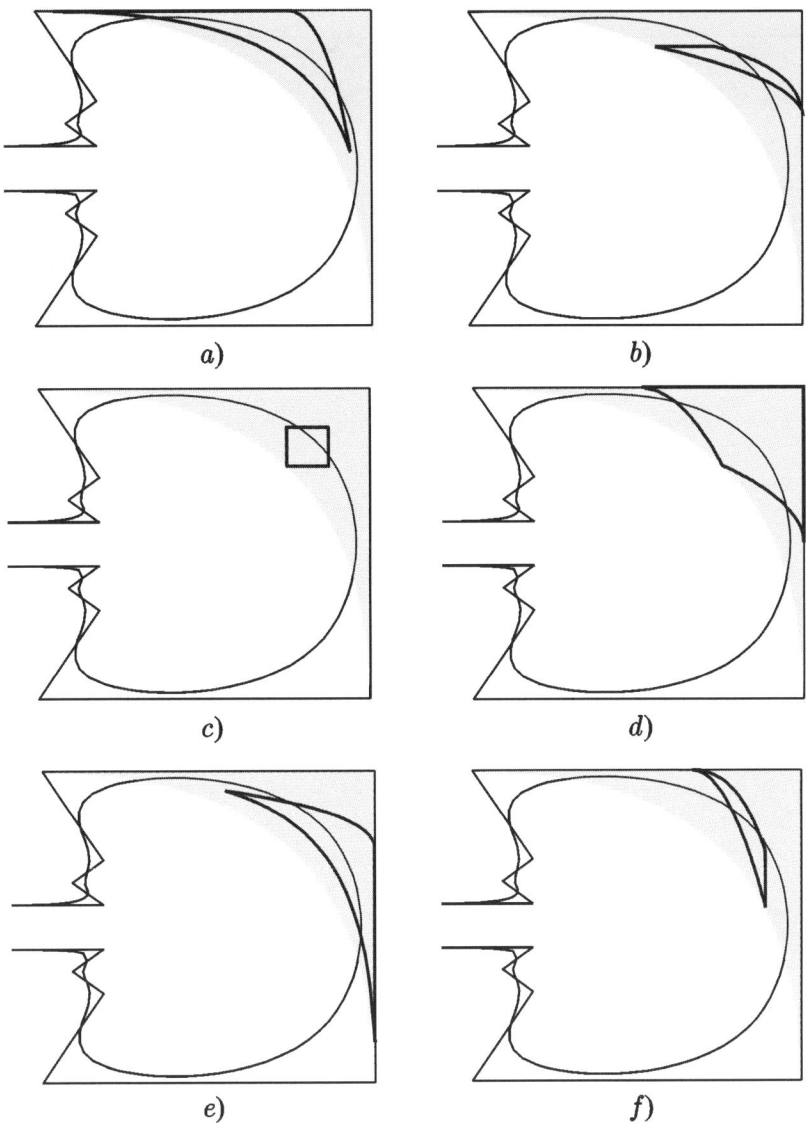

Figure 22.3 A cubic B-spline curve with the knot vector $\{0, 0, 0, 0, 0.1, 0.2, \ldots, 0.8, 0.9, 1, 1, 1, 1\}$.
The shaded region shows the permissible positions of the point $\mathbf{p}$ in general. Bold lines
form the boundary of the permissible positions of $\mathbf{p}$ at $\tilde{u} = 0.4$ if the two unknowns of the
system are chosen as: $a)$ v_{j-1}, v_j; $b)$ v_{j-1}, v_{j+1}; $c)$ v_{j-1}, v_{j+2}; $d)$ v_j, v_{j+1}; $e)$ v_{j+1}, v_{j+2};
$f)$ v_j, v_{j+2}.

One of its points will be the point $\mathbf{p}$ at the parameter value $v_p \in (0, 1)$. For this point

$$\mathbf{p} = \mathbf{d}_{j-1} + x\mathbf{e}_1 + y\mathbf{e}_2$$

holds. The vectors $\mathbf{e}_1$ and $\mathbf{e}_2$ are linearly independent, hence from the equation $\mathbf{p} = \mathbf{c}(v_p)$

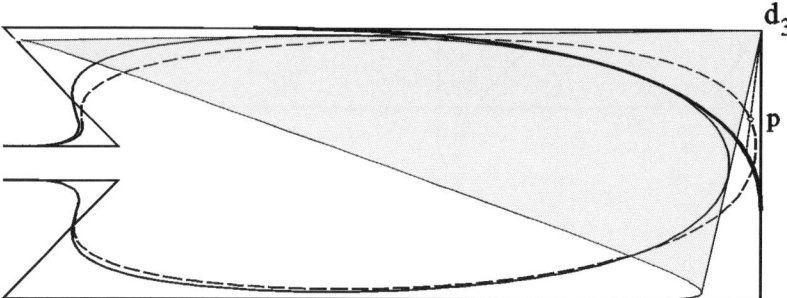

Figure 22.4   Modifying the weight w_3 and the knot value u_4 the NURBS curve passes through the given point **p** which is outside the area accessible by the only modification of w_3.

we obtain the solutions $v_p = 1 - \sqrt{x}$, $\mu = y/(1 - \sqrt{x})^2$. Figure 22.2 shows the two extreme parabolic arcs passing through **p** along with their tangent lines there. The tangent direction can be prescribed in this angular domain.

3.3. NURBS curve passing through a point

It is a well-known fact, that the modification of the weight w_j of a NURBS curve causes a perspective functional translation of points of the effected arcs, i.e., it pulls/pushes points of the curve toward/away from the control point $\mathbf{d}_j$. If a given point is on one of the line segments of the paths of this perspective change, one can easily compute the new weight value in such a way, that the new curve will pass through the given point. This point can almost be anywhere in the convex hull, but for $k > 3$ these concurrent line segments starting from $\mathbf{d}_j$ do not sweep the entire area of the triangle $\mathbf{d}_{j-1}, \mathbf{d}_j, \mathbf{d}_{j+1}$, cf. the gray area in Figure 22.4. If the given point is close to the side of the control polygon, i.e., it is out of the shaded region of Figure 22.4, the problem can only be solved with the change of two neighboring weights. Now, we give an algorithm to solve this problem with the change of one weight and one knot value.

Let $\mathbf{s}(u)$ be a cubic NURBS curve and **p** a point in the triangle $\mathbf{d}_{j-1}, \mathbf{d}_j, \mathbf{d}_{j+1}$. Consider the quadratic envelope $\mathbf{b}(v)$ of the family of NURBS curves $\mathbf{s}(u, u_{j+1})$. This parabolic arc intersects all the lines in this triangle starting from $\mathbf{d}_j$, thus suitably changing the weight w_j, there will be a parameter value $\tilde{v}$, for which $\mathbf{b}(\tilde{v}) = \mathbf{p}$. If we modify the knot value u_{j+1} of the cubic curve to be $u_{j+1} = \tilde{v}$, the cubic curve will also pass through the point **p**. This type of shape modification is illustrated in Figure 22.4.

3.4. Simultaneous modification of two knots

In the previous subsection the quadratic envelope has been modified by a weight, where the points of the curve moves along straight lines toward a control point. Similar effect, however, can be achieved in terms of non-rational quadratic B-spline curves by appropriate simultaneous modification of two knot values. More precisely, from the definition of the B-spline functions and the Corollary of Theorem 1, we can prove the following property:

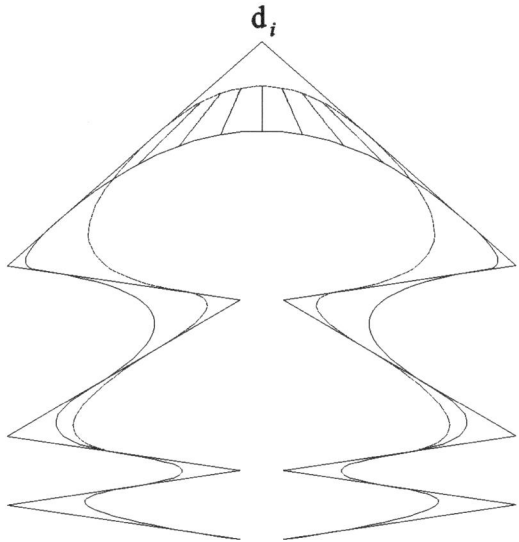

$$\mathbf{d}_i$$

Figure 22.5 Simultaneous modification of two knot values yields a perspective change of the span of a non-rational quadratic B-spline curve.

Theorem 3. *Consider the quadratic non-rational B-spline curve* $\mathbf{s}(u)$, *and simultaneously modify its knots* u_i *and* u_{i+3} *in an equal manner, i.e., let* $u_i = u_i + \lambda$, $u_{i+3} = u_{i+3} - \lambda$. *As a result of this modification, points of the span* $\mathbf{s}_{i+1}(u)$ *move along concurrent straight lines, if and only if,*

$$u_{i+2} - u_i = u_{i+3} - u_{i+1},$$

holds. The common point of these straight lines is $\mathbf{d}_i$ *(see Figure 22.5).*

Proof. As we have seen above, the span $\mathbf{s}_{i+1}(u)$ can be written in the form

$$\mathbf{s}_{i+1}(u) = \mathbf{d}_i + N_{i-1}^3(u)(\mathbf{d}_{i-1} - \mathbf{d}_i) + N_{i+1}^3(u)(\mathbf{d}_{i+1} - \mathbf{d}_i).$$

Applying the knot modification of the theorem, we obtain the family of curves

$$\mathbf{s}_{i+1}(u, \lambda) = \mathbf{d}_i + \frac{u_{i+2} - u}{u_{i+2} - u_i - \lambda} N_i^2(u)(\mathbf{d}_{i-1} - \mathbf{d}_i)$$
$$+ \frac{u - u_{i+1}}{u_{i+3} - \lambda - u_{i+1}} N_{i+1}^2(u)(\mathbf{d}_{i+1} - \mathbf{d}_i).$$

Assuming the equality $\delta = u_{i+2} - u_i = u_{i+3} - u_{i+1}$, we can factor out $1/(\delta - \lambda)$ and we obtain

$$\mathbf{s}_{i+1}(u, \lambda) = \mathbf{d}_i + \frac{1}{\delta - \lambda}\Big((u_{i+2} - u)N_i^2(u)(\mathbf{d}_{i-1} - \mathbf{d}_i)$$
$$+ (u - u_{i+1})N_{i+1}^2(u)(\mathbf{d}_{i+1} - \mathbf{d}_i)\Big)$$

which is a family of straight line segments and the pencil of lines determined by them has the center $\mathbf{d}_i$.

Conversely, if $u_{i+2} - u_i \neq u_{i+3} - u_{i+1}$ then the rational curves described above have two points at infinity (one at $\lambda = u_{i+2} - u_i$ and another at $\lambda = u_{i+3} - u_{i+1}$), therefore they can not be straight lines.

The modification of these two knot values, of course, is not so effective, than that of a weight, because the region of change is greater in the latter case while the number of changing spans is fewer (7 for the two knot values and 3 for the weight), but we have to emphasize, that this theorem allows us to modify non-rational quadratic B-spline curves similarly to NURBS curves.

Similar theorem holds for higher order B-spline curves. The generalization of Theorem 3 for arbitrary order is the following:

Theorem 4. *The points of the arc* $\mathbf{s}_{i+k-2}(u)$ *of a* k^{th} *order B-spline curve move along straight line segments when the knots* u_i *and* u_{i+2k-3} *are simultaneously and equally modified, if and only if, the equality* $u_{i+k-1} - u_i = u_{i+2k-3} - u_{i+k-2}$ *is satisfied.*

As we have seen, these straight lines are concurrent in the case of $k = 3$. If $k = 4$, the span $\mathbf{s}_{i+2}(u, \lambda)$ is of interest which has the form

$$\mathbf{s}_{i+2}(u, \lambda) = \left(\frac{u_{i+4} - u}{u_{i+4} - u_{i+1}} N_{i+1}^3(u) + N_i^3(u) \right) \mathbf{d}_i$$

$$+ \left(\frac{u - u_{i+1}}{u_{i+4} - u_{i+1}} N_{i+1}^3(u) + N_{i+2}^3(u) \right) \mathbf{d}_{i+1}$$

$$+ \frac{u_{i+3} - u}{u_{i+3} - u_i - \lambda} N_i^3(u) (\mathbf{d}_{i-1} - \mathbf{d}_i)$$

$$+ \frac{u - u_{i+2}}{u_{i+5} - \lambda - u_{i+2}} N_{i+2}^3(u) (\mathbf{d}_{i+2} - \mathbf{d}_{i+1}).$$

The coefficients of $\mathbf{d}_i$ and $\mathbf{d}_{i+1}$ are non-negative and sum to 1, i.e., the constant part of the sum is a convex linear combination of the control points $\mathbf{d}_i$ and $\mathbf{d}_{i+1}$. Therefore paths of the arc are straight line segments the extension of which intersect the side $\mathbf{d}_i$, $\mathbf{d}_{i+1}$ at its inner points moreover, they are parallel to the plane determined by the directions $\mathbf{d}_{i-1} - \mathbf{d}_i$ and $\mathbf{d}_{i+2} - \mathbf{d}_{i+1}$, cf. Figure 22.6.

3.5. Modifying two weights and a knot of a NURBS curve

If we modify two neighboring weights w_j, w_{j+1} of a NURBS curve the points of the curve move along straight lines toward or away from the leg $\mathbf{d}_j$, $\mathbf{d}_{j+1}$ of the control polygon. This translation is neither perspective nor parallel. This property can be made more intuitive geometrically by the modification of a knot value in addition. Thus we can achieve that the points of a span of the curve will move along concurrent lines passing through any given point of the line $\mathbf{d}_j$, $\mathbf{d}_{j+1}$ except the inner points of the leg. As we have mentioned in the preceding section, modifying a knot value u_j of a cubic NURBS curve the points of the spans $\mathbf{s}_{j-3}(u)$, $\mathbf{s}_{j+2}(u)$ will move along a family of concurrent straight lines each. Considering the span $\mathbf{s}_{j-3}(u)$ and assuming

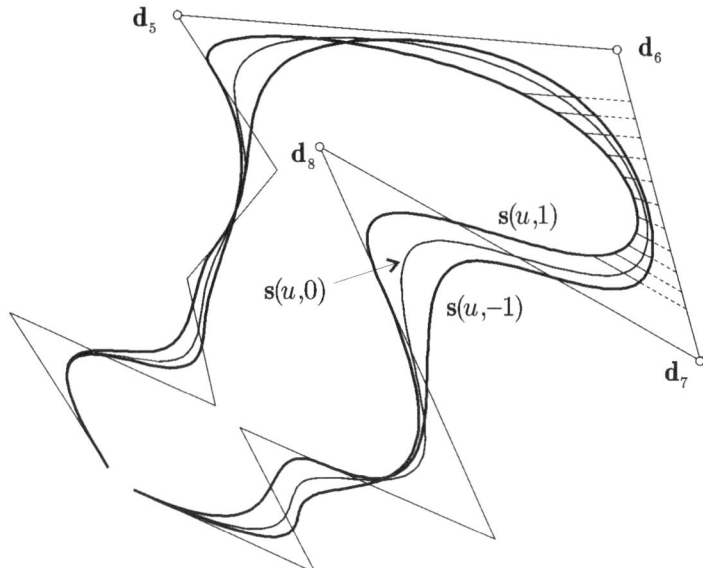

Figure 22.6 Shape modification of a cubic ($k = 4$) B-spline curve by means of symmetric alteration of knots u_6 and u_{11}. Paths of points of the arc $\mathbf{s}_8(u)$ are also shown along with their extensions which intersect the side $\mathbf{d}_6$, $\mathbf{d}_7$ of the control polygon.

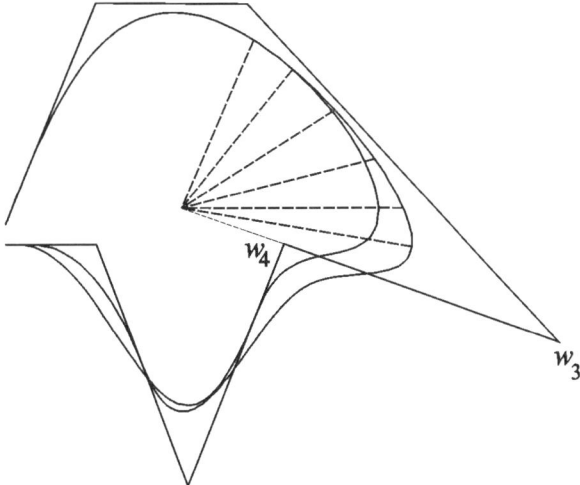

Figure 22.7 Modifying the knot u_7, points of the span $\mathbf{s}_4(u)$ move along concurrent straight lines the center of which depends on w_3 and w_4 and can be arbitrary located on the line of $\mathbf{d}_3$, $\mathbf{d}_4$, except the inner points of the segment.

that $w_{j-4} \neq w_{j-3}$ the following result can be achieved: modifying the knot value u_j the points of this span move along concurrent lines the center of which is on the line $\mathbf{d}_j, \mathbf{d}_{j+1}$ and its barycentric coordinates are

$$\left(\frac{w_{j-4}}{w_{j-4} - w_{j-3}}, \ 1 - \frac{w_{j-4}}{w_{j-4} - w_{j-3}} \right).$$

One can easily see, that one of its coordinates must be negative with the usual assumption $w_j \geq 0$, $\forall j$. Hence this center cannot be on the leg $\mathbf{d}_j, \mathbf{d}_{j+1}$ but on the rest of the line. Figure 22.7 shows a case of such a modification.

4. Summary

This chapter has been devoted to the shape control of cubic B-spline and NURBS curves. These curves can be uniquely defined by their degree, control points, weights and knot vector. While the effect of the modification of the preceding data has been widely published and used, the geometric effect of the change of the knot vector has not been studied yet. In the second section some theoretical results have been presented in terms of the paths of the points of the curve when one of its knot values is modified, and the existence of an envelope of the resulted family of curves. Applying these results, some shape control methods have been presented in the third section. Modifying one or more of the knot values of a non-rational B-spline, one can achieve constraint-based modification, such as obtaining a curve passing through a given point, or a shape modification which is similar to the effect of the modification of a weight in the rational case. For NURBS curves, the simultaneous change of one or two weights and knot values have been presented, the result of which is a NURBS curve passing through a given point or a geometrically simple perspective shape modification.

The objective of further research, besides the knot-based constrained shape modification of curves of arbitrary order k, is the study of the theoretical aspects of knot modifications for surfaces, which will hopefully generate some shape control methods both for B-spline and NURBS surfaces. One can also think to extend the theory for those curve schemes [9–19], which have ideal geometric features together with shape control characteristics.

References

[1] Au, C. K., Yuen, M. M. F. (1995) Unified approach to NURBS curve shape modification. Computer-Aided Design, 27, 85–93.

[2] Fowler B, Bartels R. (1993) Constrained-based curve manipulation. IEEE Computer Graphics and Applications, 13(5), 43–49.

[3] Juhász, I. (1999) Weight-based shape modification of NURBS curves. Computer Aided Geometric Design, 16, 377–383.

[4] Juhász I, Hoffmann M. (2001) The effect of knot modifications on the shape of B-spline curves. Journal for Geometry and Graphics, 5(2), 111–119.

[5] Piegl, L. (1989) Modifying the shape of rational B-splines. Part 1: curves. Computer-Aided Design, 21, 509–518.

[6] Piegl, L., Tiller, W. (1995) The NURBS book. Springer-Verlag, Berlin

[7] Sánchez-Reyes, J. (1997) A simple technique for NURBS shape modification. IEEE Computer Graphics and Applications, 17, 52–59.

[8] Sarfraz, M, (2003), Weighted Nu Splines: An Alternative to NURBS, Advances in Geometric Modeling, Ed.: M. Sarfraz, John Wiley, 81–95.

[9] Sarfraz, M. (2003), Optimal Curve Fitting to Digital Data, International Journal of WSCG, Vol 11(1), 128–135.

[10] Sarfraz, M. (2003), Curve Fitting for Large Data using Rational Cubic Splines, International Journal of Computers and Their Applications, Vol 10(3).

[11] Sarfraz, M., and Razzak, M. F. A., (2003), A Web Based System to Capture Outlines of Arabic Fonts, International Journal of Information Sciences, Elsevier Science Inc., Vol. 150(3–4), 177–193.

[12] Sarfraz, M., and Razzak, M. F. A., (2002), An Algorithm for Automatic Capturing of Font Outlines, International Journal of Computers & Graphics, Elsevier Science, Vol. 26(5), 795–804.

[13] Sarfraz M. (2002) Fitting curves to planar digital data. Proceedings of IEEE International Conference on Information Visualization IV'02-UK: IEEE Computer Society Press, USA, 633–638.

[14] Sarfraz, M. (1992) A C^2 Rational Cubic Alternative to the NURBS, Computers and Graphics 16(1), 69–77.

[15] Sarfraz, M. (1992) Interpolatory rational cubic spline with biased, point and interval tension. Computers and Graphics 16(4), 427–430.

[16] Sarfraz, M. (1993) Designing of Curves and Surfaces using Rational Cubics. Computers and Graphics 17(5), 529–538.

[17] Sarfraz, M. (1995) Curves and Surfaces for CAD using C^2 Rational Cubic Splines, Engineering with Computers, 11(2), 94–102.

[18] Gregory, J. A., Sarfraz, M., Yuen, P. K. (1994) Interactive Curve Design using C^2 Rational Splines, Computers and Graphics 18(2), 153–159.

[19] Sarfraz, M. (1994) Cubic Spline Curves with Shape Control, Computers and Graphics 18(4), 707–713.

Index of Authors

Aoyama, H.
Azariadis, P. N.
Brunnett, G.
Bultheel, A.
Chan K Y, T.
Dafas, P. A.
Delgado, J.
Dierckx, P.
Giraldi, G. A.
Habib, Z.
Häfele, K.-H.
Hui, K. C.
Hussain, M.
Iglesias, A.
Jiménez, W. H.
Kamiya, J.
Kim, D-S.
Kompatsiaris, I.
Li, C. L.
Li, L.
Ma, L
Ma, W.
Müller-Wittig, W.
Nasri, A. H.

Niijima, K.
Okada, Y.
Oliveira, A. A. F.
Peña, J. M.
Poliakoff, J. F.
Revesz, P.
Ryu, J.
Sacchi, R.
Sakai, M.
Sapidis, N. S.
Sarfraz, M
Schädlich, T.
Siddiqui, M. A.
Silva, R. L. S.
Strauss, E.
Strintzis, M. G.
Thomas, P. D.
Vanco, M.
Vanraes, E.
Wang, Q
Zhang, J. J.
Zheng, J. J.
Zhong, Y.

Advances in Geometric Modeling. Edited by M Sarfraz
© 2003 John Wiley & Sons, Ltd   ISBN: 0-470-85937-7

Index

Advances in Geometric Modeling. Edited by M. Sarfraz
© 2003 John Wiley & Sons, Ltd   ISBN: 0-470-85937-7